Advanced Sparsity-Driven Models and Methods for Radar Applications

Related titles on radar

Advances in Bistatic Radar Willis and Griffiths
Airborne Early Warning System Concepts, 3rd Edition Long
Bistatic Radar, 2nd Edition Willis
Design of Multi-Frequency CW Radars Jankiraman
Digital Techniques for Wideband Receivers, 2nd Edition Tsui
Electronic Warfare Pocket Guide Adamy
Foliage Penetration Radar: Detection and characterisation of objects under trees Davis
Fundamentals of Ground Radar for Air Traffic Control Engineers and Technicians Bouwman
Fundamentals of Systems Engineering and Defense Systems Applications Jeffrey
Introduction to Electronic Warfare Modeling and Simulation Adamy
Introduction to Electronic Defense Systems Neri
Introduction to Sensors for Ranging and Imaging Brooker
Microwave Passive Direction Finding Lipsky
Microwave Receivers with Electronic Warfare Applications Tsui
Phased-Array Radar Design: Application of radar fundamentals Jeffrey
Pocket Radar Guide: Key facts, equations, and data Curry
Principles of Modern Radar, Volume 1: Basic principles Richards, Scheer and Holm
Principles of Modern Radar, Volume 2: Advanced techniques Melvin and Scheer
Principles of Modern Radar, Volume 3: Applications Scheer and Melvin
Principles of Waveform Diversity and Design Wicks *et al.*
Pulse Doppler Radar Alabaster
Radar Cross Section Measurements Knott
Radar Cross Section, 2nd Edition Knott *et al.*
Radar Design Principles: Signal processing and the environment, 2nd Edition Nathanson *et al.*
Radar Detection DiFranco and Rubin
Radar Essentials: A concise handbook for radar design and performance Curry
Radar Foundations for Imaging and Advanced Concepts Sullivan
Radar Principles for the Non-Specialist, 3rd Edition Toomay and Hannan
Test and Evaluation of Aircraft Avionics and Weapons Systems McShea
Understanding Radar Systems Kingsley and Quegan
Understanding Synthetic Aperture Radar Images Oliver and Quegan
Radar and Electronic Warfare Principles for the Non-specialist, 4th Edition Hannen
Inverse Synthetic Aperture Radar Imaging: Principles, algorithms and applications Chen and Martorella
Stimson's Introduction to Airborne Radar, 3rd Edition Baker, Griffiths and Adamy
Test and Evaluation of Avionics and Weapon Systems, 2nd Edition McShea
Angle-of-Arrival Estimation Using Radar Interferometry: Methods and applications Holder
Biologically-Inspired Radar and Sonar: Lessons from nature Balleri, Griffiths and Baker
The Impact of Cognition on Radar Technology Farina, De Maio and Haykin
Novel Radar Techniques and Applications, Volume 1: Real aperture array radar, imaging radar, and passive and multistatic radar Klemm, Nickel, Gierull, Lombardo, Griffiths and Koch
Novel Radar Techniques and Applications, Volume 2: Waveform diversity and cognitive radar, and target tracking and data fusion Klemm, Nickel, Gierull, Lombardo, Griffiths and Koch
Radar and Communication Spectrum Sharing Blunt and Perrins
Systems Engineering for Ethical Autonomous Systems Gillespie
Shadowing Function from Randomly Rough Surfaces: Derivation and applications Bourlier and Li
Photo for Radar Networks and Electronic Warfare Systems Bogoni, Laghezza and Ghelfi
Multidimensional Radar Imaging Martorella
Radar Waveform Design Based on Optimization Theory Cui, De Maio, Farina and Li
Micro-Doppler Radar and its Applications Fioranelli, Griffiths, Ritchie and Balleri
Maritime Surveillance with Synthetic Aperture Radar Di Martino and Antonio Iodice
Electronic Scanned Array Design Williams

Advanced Sparsity-Driven Models and Methods for Radar Applications

Gang Li

The Institution of Engineering and Technology

Published by SciTech Publishing, an imprint of The Institution of Engineering and Technology, London, United Kingdom

The Institution of Engineering and Technology is registered as a Charity in England & Wales (no. 211014) and Scotland (no. SC038698).

© The Institution of Engineering and Technology 2021

First published 2020

The Institution of Engineering and Technology
Michael Faraday House
Six Hills Way, Stevenage
Herts, SG1 2AY, United Kingdom

www.theiet.org

British Library Cataloguing in Publication Data
A catalogue record for this product is available from the British Library

ISBN 978-1-83953-075-3 (hardback)
ISBN 978-1-83953-076-0 (PDF)

Typeset in India by MPS Limited

Contents

About the author

Gang Li is a Professor at the Department of Electronic Engineering, Tsinghua University, China. His research interests include radar signal processing, remote sensing, distributed signal processing, and information fusion. He has published over 150 papers on these subjects. He is a recipient of the National Science Fund for Distinguished Young Scholars of China and the Royal Society Newton Advanced Fellowship of United Kingdom. He is a Senior Member of the IEEE.

Preface

Compressed sensing (CS) has been one of the most active topics in signal processing area. By exploiting the sparsity of the signals, CS offers a prospective way for reducing data amount without compromising the performance of signal recovery or enhancing resolution without increasing the number of measurements. The signals in many radar applications are sparse or compressible, so the radar systems may benefit from the sparsity-driven models and methods in terms of reducing observation duration, simplifying hardware, and enhancing performance. However, in practical radar applications, it is found that directly applying the basic CS models and algorithms to radar data may be less than optimal and even unsatisfactory. Thus, it is necessary to develop advanced sparsity-based models and algorithms to fit various radar tasks, which has become a fast-growing branch of radar signal processing in recent years.

The objective of this book is to introduce more recent developments on advanced sparsity-driven models and methods that are designed for radar tasks including clutter suppression, signal detection, radar imaging, target parameter estimation, and target recognition, mainly based on my publications in the last decade. Besides the theoretical analysis, numerous simulation examples and experiments on real radar data are presented throughout the book. The material presented in this book can be understood by readers who have a fundamental knowledge of radar signal processing. The book can serve as a reference book for academic researchers, practicing engineers, and graduate students.

The outline of this book is as follows. Before introducing the advanced sparsity-driven models and methods designed for radar tasks, the fundamentals of CS are briefly reviewed in Chapter 1. In Chapter 2, the hybrid greedy pursuit algorithms are presented for enhancing radar imaging quality. In Chapter 3, the two-level block sparsity model is introduced to promoting the sparsity of signals of multichannel radar systems. In Chapter 4, the parametric sparse representation is studied to deal with model uncertainty during the radar data collection. Chapter 5 investigates how to simultaneously achieve high-resolution and wide-swath in single-channel synthetic aperture radar (SAR) imaging by utilizing the Poisson disk sampling. Chapter 6 concentrates on the sparsity-driven algorithms of radar image formation from coarsely quantized data. Chapter 7 is concerned with sparsity aware radar micro-Doppler analysis for micromotion parameter estimation and target recognition. Chapter 8 is devoted to the distributed detection of sparse signals with radar networks. Chapter 9 summarizes the book and discusses some perspectives.

Acknowledgments

I would like to express my sincere appreciation to Prof. Ying-Ning Peng and Prof. Xiang-Gen Xia for their continued support. I would like to thank Prof. Chris Baker for hosting my visit at The Ohio State University and Prof. Pramod K. Varshney for hosting my visit at Syracuse University. My gratitude also goes to my collaborators for their valuable suggestions and discussions. I gratefully acknowledge my students, Wei Rao, Rui Zhang, Yichang Chen, Xueqian Wang, Xiaoyu Yang, Jianghong Han, and Zhizhuo Jiang, for their invaluable contribution to the contents contained in this book. I am also grateful to The National Natural Science Foundation of China, The Chang Jiang Scholars Program of Ministry of Education of China, The National Ten Thousand Talent Program of China, The National Basic Research Program of China, and The Royal Society of United Kingdom for sponsoring my research. I also wish to express my thanks to the reviewers of this book for the constructive comments and to the staff of IET for the interest and efforts in the publication of this book. Special thanks must go to my family for their understanding and forbearance. Before I started writing this book, I had no idea how it would take my time and energy. This book would never have been completed without the continual encouragement and support of my family.

Notation

	All matrices and vectors are denoted by boldface letters.
$(\cdot)^T$	denotes the transpose.
$(\cdot)^*$	denotes the conjugate.
$(\cdot)^H$	denotes the Hermitian transpose.
$(\cdot)^{-1}$	denotes the inverse of a matrix.
$(\cdot)^\dagger = \left[(\cdot)^H(\cdot)\right]^{-1}(\cdot)^H$	denotes the pseudo inverse of a matrix.
$\odot$	denotes the Hadamard product.
$\otimes$	denotes the Kronecker product.
$\lvert \cdot \rvert$	denotes the absolute value.
$\lVert \cdot \rVert_0$	denotes the L_0-norm of a vector. $\lVert \mathbf{x} \rVert_0$ counts the nonzero elements in $\mathbf{x}$.
$\lVert \cdot \rVert_1$	denotes the L_1-norm of a vector. $\lVert \mathbf{x} \rVert_1 = \sum_{i=1}^{N} \lvert x_i \rvert$ for $\mathbf{x} \in \mathbb{C}^{N \times 1}$.
$\lVert \cdot \rVert_2$	denotes the L_2-norm of a vector. $\lVert \mathbf{x} \rVert_2 = \sqrt{\sum_{i=1}^{N} \lvert x_i \rvert^2}$ for $\mathbf{x} \in \mathbb{C}^{N \times 1}$.
$\lVert \cdot \rVert_F$	denotes the Frobenius norm of a matrix. $\lVert \mathbf{X} \rVert_2 = \sqrt{\sum_{m=1}^{M} \sum_{n=1}^{N} \lvert X_{m,n} \rvert^2}$ for $\mathbf{X} \in \mathbb{C}^{M \times N}$.
$\langle \cdot, \cdot \rangle$	denotes the inner product of two matrices or vectors.
$\mathrm{rank}(\cdot)$	denotes the rank of a matrix.
$\mathrm{tr}(\cdot)$	denotes the trace of a matrix.
$\nabla(\cdot)$	denotes the gradient operator.
$E(\cdot)$	denotes the expected value.
$\mathrm{diag}(\cdot)$	denotes the diagonal matrix with a vector on its diagonal.
$\mathrm{mean}(\cdot)$	denotes the averaging operator.
$\mathrm{card}(\cdot)$	denotes the cardinality of a set.

Chapter 1

Introduction

1.1 Sparsity of radar signals

Radar plays an important role in military and civil applications thanks to its day-and-night and all-weather capability. In many radar applications, the received signals are compressible or sparse, since the intrinsic information of high-dimensional signals may be fully described by a small number of parameters. Some examples of sparse radar signals are provided below:

- In air surveillance with inverse synthetic aperture radar (ISAR), the aircrafts to be imaged are sparsely distributed in the air space.
- In ocean surveillance with synthetic aperture radar (SAR), the echoes reflected from the sparsely distributed ships to be imaged are dominant, while the echoes reflected from the sea surface are much weaker.
- In the detection of vital signs behind a wall with through-the-wall radar for rescue, the number of potential objects is usually assumed to be small.
- In space-time-adaptive-processing (STAP) for detection of ground moving targets with airborne array radar, the target and the clutter occupy a small part of the angle-Doppler domain.
- The radar signals reflected from human gait and hand gestures show a sparse pattern in the time-frequency domain.

Mathematically, a signal is called sparse if only a few entries are nonzero. Because the noise always exists in real radar signals, the sparsity of radar signals means that only a few entries have dominant magnitudes in certain domains. Radar systems may benefit from exploiting the sparsity as a priori knowledge in the following aspects.

- Supper resolution. Two closely located sources that are difficult to be distinguished by the traditional signal processing methods (i.e., matched filtering) may be separable in the solution of sparse signal recovery.
- Less signal samples. Sparse signal processing provides a way to decrease the amount of data without sacrificing signal recovery accuracy and signal detection performance, which can reduce the cost of data acquiring, storage and transmission.
- Inexpensive analog-to-digital converter (ADC). The sparse signals can be reconstructed or detected from coarsely quantized data even one-bit data, which may significantly simplify the hardware implementation.

Note that the traditional principle component analysis (PCA) and the related subspace analysis can also be viewed as sparsity-promoting algorithms. When the signal of interest is stationary, that is, the locations/frequencies of the dominant components are constant during the observation time interval, and the number of snapshots is sufficiently large, the signal subspace spanned by the dominant components and the noise subspace can be distinguished by PCA. In comparison with PCA, sparse signal processing does not require the stationarity of the signals and the sufficient number of snapshots. That is, the sparse signal processing can retrieve the dominant components from even single snapshot and accordingly works well when the locations/frequencies of dominant components are time-varying. This implies that the sparse signal processing is suitable for a wider range of scenarios.

Although the sparse signal processing has been studied for several decades, it is found that the direct applications of the basic models and algorithms based on the sparsity of signals to radar data may be less than optimal and even unsatisfactory. The reason is that radar signal processing algorithms are generally application-dependent. We will first review the fundamentals of sparse signal recovery in this chapter and then present some advanced sparsity-driven models and algorithms specially designed for radar tasks in the rest of this book. The radar tasks discussed in this book include clutter suppression, signal detection, radar imaging, target parameter estimation, and target recognition.

1.2   Fundamentals of sparse signal recovery

The sparse signal recovery is also commonly known as compressed sensing (CS). In this section, the fundamentals of CS, including the basic signal model and the typical algorithms, are briefly reviewed.

1.2.1   Signal model

Consider the following problem:

$$\mathbf{y} = \mathbf{\Phi}\mathbf{x} + \mathbf{w} \tag{1.1}$$

where $\mathbf{y} \in \mathbb{C}^{M \times 1}$ is the measurement vector, $\mathbf{\Phi} \in \mathbb{C}^{M \times N}$ is a dictionary matrix, $\mathbf{x} \in \mathbb{C}^{N \times 1}$ is the vector to be solved, and $\mathbf{w} \in \mathbb{C}^{M \times 1}$ is the additive noise. Assume that $M < N$. In this case, the problem in (1.1) appears ill-conditioned and some elaborate algorithms are required to solve it.

Especially, here we focus on the sparse signals. The signal $\mathbf{x}$ is said to be K-sparse if the number of nonzero elements in $\mathbf{x}$ is smaller than or equal to K, where $K \ll N$ and $K < M$. The value of K is called the sparsity level of the signal. The set of the locations of nonzero elements in $\mathbf{x}$ is referred to as the support set, defined by

$$\Lambda(\mathbf{x}) = \{i | x_i \neq 0, i = 1, 2, \cdots, N\} \tag{1.2}$$

where x_i is the ith entry of $\mathbf{x}$.

Given the received signal **y**, it is obvious that the sparse solution **x** depends on the dictionary **Φ**. The columns of the dictionary **Φ**, which are called atoms or basis-signals, are expected to be capable of representing the received signal **y** in a sparse fashion. The choice of **Φ** is application-dependent. Some widely used dictionaries in radar tasks are introduced as follows:

1. The Fourier dictionary [1].

    The entry at the mth row and the nth column of the Fourier dictionary is formulated as

$$\Phi(m, n) = \exp\{j2\pi f_n t_m\} \tag{1.3}$$

    where t_m is the mth sampling time, f_n is the nth frequency point, $m = 1, 2, \cdots, M$, $n = 1, 2, \cdots, N$. In this case, **x** denotes the frequency-domain coefficients. This dictionary works well when only a few frequency-domain coefficients are dominant.

2. The dictionary consisting of array steering vectors [1].

    The entry of the dictionary consisting of array steering vectors is formulated as

$$\Phi(m, n) = \exp\left\{-j2\pi \frac{d_m \sin(\theta_n)}{\lambda}\right\} \tag{1.4}$$

    where d_m denotes the position of the mth antenna, θ_n is the nth spatial direction grid, and λ is the radar wavelength. In this case, **x** indicates the strengths of the sources at all the possible direction of arrivals (DOA). This dictionary is suitable for the case where the spatial distribution of the sources is sparse.

3. The dictionary consisting of ambiguity functions [1].

    The entry of the dictionary consisting of ambiguity functions is expressed as

$$\Phi(m, n) = s_T\left(t_m - \frac{2r_n}{c}\right)\exp\left(-j\frac{4\pi(r_n - v_n t_m)}{\lambda}\right) \tag{1.5}$$

    where t_m is the mth sampling time, $s_T(\cdot)$ denotes the waveform of the transmitted signal, (r_n, v_n) denotes the nth discretized grid in the range-velocity domain, and c is the light speed. In this case, **x** denotes the high-resolution profile of the ambiguity range-velocity domain. This dictionary is suitable for range and velocity estimation when there are a few dominant scatterers in the observed scene.

4. The Gabor dictionary [2].

    The entry of the Gabor dictionary is expressed as

$$\Phi(m, n) = \frac{1}{\pi^{1/4}\sqrt{\tau}}\exp\left(-\frac{(t_m - T_n)^2}{2\tau^2}\right)\exp(j2\pi t_m F_n) \tag{1.6}$$

    where (T_n, F_n) denotes the nth discretized position in the time-frequency domain and τ indicates the width of the Gaussian window. When this dictionary is used, $\mathbf{x}$ denotes the complex amplitude distribution in the time-frequency domain. This dictionary is appropriate for describing the signals with sparse time-frequency distributions.

5.   The chirp dictionary [3].

The entry of the chirp dictionary is expressed as

$$\Phi(m,n) = \exp\left[j\left(2\pi F_n t_m + \pi a_n t_m^2\right)\right] \tag{1.7}$$

where (F_n, a_n) denotes the nth discretized position of the frequency-chirp-rate domain. When this dictionary is used, $\mathbf{x}$ denotes the coefficients in the frequency-chirp-rate domain. This dictionary is appropriate for decomposing chirp signals in a sparse fashion.

1.2.2   Typical algorithms for sparse signal recovery

A natural idea of finding the sparse solution to (1.1) is to seek the sparsest solution at a given tolerance of the recovery error level η:

$$\widehat{\mathbf{x}} = \arg\min_{\mathbf{x}} \|\mathbf{x}\|_0, \quad \text{s.t.} \quad \|\mathbf{y} - \Phi\mathbf{x}\|_2^2 \leq \eta \tag{1.8}$$

However, solving (1.8) requires an exhaustive enumeration of all possible positions of the nonzero entries in $\mathbf{x}$, which is referred to as an NP-hard problem. The commonly used algorithms for finding the approximate solution include regularization methods, greedy algorithms, iterative thresholding algorithms, and Bayesian algorithms. These algorithms offer stable recovery of K-sparse N-dimensional signals from only $M = O(K \log(N/K))$ measurements. That is to say, the number of measurements is significantly smaller than the dimension of the signal to be solved and greater than the sparsity level by a small logarithmic factor, which is the cost for not knowing the locations of the nonzero coefficients in advance. Some typical CS algorithms will be introduced in what follows.

1.2.2.1   Regularization methods

The regularization methods are commonly used to stabilize the solution of the ill-conditioned problem in (1.1). The idea of the regularization method is to replace the original ill-conditioned problem with another problem whose stable solution can approximate the desired solution. A popular method is the Tikhonov regularization [4], which gives the approximated solution to (1.1) by

$$\widehat{\mathbf{x}} = \arg\min_{\mathbf{x}} \left(\|\mathbf{y} - \Phi\mathbf{x}\|_2^2 + \gamma \cdot p(\mathbf{x}) \right) \tag{1.9}$$

where the first term of the objective function represents the recovery error or the data fidelity, $p(\mathbf{x})$ in the second term of the objective function is a regularization term that stands for prior or penalty, and γ is the regularization parameter that balances the above two terms. Some common choices of the regularization term are listed as follows:

1. $p(\mathbf{x}) = \|\mathbf{x}\|_2^2$ is the energy of the signal.
2. $p(\mathbf{x}) = \|\Gamma\mathbf{x}\|_2^2$ is the energy of the signal through a linear transformation Γ.
3. $p(\mathbf{x}) = \|\nabla\mathbf{x}\|_1$ means the total variation (TV) that tends to preserve edges in an image.
4. $p(\mathbf{x}) = \|\mathbf{x}\|_1$ induces the sparsity in the solution.
5. $p(\mathbf{x}) = \|\mathbf{H}\mathbf{x}\|_1$ promotes the sparsity of the coefficients in a specific domain through a linear transformation $\mathbf{H}$.

The typical sparse signal recovery problem corresponds to the case that $p(\mathbf{x}) = \|\mathbf{x}\|_1$, that is, (1.9) becomes:

$$\widehat{\mathbf{x}} = \arg \min_{\mathbf{x}} \left\{ \|\mathbf{y} - \Phi\mathbf{x}\|_2^2 + \gamma\|\mathbf{x}\|_1 \right\} \tag{1.10}$$

This optimization problem is also referred as to the basis pursuit denoising (BPDN) [5] or the least absolute shrinkage and selection operator (Lasso) [6]. The problem in (1.10) is a second-order cone programming problem and thus can be solved by the interior point methods [7]. A number of toolboxes of convex optimization are available for solving (1.10), such as ℓ_1-Magic [8] and CVX [9], etc. The computational complexity of solving (1.10) via convex optimization is $O(N^3)$ [7]. When the problem is large scale, for example, there are millions of pixels or resolution cells to be solved in SAR imaging applications, the computational burden of the convex optimization methods is considerably high, which may preclude the practical use of this kind of methods.

There are a number of studies on the performance guarantee for sparse signal recovery. A well-known sufficient condition for the stable recovery of sparse signals is the restricted isometry property (RIP).

Definition 1.1 [10] The dictionary Φ is said to satisfy the restricted RIP of order K if there exists a constant $\delta_K \in (0, 1)$ such that:

$$(1 - \delta_K)\|\mathbf{x}\|_2^2 \leq \|\Phi\mathbf{x}\|_2^2 \leq (1 + \delta_K)\|\mathbf{x}\|_2^2 \tag{1.11}$$

holds for all the K-sparse signals.

Theorem 1.1 *[11,12] Suppose Φ satisfies the RIP of order 2K with $\delta_{2K} < 0.4651$. If $\eta \geq \|\mathbf{w}\|_2$, then the solution to (1.10) will obey:*

$$\|\mathbf{x} - \widehat{\mathbf{x}}\|_2 \leq C_1 \frac{\|\mathbf{x} - \mathbf{x}_K\|_1}{\sqrt{K}} + C_2\eta \tag{1.12}$$

where C_1 and C_2 depends only on δ_{2K}.

Motivated by the practical needs, more than one regularization terms can be also added to the objective function in (1.9), which is referred to as multiparameter regularization or multiparameter Tikhonov regularization or compound regularization [13–16]. For example, in image processing applications, the sparsity of the coefficients in wavelet domain and the edges in the image may be required to be preserved simultaneously, which leads to both L_1 and TV regularization terms in the objective function.

1.2.2.2   Greedy algorithms

From (1.1), we can see that the measurement vector $\mathbf{y}$ can be viewed as the sum of several basis-signals (i.e., the columns of the dictionary $\mathbf{\Phi}$) whose indices are recorded in the support set $\Lambda(\mathbf{x})$. The greedy algorithms aim to recover the support set $\Lambda(\mathbf{x})$, which directly represents the sparse nature of the signals. Thus the goal of greedy algorithms can be regarded as the direct pursuit of L_0 minimization under the recovery error constraint. Once $\Lambda(\mathbf{x})$ consisting of K indices is determined, the nonzero coefficients indexed by $\Lambda(\mathbf{x})$ can be easily estimated by using the least squares estimation and other coefficients are set to zero.

In what follows, two typical greedy algorithms are introduced in detail. The input variables of these greedy algorithms include the measurement vector $\mathbf{y}$, the dictionary $\mathbf{\Phi}$, and the sparsity K, while the output of these algorithms is the sparse solution estimate $\widehat{\mathbf{x}}$. The steps of greedy algorithms are composed of matrix and vector operations, which are much easier than the calculations of the interior point methods. Thus, generally speaking, the greedy algorithms are more computationally efficient than the optimization-based algorithms. This is the reason why the greedy algorithms are usually preferred in real-time processing systems.

1. Orthogonal matching pursuit (OMP) [17].

The OMP algorithm operates in an iterative manner, as summarized in Table 1.1. At the initialization phase, the recovery residual $\mathbf{r}^{(0)}$ is set to be equal to

Table 1.1   The OMP algorithm

Input:
   The measurement vector $\mathbf{y}$, the dictionary $\mathbf{\Phi}$, and the sparsity K.
Initialization:
   $\mathbf{r}^{(0)} = \mathbf{y}$, $\widehat{\mathbf{x}}^{(0)} = \mathbf{0}^{N \times 1}$, $\Lambda^{(0)} = \varnothing$, $k = 1$.
Iteration:

1. $\mathbf{g}^{(k)} = \mathbf{\Phi}^H \mathbf{r}^{(k-1)}$.

2. $\Lambda^{(k)} = \Lambda^{(k-1)} \cup \left\{ \underset{i \in \{1,2,\cdots,N\}}{\arg\max} \left| g_i^{(k)} \right| \right\}$.

3. $\widehat{\mathbf{x}}_{\Lambda^{(k)}}^{(k)} = \mathbf{\Phi}_{\Lambda^{(k)}}^\dagger \mathbf{y}$, $\widehat{\mathbf{x}}_{\{1,2,\cdots,N\}-\Lambda^{(k)}}^{(k)} = \mathbf{0}$.

4. $\mathbf{r}^{(k)} = \mathbf{y} - \mathbf{\Phi}\widehat{\mathbf{x}}^{(k)}$

5. If $k = K$, quit the iteration; else let $k = k + 1$ and go to step 1.

Output:
   $\widehat{\mathbf{x}}^{(k)}$.

---

the measurement vector **y**, and the solution $\widehat{\mathbf{x}}^{(0)}$ and the support set $\Lambda^{(0)}$ are set to be zero and empty, respectively, where the superscript denotes the iteration index. At each iteration, a column of $\mathbf{\Phi}$, which is most correlated to the recovery residual, is chosen and merged into the set containing all the basis-signals selected before, and the residual is updated by projecting the measurement vector onto the orthogonal complement of the subspace spanned by all the selected basis-signals. The iteration is terminated when the number of the selected basis-signals reaches K. The computational complexity of the OMP algorithm is $O(KMN)$.

An alternative criterion of terminating OMP is to compare the residual with an error level. That is to say, step 5 in Table 1.1 can be replaced with "if $\|\mathbf{r}^{(k)}\|_2^2 \leq \varepsilon$, quit the iteration; else let $k = k + 1$ and go to step 1", where ε is the user-defined parameter denoting the error tolerance. For example, in some radar applications, $\varepsilon = 0.05\|\mathbf{y}\|_2^2$ is a reasonable value, which means that the relative recover error is smaller than 5%. If this criterion is used, the error tolerance parameter ε should also be one of the input variables instead of the sparsity K.

2. Subspace pursuit (SP) [18].

The detailed steps of the SP algorithm are given in Table 1.2. Similar to OMP, the basic idea of SP is also to iteratively test subsets of the basis-signals, for the purpose of finding the correct signal support set. At each iteration of SP, the previous support set estimate is merged with K newly added indices corresponding to the largest correlation coefficients between the residual and the basis-signals. This enlarges the number of basis-signal candidates to $2K$. Then the measurement vector is projected onto the subspace spanned by the $2K$ basis-signal candidates, and the K indices corresponding

Table 1.2   The SP algorithm

Input:
 The measurement vector **y**, the dictionary $\mathbf{\Phi}$, and the sparsity K.
Initialization:
 $\Lambda^{(0)} = \{K$ indices corresponding to the largest magnitude entries in the vector $\mathbf{\Phi}^H\mathbf{y}\}$,
 $\widehat{\mathbf{x}}_{\Lambda^{(0)}}^{(0)} = \mathbf{\Phi}_{\Lambda^{(0)}}^\dagger\mathbf{y}$, $\widehat{\mathbf{x}}_{\{1,2,\cdots,N\}-\Lambda^{(0)}}^{(0)} = \mathbf{0}$, $\mathbf{r}^{(0)} = \mathbf{y} - \mathbf{\Phi}\widehat{\mathbf{x}}^{(0)}$, $k = 1$.
Iteration:

1. $\widetilde{\Lambda}^{(k)} = \Lambda^{(k-1)}\cup\{K$ indices corresponding to the largest magnitude entries in the vector $\mathbf{\Phi}^H\mathbf{r}^{(k-1)}\}$.

2. Let $\widetilde{\mathbf{x}} = \mathbf{\Phi}_{\widetilde{\Lambda}^{(k)}}^\dagger\mathbf{y}$.

3. $\Lambda^{(k)} = \{K$ indices corresponding to the largest magnitude entries in the vector $\widetilde{\mathbf{x}}\}$.

4. $\widehat{\mathbf{x}}_{\Lambda^{(k)}}^{(k)} = \mathbf{\Phi}_{\Lambda^{(k)}}^\dagger\mathbf{y}$, $\widehat{\mathbf{x}}_{\{1,2,\cdots,N\}-\Lambda^{(k)}}^{(k)} = \mathbf{0}$, $\mathbf{r}^{(k)} = \mathbf{y} - \mathbf{\Phi}\widehat{\mathbf{x}}^{(k)}$.

5. If $\|\mathbf{r}^{(k)}\|_2 > \|\mathbf{r}^{(k-1)}\|_2$, let $\widehat{\mathbf{x}}_{\Lambda^{(k-1)}}^{(k)} = \mathbf{\Phi}_{\Lambda^{(k-1)}}^\dagger\mathbf{y}$ and $\widehat{\mathbf{x}}_{\{1,2,\cdots,N\}-\Lambda^{(k-1)}}^{(k)} = \mathbf{0}$ and quit the iteration; otherwise, let $k = k+1$ and go to step 1.

Output:
 $\widehat{\mathbf{x}}^{(k)}$.

---

to the largest projection coefficients are kept and other indices are abandoned. The main difference between OMP and SP lies in the following aspects: (1) OMP selects only one basis-signal per iteration, while SP selects multiple basis-signal candidates in a group per iteration; (2) OMP never removes any indices from the support set estimated in previous iterations, while SP updates the support set by adding and removing its elements at each iteration. SP benefits from the adding-removing operation, since those elements of the support set, which are regarded reliable in previous iterations but actually wrong, can be excluded by re-evaluating the reliability of support set at each iteration. The SP algorithm is terminated when the minimum residual is reached. The computational complexity of the SP algorithm is upper-bounded by $O(KMN)$ and can be further reduced to $O(MN \log K)$ when the nonzero entries of the sparse signal decay slowly. By the way, the pseudo-inverse operation with the temporary set $\widetilde{\Lambda}^{(k)}$ in step 2 of the SP algorithm requires $M > 2K$.

Another algorithm that is similar to SP is the compressive sampling matching pursuit (CoSaMP) [19]. The only difference between SP and CoSaMP lies in that, in step 1 of Table 1.2, the former adds K new candidates into the temporary set $\widetilde{\Lambda}^{(k)}$ while the latter adds $2K$ new candidates. These two algorithms were independently developed by different groups of authors at almost the same time.

1.2.2.3   Iterative thresholding algorithms

The iterative thresholding algorithms aim to solve the problem of sparse signal recovery from the viewpoint of gradient-based updating. If we only consider the minimization of the recovery error $\|\mathbf{y} - \mathbf{\Phi x}\|_2^2$, the gradient algorithm directly generates a sequence $\{\widehat{\mathbf{x}}^{(k)}\}$ via:

$$
\begin{aligned}
\widehat{\mathbf{x}}^{(k)} &= \widehat{\mathbf{x}}^{(k-1)} - \rho_k \cdot \nabla_{\mathbf{x}^*}\left(\|\mathbf{y} - \mathbf{\Phi x}\|_2^2\right)\Big|_{\mathbf{x} = \widehat{\mathbf{x}}^{(k-1)}} \\
&= \widehat{\mathbf{x}}^{(k-1)} + \rho_k \cdot \mathbf{\Phi}^H\left(\mathbf{y} - \mathbf{\Phi}\widehat{\mathbf{x}}^{(k-1)}\right)
\end{aligned}
\tag{1.13}
$$

where k denotes the iteration index and ρ_k is the stepsize. Further considering the sparsity constraint, some thresholding strategies can be imposed on the temporary solution, that is,

$$
\widehat{\mathbf{x}}(k) = \mathrm{Th}\left[\widehat{\mathbf{x}}^{(k-1)} + \rho_k \cdot \mathbf{\Phi}^H\left(\mathbf{y} - \mathbf{\Phi}\widehat{\mathbf{x}}^{(k-1)}\right)\right]
\tag{1.14}
$$

where $\mathrm{Th}(\cdot)$ denotes a thresholding operation.

Take the iterative hard thresholding (IHT) algorithm [20] as an example. The thresholding operation $\mathrm{Th}(\cdot)$ in IHT preserves the K largest entries and sets other entries to zero at each iteration. The steps of IHT are summarized in Table 1.3. At each iteration of IHT, the K largest correlation coefficients between the residual and the basis-signals are added to the previous estimate of the sparse solution, and then the thresholding is carried out to keep only the K largest entries. If there is no unique such set, that is, some entries have the same magnitude, the random selection from the entries with the same magnitude can be taken. The steps 1 and 2 in Table 1.3 are

Table 1.3   The IHT algorithm

Input:
 The measurement vector $\mathbf{y}$, the dictionary $\mathbf{\Phi}$, the sparsity K, the stepsize parameter ρ, and the error tolerance ε.
Initialization:
 $\widehat{\mathbf{x}}^{(0)} = \mathbf{0}^{N \times 1}$, $k = 1$.
Iteration:

1. $\widetilde{\mathbf{x}} = \widehat{\mathbf{x}}^{(k-1)} + \rho \cdot \mathbf{\Phi}^H \left(\mathbf{y} - \mathbf{\Phi}\widehat{\mathbf{x}}^{(k-1)}\right).$
2. $\Lambda^{(k)} = \{K$ indices corresponding to the largest magnitude entries in the vector $\widetilde{\mathbf{x}}\}.$
3. $\widehat{\mathbf{x}}^{(k)}_{\Lambda^{(k)}} = \widetilde{\mathbf{x}}_{\Lambda^{(k)}}$ and $\widehat{\mathbf{x}}^{(k)}_{\{1,2,\cdots,N\}-\Lambda^{(k)}} = \mathbf{0}.$
4. If $\|\mathbf{y} - \mathbf{\Phi}\widehat{\mathbf{x}}^{(k)}\|_2^2 \leq \varepsilon$, terminate the iteration; otherwise, let $k = k + 1$ and go to step 1.

Output:
 $\widehat{\mathbf{x}}^{(k)}$.

---

also capable of re-evaluating the sparse solution estimated at previous iterations, similar to the adding-removing operation in SP. The stepsize parameter ρ, which is assumed to be constant for all the iterations, balances the newly added entries and the sparse solution obtained previously. The iterative process of IHT is stopped when the strength of the residual is smaller than a predefined error tolerance level ε. The computational complexity of the IHT algorithm is $O(KMN)$.

The thresholding operation $\mathrm{Th}(\cdot)$ varies in different algorithms. In the iterative shrinkage thresholding (IST) algorithm [21], the thresholding operation is given by

$$\widehat{\mathbf{x}}^{(k)} = \mathrm{Th}_{\gamma\rho_k}\left[\widehat{\mathbf{x}}^{(k-1)} + \rho_k \cdot \mathbf{\Phi}^H \left(\mathbf{y} - \mathbf{\Phi}\widehat{\mathbf{x}}^{(k-1)}\right)\right] \tag{1.15}$$

where

$$\mathrm{Th}_{\gamma\rho_k}(z) = \begin{cases} \mathrm{sgn}(z)(|z| - \gamma\rho_k), & |z| \geq \gamma\rho_k \\ 0, & |z| < \gamma\rho_k \end{cases} \tag{1.16}$$

is carried out for each element of the vector, where $\mathrm{sgn}(\cdot)$ is the sign function. Various versions of the iterative thresholding algorithms have been independently proposed by different groups of authors in [22–27].

1.2.2.4 Bayesian algorithms

The above three kinds of algorithms are designed to recover the deterministic signals. Another way to formulate the sparse signals is the Bayesian framework, that is, to formulate the sparse signals by probabilistic models. The probability density function (PDF) of the stochastic sparse signal $\mathbf{x}$ in (1.1) is assumed to be known a priori. The sparse feature of the signal is encouraged by the PDF with a sharp peak at zero. That is to say, the entries of $\mathbf{x}$ are close to zero with a large probability. Then the sparse signal recovery problem is converted into the problem of finding the maximum a posterior (MAP) solution.

Suppose that **w** in (1.1) is the zero-mean Gaussian white noise with variance σ^2. Then the likelihood function can be formulated as

$$f(\mathbf{y}|\mathbf{x}) = \left(2\pi\sigma^2\right)^{-M/2} \exp\left(-\frac{1}{2\sigma^2}\|\mathbf{y} - \mathbf{\Phi}\mathbf{x}\|_2^2\right) \tag{1.17}$$

and the MAP solution to (1.1) is given by

$$\hat{\mathbf{x}} = \arg\min_{\mathbf{x}} \left\{ \|\mathbf{y} - \mathbf{\Phi}\mathbf{x}\|_2^2 - 2\sigma^2 \log f(\mathbf{x}) \right\} \tag{1.18}$$

where $f(\mathbf{x})$ is the prior PDF of **x**. A widely used sparsity-promoting prior on **x** is the Laplace distribution [28]:

$$f(\mathbf{x}) = \left(\frac{\beta}{2}\right)^N \exp\left(-\beta \sum_{i=1}^N |x_i|\right) \tag{1.19}$$

where β is the model parameter. Submitting (1.19) into (1.18) yields the MAP solution that has a similar form of (1.10). More sophisticated prior distributions of **x** have been studied in [29–31] for performance enhancement. However, in practice, the exact knowledge about the prior distributions of **x** may be unavailable. A practical way of imposing a proper prior on **x** is to empirically select the probabilistic models and corresponding parameters based on analysis of the data previously collected from similar scenes.

1.2.3   Beyond the standard sparsity

The standard sparsity discussed above assumes nothing about the inter-relation between the coefficients of the sparse signal. In fact, in a number of applications, we have knowledge about the structure of the coefficients in addition to the sparsity. An example is the group sparsity (also called block sparsity or clustered sparsity). In high-resolution images, a target may occupy multiple pixel/resolution cells. In this case, all the coefficients in a group tend to be zero or nonzero. Another example is the tree-structured sparsity. The coefficients of a signal that is sparse in the wavelet basis lie on a connected tree. Some local structures in an image, for example, the edges, textures, and smooth shades, may be further emphasized to enhance the inter-relation between the coefficients of the sparse signal.

The structured sparse models give rise to develop more sophisticated algorithms. The convex optimization methods based on the group sparsity have been proposed in [32,33]. In [34,35], the original greedy algorithms have been extended to some versions based on the tree-structured sparsity. The self-similarity of the signal has been formulated as a kind of local structures in [36].

The purpose of developing the models beyond the standard sparsity and the corresponding algorithms is a further reduction of the number of measurement or performance enhancement of signal recovery. For instance, they offer robust recovery of K-sparse signals from just $M = O(K)$ measurements [35,37] and significantly improve the signal reconstruction quality [32,34,36]. More details of the general structure-based or model-assisted CS algorithms are beyond the scope of this book.

This book focuses on the application of sparsity-driven signal processing in radar tasks. In the rest of this book, we will present some advanced sparsity-driven models and methods in specific radar scenarios, instead of discussing the general case and the applications in other areas. For more details of the fundamentals of CS, we refer the readers to [38–41].

References

[1]   Ender J. H. G. "On compressive sensing applied to radar." *Signal Processing.* 2010; (90): 1402–1414.

[2]   Li G., Zhang R., Ritchie M., and Griffiths H. "Sparsity-driven micro-doppler feature extraction for dynamic hand gesture recognition." *IEEE Transactions on Aerospace and Electronic Systems.* 2018; 54(2): 655–665.

[3]   Amin M.G., Jokanovic B., Zhang Y. D., and Ahmad F. "A sparsity-perspective to quadratic time-frequency distributions." *Digital Signal Processing.* 2015; 46: 175–190.

[4]   Tikhonov A. N. and Arsenin V. Y. *Solution of ill-posed problems.* Washington, DC: V. H. Winston; 1977.

[5]   Chen S.S., Donoho D.L., and Saunders M.A. "Atomic decomposition by basis pursuit." *SIAM Review.* 2001; 43(1): 129–159.

[6]   R. Tibshirani. "Regression shrinkage and selection via the Lasso." *Journal of the Royal Statistical Society: Series B.* 1996; 58(1): 267–288.

[7]   Boyd S. and Vandenberghe L. *Convex optimization.* Cambridge, UK: Cambridge University Press: 2004.

[8]   Candes E.J. and Romberg J. "L1-magic: Recovery of sparse signals via convex programming." http://brainimaging.waisman.wisc.edu/~chung/BIA/download/matlab.v1/l1magic-1.1/l1magic_notes.pdf

[9]   Grant M., Boyd S., and Ye Y. "CVX: MATLAB software for disciplined convex programming." http://cvxr.com/cvx/

[10]   Candes E.J. and Tao T. "Decoding by linear programming." *IEEE Transactions on Information Theory.* 2005; 51(12): 4203–4215.

[11]   Foucart S. "A note on guaranteed sparse recovery via L_1-minimization." *Applied and Computational Harmonic Analysis.* 2010; 29(1): 97–103.

[12]   Candès E. J. "The restricted isometry property and its implications for compressed sensing." *Comptes Rendus Mathematique.* 2008; 346 (9):589–592.

[13]   Ito K., Jin B., and Takeuchi T. "Multi-parameter Tikhonov regularization." arXiv preprint arXiv:1102.1173, 2011.

[14]   Brezinski1 C., Redivo-Zaglia M., Rodriguez G., and Seatzu S. "Multi-parameter regularization techniques for ill-conditioned linear systems." *Numerische Mathematik.* 2003; 94: 203–228.

[15] Lu S. and Pereverzev, S. V. "Multi-parameter regularization and its numerical realization." *Numerische Mathematik.* (2011) 118:1–31.

[16] Aghamiry H. S., Gholami A., and Operto S. "Compound regularization of full-waveform inversion for imaging piecewise media." *IEEE Transactions on Geoscience and Remote Sensing.* 2020; 58(2): 1192 – 1204.

[17] Tropp J. and Gilbert A.C. "Signal recovery from partial information via orthogonal matching pursuit." *IEEE Transactions on Information Theory.* 2007; 53(12): 4655–4666.

[18] Dai W. and Milenkovic O. "Subspace pursuit for compressive sensing signal reconstruction." *IEEE Transactions on Information Theory.* 2009; 55(5): 2230–2249.

[19] Needell D. and Tropp J.A. "CoSaMP: Iterative signal recovery from incomplete and inaccurate samples." *Applied and Computational Harmonic Analysis.* 2009; 26(3): 301–321.

[20] Blumensath T. and Davies M. "Iterative hard thresholding for compressed sensing." *Applied and Computational Harmonic Analysis.* 2009; 27(3): 265–274.

[21] Beck A. and Teboulle M. "A fast iterative shrinkage-thresholding algorithm for linear inverse problems." *SIAM Journal on Imaging Sciences.* 2009; 2(1): 183–202.

[22] Blumensath T. and Davies M. "Iterative thresholding for sparse approximations." *Journal of Fourier Analysis and Applications.* 2008; 14: 629–654.

[23] Bayram I. and Selesnick I. W. "A subband adaptive iterative shrinkage/thresholding algorithm." *IEEE Transactions on Signal Processing.* 2010; 58 (3): 1131 – 1143.

[24] Zulfiquar Ali Bhotto Md., Omair Ahmad M., and Swamy M. N. S. "An improved fast iterative shrinkage thresholding algorithm for image deblurring." *SIAM Journal on Imaging Sciences.* 2015; 8(3): 1640–1657.

[25] Nesterov Y. E. "Gradient methods for minimizing composite objective function." CORE Report, 2007. Available at http://www.ecore.be/DPs/dp 1191313936.pdf.

[26] Blumensath T. and Davies M. E. "Normalized iterative hard thresholding: Guaranteed stability and performance." *IEEE Journal of Selected Topics in Signal Processing.* 2010; 4(2): 298–309.

[27] Elad M. "Why simple shrinkage is still relevant for redundant representations? ." *IEEE Transactions on Information Theory.* 2006; 52(12): 5559–5569.

[28] Ji S., Xue Y. and Carin L. "Bayesian compressive sensing." *IEEE Transactions on Signal Processing.* 2008; 56(6): 2346- 2356.

[29] He L. and Carin, L. "Exploiting structure in wavelet-based Bayesian compressive sensing." *IEEE Transactions on Signal processing.* 2008; 57 (9): 3488–3497.

[30] Baraniuk R.G., Cevher V., and Wakin M.B. "Low-dimensional models for dimensionality reduction and signal recovery: A geometric perspective." *Proceedings of the IEEE.* 2010; 98(6): 959–971.

[31] Wipf D.P. and Rao B.D. "Sparse Bayesian learning for basis selection." *IEEE Transactions on Signal processing.* 2004; 52(8): 2153–2164.

[32]   Yuan M. and Lin Y. "Model selection and estimation in regression with grouped variables." *Journal of The Royal Statistical Society Series B.* 2006; 68(1):49–67.

[33]   Elhamifar E. and Vidal R. "Block-sparse recovery via convex optimization." *IEEE Transactions on Signal Processing.* 2012; 60(8): 4094 – 4107.

[34]   Huang J., Zhang T., and Metaxas D. "Learning with structured sparsity." *Journal of Machine Learning Research.* 2011; 12: 3371–3412

[35]   Hegde C., Duarte M. F., and Cevher V. "Compressive sensing recovery of spike trains using a structured sparsity model." *Proceedings of Signal Processing with Adaptive Sparse Structured Representations.* Saint-Malo, France. 2009; 1–6.

[36]   Wu X., Dong W., Zhang X., and Shi G. "Model-assisted adaptive recovery of compressed sensing with imaging applications." *IEEE Transactions on Image Processing.* 2012; 21(2): 451–458.

[37]   Baraniuk R., Cevher V., Duarte M. F., and Hegde C. "Model based compressive sensing." *IEEE Transactions on Information Theory.* 2010; 56(4): 1982–2001.

[38]   Yuan G. X., Chang K.W., Hsieh C. J., and Lin C. J. "A comparison of optimization methods and software for large-scale L1-regularized linear classification." *The Journal of Machine Learning Research.* 2010; 11:3183–3234.

[39]   Tropp J. A. and Wright S. J. "Computational methods for sparse solution of linear inverse problems." *Proceedings of the IEEE.* 2010; 98(6):948–958.

[40]   Eldar Y.C. and Kutyniok G. *Compressed sensing: Theory and applications.* New York, USA: Cambridge University Press: 2012.

[41]   Elad M. *Sparse and redundant representations: From theory to applications in signal and image processing.* New York, USA: Springer: 2010.

Chapter 2

Hybrid greedy pursuit algorithms for enhancing radar imaging quality

2.1 Introduction

Compared to convex optimization methods, the greedy algorithms are more computationally efficient and therefore more suitable for real-time radar processing. In this chapter, we will analyze the strengths and limitations of typical greedy algorithms such as orthogonal matching pursuit (OMP) [1] and subspace pursuit (SP) [2] and then present two advanced greedy algorithms, that is, the hybrid matching pursuit (HMP) [3] and the look-ahead hybrid matching pursuit (LAHMP) [4], for enhancing the quality of radar imaging. By combining the strengths of OMP and SP, the HMP algorithm ensures the orthogonality among the selected basis-signals like OMP and keeps the reevaluation of the selected basis-signals like SP. The LAHMP algorithm is an extension of HMP, based on embedding the look-ahead operation into HMP. The basis-signal selection strategy is refined in LAHMP by evaluating the effect of the basis signal selection at each iteration on the recovery error at subsequent iterations. Experimental results on real radar data demonstrate that these two advanced greedy algorithms can improve the concentration of dominant coefficients in the imaging result around the true target locations and the suppression of clutter and artifacts outside the target areas, at the cost of increased computational complexity.

2.2 Radar imaging with multiple measurement vectors

Here the through-wall radar imaging (TWRI) scenario is taken as an example. In this section, the background of TWRI with multiple measurement vectors (MMV) and the corresponding extensions of OMP and SP are reviewed.

2.2.1 Signal model

Consider an antenna array radar with Q receiving channels. At the qth receiving channel, the data are collected with $M^{(q)}$ equivalent antenna phase centers located at $\{\mathbf{r}_1^{(q)}, \mathbf{r}_2^{(q)}, \cdots, \mathbf{r}_{M^{(q)}}^{(q)}\}$ at $L^{(q)}$ frequency points $\{f_1^{(q)}, f_2^{(q)}, \cdots, f_{L^{(q)}}^{(q)}\}$, $q = 1, 2, \cdots, Q$. Consider a three-dimensional scene occupying a volume V, where the location of a voxel is denoted by vector **r**. At the qth receiving channel, the signal received by the mth phase center at the lth frequency point can be expressed as [5]

$$y_{m,l}^{(q)} = \int_V x^{(q)}(\mathbf{r}) \exp\left[-j\frac{4\pi}{c} f_l\left(\left\|\mathbf{r} - \mathbf{r}_m^{(q)}\right\|_2\right)\right] + w_{m,l}^{(q)} \tag{2.1}$$

where $x^{(q)}(\mathbf{r})$ is the reflectivity of the voxel $\mathbf{r}$, and c is the propagation speed in free-space, and $w_{m,l}^{(q)}$ is the additive noise. The amplitude spread of each scatterer in the scene and the effect of the antenna pattern have been included in $x^{(q)}(\mathbf{r})$. The frequency response characteristic of each scatterer in the scene is assumed to be constant over the frequency band at the qth receiving channel. It should be mentioned that the mismatch between (2.1) and the actual propagation through the wall may cause some image distortion and displacement, and more sophisticated through-wall models [6,7] can be used to correct this effect.

By discretizing the observed volume V as $N_x \times N_y \times N_z$ voxels, (2.1) can be rewritten as

$$\mathbf{y}^{(q)} = \mathbf{\Phi}^{(q)} \mathbf{x}^{(q)} + \mathbf{w}^{(q)} \tag{2.2}$$

where $\mathbf{x}^{(q)} = [x^{(q)}(\mathbf{r}_{(1,1,1)}), \cdots, x^{(q)}(\mathbf{r}_{(n_x,n_y,n_z)}), \cdots, x^{(q)}(\mathbf{r}_{(N_x,N_y,N_z)})]^T \in \mathbb{C}^{N \times 1}$ denotes the discrete reflectivity of the observed scene, $N \overset{\Delta}{=} N_x \times N_y \times N_z$, $\mathbf{y}^{(q)} = [y_{1,1}^{(q)}, \cdots, y_{m,l}^{(q)}, \cdots, y_{M^{(q)},L^{(q)}}^{(q)}]^T \in \mathbb{C}^{M^{(q)} L^{(q)} \times 1}$ is the measured vector at the qth receiving channel, $\mathbf{w}^{(q)}$ represents the additive noise, the dictionary $\mathbf{\Phi}^{(q)} \in \mathbb{C}^{M^{(q)} L^{(q)} \times N}$ and its elements are given by

$$\mathbf{\Phi}^{(q)}\left(l + (m-1)L^{(q)}, n_x + (n_y - 1)N_x + (n_z - 1)N_x N_y\right)$$
$$= \exp\left(-j\frac{4\pi}{c} f_l\left(\left\|\mathbf{r}_{(n_x,n_y,n_z)} - \mathbf{r}_m^{(q)}\right\|_2\right)\right) \tag{2.3}$$

where $l = 1, 2, \cdots, L^{(q)}$, $m = 1, 2, \cdots, M^{(q)}$, $n_x = 1, 2, \cdots, N_x$, $n_y = 1, 2, \cdots, N_y$, $n_z = 1, 2, \cdots, N_z$, and the position of the voxel with coordinates (n_x, n_y, n_z) is denoted by $\mathbf{r}_{(n_x,n_y,n_z)}$. The model in (2.2) is referred as to the multiple measurement vectors (MMV) with joint sparsity, that is, each one of $\{\mathbf{x}^{(1)}, \mathbf{x}^{(2)}, \cdots, \mathbf{x}^{(Q)}\}$ has only a few nonzero entries and they share a common support set indicting the locations of the nonzero entries. This holds because the radar signals at all the receiving channels are reflected from the same scene and the same discretization of the imaging space is taken at all the receiving channels. The sparsity level of $\mathbf{x}^{(q)}$ is denoted as K, and the common support set indicating the positions of the nonzero coefficients is denoted as Λ. Assume $2K < M^{(q)} L^{(q)} < N$ for $q = 1, 2, \cdots, Q$. Generally speaking, the dimension of $\mathbf{x}^{(q)}$ is required to be unique for all the channels, but the length of $\mathbf{y}^{(q)}$ for different channel index q is not required to be the same. The values of the nonzero coefficients in $\mathbf{x}^{(q)}$ may vary with the channel index q because of possible polarization diversity, spatial diversity, and frequency diversity from different receiving channels.

2.2.2 Extended OMP and SP for MMV

The original OMP and SP algorithms have been described in Chapter 1. Their extensions to the case of MMV are denoted as the simultaneous orthogonal matching pursuit (SOMP) algorithm [8,9] and the simultaneous subspace pursuit (SSP) algorithm [10], respectively. These two algorithms are detailed in Table 2.1 and Table 2.2, respectively, where:

$$\text{max\_ind}(\mathbf{x}, K) = \{\text{indices of the } K \text{ largest magnitude entries in } \mathbf{x}\} \qquad (2.4)$$

The strengths and limitations of SOMP and SSP are similar to that of OMP and SP. When the number of channels $Q = 1$, (2.2) degenerates to the model of single measurement vector (SMV), and accordingly SOMP and SSP degenerate to the original OMP and SP algorithms, respectively. In SOMP and SSP, the support set is always assumed to be common for all the channels under the joint sparsity constraint, and its estimation is based on sum fusion of the coefficients calculated at all the channels.

2.3 Hybrid matching pursuit algorithm

In this section, the strengths and limitations of OMP and SP are analyzed and the HMP algorithm [3] is presented based on the combination of the index selection strategy of OMP and the index reevaluation strategy of SP.

2.3.1 Analysis of strengths and limitations of OMP and SP

In a number of radar imaging scenarios, the basis-signals are generally Fourier-like harmonic functions, which lead to the fact that the adjacent basis-signals (i.e., the columns of the dictionary with adjacent indices) are strongly correlated. As reported in [11], with such a kind of dictionary, OMP is superior to SP in terms of

Table 2.1   The SOMP algorithm

Input:
    The measurement vectors $\mathbf{y}^{(q)}$, the dictionary $\mathbf{\Phi}^{(q)}$, for $q = 1, 2, \ldots, Q$, and the sparsity K.
Initialization:
    $\mathbf{r}^{(q)} = \mathbf{y}^{(q)}$, $\widehat{\mathbf{x}}^{(q)} = \mathbf{0}^{N \times 1}$, $\Lambda = \emptyset$, $k = 1$.
Iteration:

1. $\widetilde{i} = \underset{i \in \{1,2,\cdots,N\}}{\arg\max} \sum_{q=1}^{Q} \left| \left(\mathbf{\Phi}_i^{(q)} \right)^H \mathbf{r}^{(q)} \right|$;

2. $\Lambda = \Lambda \cup \{\widetilde{i}\}$;

3. $\widehat{\mathbf{x}}_\Lambda^{(q)} = \mathbf{\Phi}_\Lambda^\dagger \mathbf{y}^{(q)}$ and $\widehat{\mathbf{x}}_{\{1,2,\cdots,N\}-\Lambda}^{(q)} = \mathbf{0}$, for $q = 1, 2, \ldots, Q$;

4. $\mathbf{r}^{(q)} = \mathbf{y}^{(q)} - \mathbf{\Phi}\widehat{\mathbf{x}}^{(q)}$, for $q = 1, 2, \ldots, Q$

5.   If $k = K$, quit the iteration; else let $k = k + 1$ and go to step 1.

Output:
    Λ and $\widehat{\mathbf{x}}^{(q)}$, for $q = 1, 2, \ldots, Q$.

Table 2.2   The SSP algorithm

Input:
   The measurement vectors $\mathbf{y}^{(q)}$, the dictionary $\boldsymbol{\Phi}^{(q)}$, for $q = 1, 2, \ldots, Q$, and the sparsity K.

Initialization:
$$\Lambda_{\text{old}} = \text{max\_ind}\left(\sum_{q=1}^{Q}\left|(\boldsymbol{\Phi}^{(q)})^H \mathbf{y}^{(q)}\right|, K\right), \widehat{\mathbf{x}}_{\Lambda_{\text{old}}}^{(q)} = (\boldsymbol{\Phi}_{\Lambda_{\text{old}}}^{(q)})^\dagger \mathbf{y}^{(q)},$$

$$\widehat{\mathbf{x}}_{\{1,2,\cdots,N\}-\Lambda_{\text{old}}}^{(q)} = \mathbf{0}; \mathbf{r}_{\text{old}}^{(q)} = \mathbf{y}^{(q)} - \boldsymbol{\Phi}^{(q)}\widehat{\mathbf{x}}^{(q)}, \text{ for } q = 1, 2, \ldots, Q.$$

Iteration:

1. $\Lambda_{\text{temp}} = \Lambda_{\text{old}} \cup \text{max\_ind}\left(\sum_{q=1}^{Q}\left|\left(\boldsymbol{\Phi}^{(q)}\right)^H \mathbf{r}_{\text{old}}^{(q)}\right|, K\right);$

2. $\Lambda_{\text{new}} = \text{max\_ind}\left(\sum_{q=1}^{Q}\left|\left(\boldsymbol{\Phi}_{\Lambda_{\text{temp}}}^{(q)}\right)^\dagger \mathbf{y}^{(q)}\right|, K\right);$

3. $\widehat{\mathbf{x}}_{\Lambda_{\text{new}}}^{(q)} = \left(\boldsymbol{\Phi}_{\Lambda_{\text{new}}}^{(q)}\right)^\dagger \mathbf{y}^{(q)}, \widehat{\mathbf{x}}_{\{1,2,\cdots,N\}-\Lambda_{\text{new}}}^{(q)} = \mathbf{0}, \text{ and } \mathbf{r}_{\text{new}}^{(q)} = \mathbf{y}^{(q)} - \boldsymbol{\Phi}^{(q)}\widehat{\mathbf{x}}^{(q)}, \text{ for } q = 1, 2, \ldots, Q;$

4. if $\sum_{q=1}^{Q}\left\|\mathbf{r}_{\text{old}}^{(q)}\right\|_2^2 > \sum_{q=1}^{Q}\left\|\mathbf{r}_{\text{new}}^{(q)}\right\|_2^2$, let $\Lambda_{\text{old}} = \Lambda_{\text{new}}$ and $\mathbf{r}_{\text{old}}^{(q)} = \mathbf{r}_{\text{new}}^{(q)}$, and return to step 1;

    otherwise, let $\widehat{\mathbf{x}}_{\Lambda_{\text{old}}}^{(q)} = \left(\boldsymbol{\Phi}_{\Lambda_{\text{old}}}^{(q)}\right)^\dagger \mathbf{y}^{(q)}$ and $\widehat{\mathbf{x}}_{\{1,2,\cdots,N\}-\Lambda_{\text{old}}}^{(q)} = \mathbf{0}$ and quit the iteration.

Output:
   Λ_{old} and $\widehat{\mathbf{x}}^{(q)}$, for $q = 1, 2, \ldots, Q$.

---

resolution. The reason is explained as follows. At each iteration of SP, multiple basis-signals highly correlated to the measurement vector are selected simultaneously. Such a strategy may cause the loss of resolution due to the high correlation coefficients among the adjacent basis-signals. As a contrast, OMP selects only one basis-signal per iteration and ensures the orthogonality among the selected basis-signals through the whole iterative process, which is helpful to distinguish two closely spaced components.

On the other hand, OMP adds an index into the support set estimate per iteration but never removes any indices selected at past iterations. This causes a drawback that the wrong indices selected at past iterations cannot be corrected. That is to say, once an index is deemed as reliable and added to the support set estimate, there is no chance to remove it at subsequent iterations. As a result, the final radar imaging results produced by OMP may suffer from undesired artifacts outside the actual target region. As a contrast, SP is capable of correcting index selection errors at past iterations thanks to the backtracking operation at each iteration. At each iteration of SP, K indices in the previous support set estimate are merged with newly added K indices corresponding to the largest correlation coefficients between the residual and the basis-signals. This enlarges the number of basis-signal candidates to $2K$. Then the support set estimate is updated by finding K indices corresponding to the largest projection coefficients of the measurement data onto the subspace spanned by the $2K$ basis-signal candidates. By doing so, the

Table 2.3   The HMP algorithm

Input:
  The measurement vectors $\mathbf{y}^{(q)}$, the dictionary $\mathbf{\Phi}^{(q)}$, for $q = 1, 2, \ldots, Q$, and the sparsity K.

Initialization:

$\tilde{\mathbf{x}}_{omp}^{(q)} = \text{OMP}\,(\mathbf{y}^{(q)}, \mathbf{\Phi}^{(q)}, K)$, where $\tilde{\mathbf{x}}_{omp}^{(q)}$ is the output of OMP at the qth channel,

$\Lambda_{\text{old}} = \text{max\_ind}\left(\sum_{q=1}^{Q}\left|\tilde{\mathbf{x}}_{omp}^{(q)}\right|, K\right)$, $\mathbf{r}_{\text{old}}^{(q)} = \mathbf{y}^{(q)} - \left(\mathbf{\Phi}_{\Lambda_{\text{old}}}^{(q)}\right)\left(\mathbf{\Phi}_{\Lambda_{\text{old}}}^{(q)}\right)^{\dagger}\mathbf{y}^{(q)}$,   for $q = 1, 2, \cdots, Q$.

Iterations:

1. $\tilde{\mathbf{x}}_{omp}^{(q)} = \text{OMP}\,\left(\mathbf{r}_{\text{old}}^{(q)}, \mathbf{\Phi}^{(q)}, K\right)$, for $q = 1, 2, \cdots, Q$.

2. $\Lambda_{\text{temp}} = \Lambda_{\text{old}} \cup \text{max\_ind}\left(\sum_{q=1}^{Q}\left|\tilde{\mathbf{x}}_{omp}^{(q)}\right|, K\right)$.

3. $\Lambda_{\text{new}} = \text{max\_ind}\left(\sum_{q=1}^{Q}\left|(\mathbf{\Phi}_{\Lambda_{\text{temp}}}^{(q)})^{\dagger}\mathbf{y}^{(q)}\right|, K\right)$.

4. $\mathbf{r}_{\text{new}}^{(q)} = \mathbf{y}^{(q)} - \left(\mathbf{\Phi}_{\Lambda_{\text{new}}}^{(q)}\right)\left(\mathbf{\Phi}_{\Lambda_{\text{new}}}^{(q)}\right)^{\dagger}\mathbf{y}^{(q)}$, for $q = 1, 2, \cdots, Q$.

5.   if $\sum_{q=1}^{Q}\left\|\mathbf{r}_{\text{old}}^{(q)}\right\|_{2}^{2} > \sum_{q=1}^{Q}\left\|\mathbf{r}_{\text{new}}^{(q)}\right\|_{2}^{2}$, let $\Lambda_{\text{old}} = \Lambda_{\text{new}}$, $\mathbf{r}_{\text{old}}^{(q)} = \mathbf{r}_{\text{new}}^{(q)}$, and return to step 1; otherwise,

  let $\hat{\mathbf{x}}_{\Lambda_{\text{old}}}^{(q)} = (\mathbf{\Phi}_{\Lambda_{\text{old}}}^{(q)})^{\dagger}\mathbf{y}^{(q)}$ and $\hat{\mathbf{x}}_{\{1,2,\cdots,N\}-\Lambda_{\text{old}}}^{(q)} = 0$ and quit the iteration.

Output:
  Λ_{old} and $\hat{\mathbf{x}}^{(q)}$, for $q = 1, 2, \ldots, Q$.

---

reliability of the basis-signals selected at past iterations is re-evaluated, which makes it possible to remove poor index estimates from and to add new reliable index candidates into the support set estimate.

2.3.2   Algorithm description

The above analysis inspires the design of the HMP algorithm [3], which combines the index selection strategy of OMP and the index reevaluation strategy of SP. The steps of HMP are summarized in Table 2.3. Through the HMP algorithm, the OMP algorithm is used to solve for the local sparse solutions, and the global estimate of the common support set is carried out by fusion of all local solutions. In the initialization phase, Q sparse solutions are obtained by independently applying the OMP algorithm on every local data subset. Then, the global estimate of the common support set, denoted as Λ_{old}, is determined by finding K indices corresponding to the largest coefficients in the fusion result of all the local solutions. By using the common support set estimate Λ_{old}, the local residual vectors are updated by projecting the local data subset $\mathbf{y}^{(q)}$ onto the complement of the column space of $\mathbf{\Phi}^{(q)}$, for $q = 1, 2, \ldots, Q$. At each iteration, Q sparse solutions are obtained by independently applying the OMP algorithm on every local residual vector. The support set estimate is merged with K indices corresponding to the largest entries in the fusion result of all the local OMP outputs. As a result, the enlarged set Λ_{temp}

contains no more than $2K$ indices. Then, the projection coefficients of the local measurement data $\mathbf{y}^{(q)}$ onto the column subspace $\boldsymbol{\Phi}_{\Lambda_{\text{temp}}}^{(q)}$ are calculated for $q = 1, 2,$ ..., Q. The set Λ_{new}, which is composed of K indices corresponding to the largest entries in the fusion result of all the local projection coefficients, is considered as the refined estimate of the common support set. Accordingly, the residual vectors are updated by projecting the local measurement data $\mathbf{y}^{(q)}$ onto the complement of the column subspace of $\boldsymbol{\Phi}_{\Lambda_{\text{new}}}^{(q)}$, for $q = 1, 2, \ldots, Q$.

It is clear that the HMP is a combination of the operations of OMP and SP algorithms. The strategy of local index selection in HMP follows OMP, that is, the basis-signals are selected one by one. This guarantees the orthogonality among all the selected basis-signals and, therefore, the capability of distinguishing closely spaced components with a Fourier-like dictionary. The backtracking operations in HMP are similar to that in SP. That is, the support set estimate is first enlarged by adding K new candidates and then refined by finding K largest projection coefficients. This can remove poor indices chosen at past iterations and add new potential index candidates to the support set estimate. From the above observations, the performance of HMP is expected to be better than that of OMP and SP and their extended versions for MMV, that is, SOMP and SSP, in radar imaging applications.

Similar to the existing greedy algorithms, the HMP algorithm also requires the sparsity K as one of its inputs. As suggested in [12], the estimate $K \approx \max\{M^{(q)}L^{(q)}, q = 1, 2, \cdots, Q\}/(2 \log N)$ is often reasonable for the successful recovery of most sparse signals. A more accurate approach of choosing the value of K is to run a greedy algorithm using a range of sparsity levels such as $K = 1, 2, 4, \ldots, \max\{M^{(q)}L^{(q)}, q = 1, 2, \cdots, Q\}$ and to select the satisfactory sparse recovery result, at the cost that the computational complexity increases by a factor no more than $O(\log M)$ [12]. If the change of the recovered error is no longer significant with the increase of the value of K, the current value of K can be deemed sufficient. That is to say, the dimension of the determined subspace is large enough to contain the desired dominant components of the received signal.

2.3.3   Computational complexity of HMP

The comparison among Table 2.1, Table 2.2 and Table 2.3 clearly indicates that HMP is more computationally expensive than SOMP and SSP. The quantitative comparison of the computational complexities of SOMP, SSP, and HMP is provided as follows. For simplicity of expression, we assume that $M^{(q)} = M$, $L^{(q)} = L$, for $q = 1, 2, \cdots, Q$. We first look at the computational complexities of SOMP and SSP. As reported in [13], for the scale of the data collected at the qth channel, the computational complexities of OMP and SP are given by $O(KMLN)$. Compared to OMP and SP, the extra computational burdens of SOMP and SSP come from the fusion of all of the local correlation and projection coefficients. The fusion procedure at each iteration in SOMP and SSP includes two steps, that is, summation of multiple vectors and search for large coefficients, which costs $O(QN)$ floating point operations per iteration. Since OMP and SSP require K and $O(K)$ iterations, respectively, the total complexities of the fusion in SOMP and SSP are given by

$O(KQN)$. Assume $Q \ll ML$, that is, the number of the receiving channels is much smaller than the number of measurements at each channel, which is applicable in most radar applications. Accordingly, the computational burden of the fusion in SOMP and SSP are marginal in comparison with the computations on the data at all the receiving channels. Once the support set is determined, the final sparse solution based on least square estimation requires $O(K^2ML)$ floating point operations, if the modified Gram–Schmidt algorithm [14] is used to implement the QR decomposition. Therefore, the total computational complexities of SOMP and SSP are given by $O(KMLNQ)$.

Next, we analyze the computational complexity of HMP. The HMP algorithm contains four types of computations:

1. The independent implementations of OMP at all the receiving channels cost $O(KMLNQ)$ floating point operations;
2. The complexity of fusing the local correlation and projection coefficients is $O(QN)$;
3. The projections at all the receiving channels cost $O(K^2MLQ)$ when the modified Gram–Schmidt algorithm is used; and
4. After the support set is determined, the final sparse solution based on least square estimation requires $O(K^2ML)$ floating point operations.

Under the assumption $Q \ll ML$ that and $2K < ML < N$, the complexity of HMP is $(1 + T_{\mathrm{HMP}})O(KMLNQ)$, where T_{HMP} is the required number of iterations. From Table 2.3, we can see $T_{\mathrm{HMP}} \geq 1$ because at least one iteration is required for HMP to check the termination criterion. Since HMP is the combination of SOMP and SSP, the number of iterations of HMP should be of the same order as that of SOMP and SSP, that is, $T_{\mathrm{HMP}} = O(K)$. Therefore, the total computational complexity of HMP is given by $O(K^2MLNQ)$.

2.3.4 Experimental results

In this section, the HMP algorithm is evaluated by using real through-wall radar data. The data were collected with the lightweight flat-panel array [15,16] designed by the ElectroScience Laboratory at The Ohio State University. This antenna array is 55.25 cm wide, 45.72 cm high and 4.44 cm thick, and it consists of 12 dual-polarized antenna elements. Each antenna element not only operates in monostatic mode individually but also cooperates with adjacent ones in bistatic mode to form virtual elements. Thus, there are 37 equivalent phase centers (12 real and 25 virtual), as depicted in Figure 2.1. The array operates from 0.9 to 2.3 GHz with 101 equally distributed frequency points.

The observed scene consists of a volume containing two 22.86 cm trihedral targets suspended 52 cm above the ground and located approximately 3.2 m from the radar, as depicted in Figure 2.2. Background subtraction is used to remove the direct echo reflected from the wall (in the through-wall scenario) as well as background clutter from the room, by performing a difference measurement without the trihedral targets present. This allows us to focus on the isolated targets. However, as

(a)

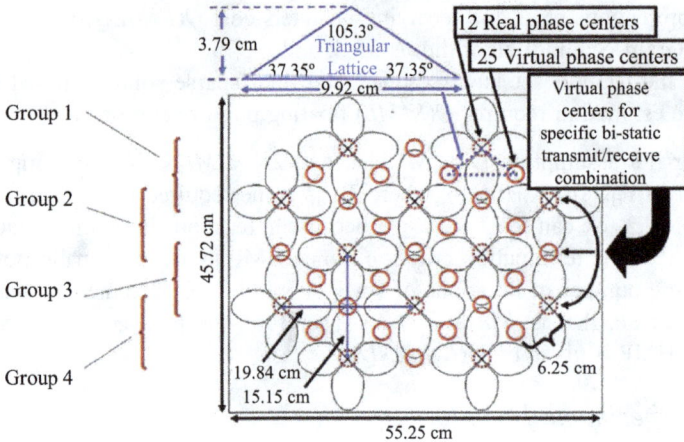

(b)

*Figure 2.1   The configuration of the antenna array radar: (a) the photograph of
the antenna array and (b) the locations of virtual (solid circles) and
real (dashed circles) array phase centers*

shown in [17], high-observable targets such as trihedrals clearly stand out in the
images even in the presence of the background. In that case, it is advisable to first
isolate the region of interest (ROI) in the image and filter out all scattering con-
tributions outside ROI. Other algorithms for wall-clutter removal such as the spatial
filtering [18] and the singular value decomposition (SVD) [19] can also be used
before the image formation. The center-to-center cross-range spacing between
these two trihedral targets is 60 cm.

Our goal is to form the composite through-wall radar image from multiple
receiving channels, for more effectively enhancing the target reflection and

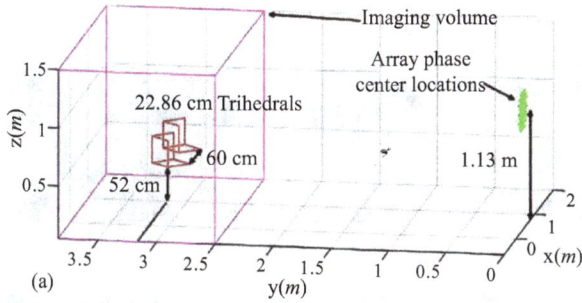

(a)

(b)

(c)

Figure 2.2   Imaging scene consisting of two closely spaced trihedral targets: (a) scene layout, (b) free-space scenario, and (c) through-wall scenario

attenuating the background clutter. Thus, the radar images shown in the rest of this chapter are actually the result of $\sum_{q=1}^{Q} |\hat{\mathbf{x}}(q)|$ produced by different algorithms. As a reference, the cross-range resolution of the traditional back projection (BP) algorithm for point-like targets 3.2 m away from the utilized radar system is approximately 76 cm. Thus, the two trihedrals are not resolvable by using the BP algorithm. All the phase centers are divided into $Q = 4$ groups with overlap in the vertical direction, as shown in Figure 2.1(b). The numbers of antenna phase centers in these four groups are 12, 14, 14, and 12, respectively. The data subsets collected from different groups of phase centers have the same sparsity pattern, which allows us to utilize the HMP algorithm and other CS algorithms based on the MMV model.

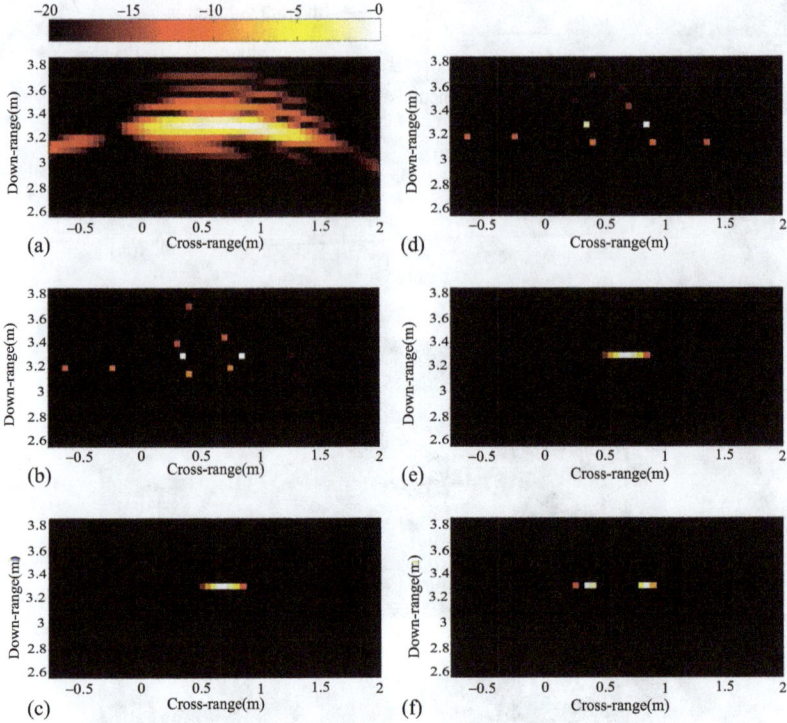

Figure 2.3   Imaging results in free-space scenario by (a) BP, (b) OMP, (c) SP,
(d) SOMP, (e) SSP, and (f) HMP

It should be mentioned that the overlapping among different groups of antenna phase centers is optional. The division of the phase centers in Figure 2.1 aims to enhance the azimuth resolution via setting several arrays with large azimuth apertures, at the cost of a reduction of the vertical resolution of each group of antennas.

The top-view imaging results of six different algorithms in the free-space scenario are provided in Figure 2.3. The volume of interest is discretized as $57 \times 27 \times 31$ voxel cells. All the measurement data collected at 37 antenna phase centers and 101 frequency points are used to form the image of the observed scene, and all the phase centers are divided into $Q = 4$ groups as described above. Here, the algorithms based on convex optimization are not considered because of their considerably high complexity for such a problem size. Figure 2.3(a)–(c) is obtained by using the SMV model, in which the full measurement data are arranged as a single vector, while Figure 2.3(d)–(f) corresponds to the MMV model, in which the data subsets are collected by four groups of antenna phase centers. As revealed in Figure 2.3(a), the BP algorithm based on matching filtering cannot distinguish the two closely spaced targets, because its cross-range resolution is larger than the actual spacing of the targets. In Figure 2.3(b)–(f), the sparsity is set to $K = 20$ for all the greedy algorithms used here. Such a value of sparsity is sufficient to image the

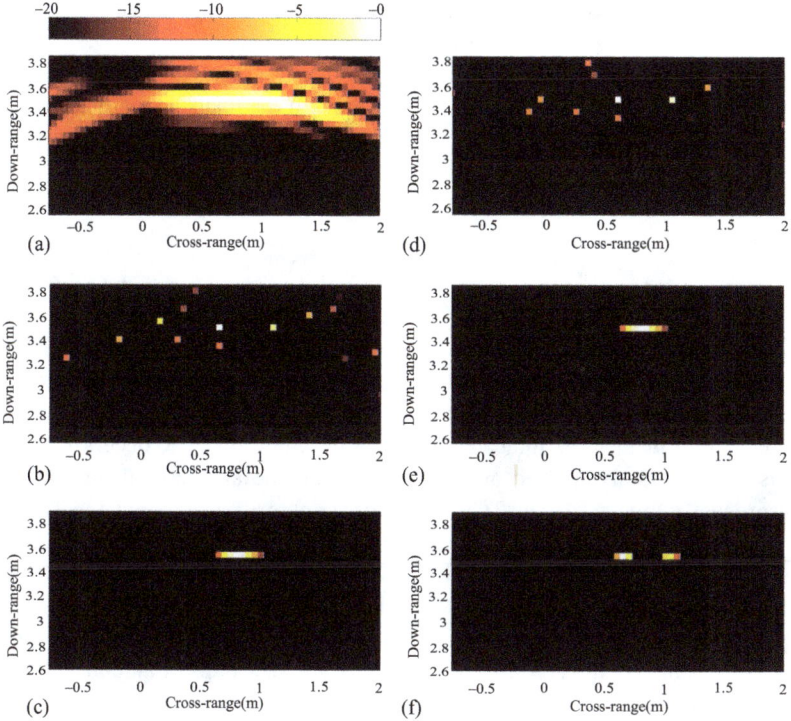

Figure 2.4   Imaging results in through-wall scenario by (a) BP, (b) OMP, (c) SP, (d) SOMP, (e) SSP, and (f) HMP

scene of interest since the number of dominant pixels in Figure 2.3 (b)–(f) is much smaller than 20. In Figure 2.3(b), OMP finds two point-like scatterers but suffers from many artifacts that appear outside the target areas. In Figure 2.3(c), SP suppresses artifacts well but fails to separate the two targets. Based on the MMV model, SOMP and SSP suffer from similar issues with OMP and SP, respectively, as shown in Figure 2.3(d) and (e). The imaging result of the HMP algorithm is given in Figure 2.3(f), where the two targets are clearly distinguished with no visible artifacts.

The above experiment is repeated for a through-wall imaging scenario with the same parameter settings. A 20.32 cm thick cinder block wall is placed between the radar and the trihedral targets, 97 cm from the radar, as shown in Figure 2.2(c). The imaging results of the six algorithms in the through-wall scenario are presented in Figure 2.4. Figure 2.4(a)–(c) is obtained by using the SMV model, while Figure 2.4(d)–(f) corresponds to the MMV model. From Figure 2.3(a) and Figure 2.4(a), we can observe the change of the clutter and the displacement of the target position due to the presence of the wall. Similar to the previous experiment, OMP and SOMP are disturbed by artifacts while SP and SSP suffer from low-resolution. Even if the presence of the wall leads to a large increase in propagation complexity, HMP is still capable of resolving the two closely spaced targets behind the wall as well as suppressing artifacts.

To quantitatively evaluate the performances of the algorithms mentioned above, the target-to-clutter ratio (TCR) is used as a measure of quality of the reconstructed images. The definition of TCR is given by the ratio between the average value of the maximal magnitudes in the two target areas and the maximal magnitude in the rest of the reconstructed image:

$$\text{TCR} = \frac{\max_{(x,y)\in R_{t,1}}\left|I(x,y)\right| + \max_{(x,y)\in R_{t,2}}\left|I(x,y)\right|}{2 \cdot \max_{(x,y)\notin R_{t,1}\cup R_{t,2}}\left|I(x,y)\right|} \tag{2.5}$$

where $I(x,y)$ is the magnitude of the pixel (x,y) in the reconstructed image, $R_{t,i}$ is the ith target area consisting of 25 pixels around the true position of the ith target, $i = 1$, 2. The definition in (2.5) indicates that a large value of TCR occurs when both targets are accurately located and the clutter and the artifacts outside the target areas are well suppressed. The TCR values of six algorithms are given in Table 2.4. The term "4 groups of antennas" refers to the case where all four groups of antenna phase centers in Figure 2.1 are active and corresponds to the imaging results in Figure 2.3 and Figure 2.4. The term "3 groups of antennas" refers to the case where the phase center group No. 4 is not activated and the other three groups of antenna phase centers in Figure 2.1 are available. The term "2 groups of antennas" refers to the case where only the phase center groups No. 1 and No. 2 in Figure 2.1 are active. For the BP algorithm, the TCR values are always around 0 dB in Table 2.4.

Table 2.4   TCR values yielded by six algorithms (in dB): (a) in free-space and (b) in through-wall case

| | **(a) Free-space** | | |
	Four groups of antennas	Three groups of antennas	Two groups of antennas
BP	−0.09	−0.10	0.14
SP	−8.36	−7.96	−8.03
SSP	−7.40	−8.44	−8.99
OMP	9.09	8.07	8.08
SOMP	9.88	9.48	8.15
HMP	33.22	20.64	9.57
	(b) Through-wall		
	Four groups of antennas	Three groups of antennas	Two groups of antennas
BP	0.20	0.35	0.32
SP	−4.23	−4.22	−4.23
SSP	−4.23	−4.21	−4.24
OMP	5.48	4.59	4.19
SOMP	5.84	4.84	4.69
HMP	37.23	27.63	8.07

This indicates that the maximal magnitudes inside and outside the target areas are comparable due to the poor resolution of BP, as observed from Figure 2.3 and Figure 2.4. Although SP and SSP suppress clutter and artifacts well, their cross-range resolutions are not sufficient to distinguish these two closely spaced targets. As seen in Figure 2.3 and Figure 2.4, SP and SSP trend to merge the images of the two targets into one, which lowers the pixel magnitudes inside the target areas and raises an undesired magnitude peak outside the target areas. This is the reason why SP and SSP always provide negative values of TCR in Table 2.4. Thanks to the orthogonality among the selected basis-signals, OMP and SOMP can accurately locate the two closely spaced targets and, therefore, provide better image quality with positive TCR values. However, OMP and SOMP suffer from strong artifacts outside the target areas, as can be seen in Figure 2.3 and Figure 2.4. This is the reason why a significant increase in the amount of measurements only brings a slight improvement of TCR for OMP and SOMP. Among these six algorithms, the HMP algorithm provides the highest value of TCR in each experimental scenario, which implies that HMP achieves more accurate localization of the targets and better suppression of the clutter and the artifacts. Moreover, the image quality generated by HMP is significantly improved as the number of data subsets increases. This characteristic of HMP is expected to be attractive when a large-scale antenna array is used.

2.4   Look-ahead hybrid matching pursuit algorithm

As demonstrated in Section 2.3, HMP combines the strengths of OMP and SP, that is, the orthogonality of the selected basis-signals and the reevaluation of the basis-signals selected at the past iterations. In the iterative process of HMP, new basis-signal candidates are selected to directly expand the previous estimate of the support set. However, such a straightforward expansion of the basis-signal candidate set cannot guarantee a decrease of the recovery error. Namely, some unreliable entries in the basis-signal candidate set may increase the recovery error at the next iteration. This inspires us to evaluate the effect of the basis-signal selection on the future recovery error before the expansion of the basis-signal candidate set so that more reliable basis-signals can be selected at each iteration. To do so, in this section, we extend HMP by embedding the look-ahead strategy into the iterative process of HMP and refer to it as the LAHMP algorithm [4].

The first application of the look-ahead strategy was presented in [20], where an optimal basis-signal is selected from the set of candidates by appraising its effect on the final performance in the sense of minimizing the recovery error at the end of all the future iterations. As reported in [20], the look-ahead strategy can improve the reliability of the selected basis-signals at the cost of considerable computational complexity. From the consideration of the tradeoff between performance and complexity, before the expansion of the previous support set, LAHMP only evaluates the effect of the basis-signal candidate selection on the recovery error at the next iteration instead of checking the end of the entire iterative process.

Table 2.5   The LAAS algorithm

Input:
    The measurement vectors $\mathbf{y}^{(q)}$, the dictionary $\mathbf{\Phi}^{(q)}$, for $q = 1, 2, \ldots, Q$, the sparsity K, previous estimate of the support set Λ_{old}, and index candidate set Λ_c

Initialization:
    $\mathbf{n} = \{n_1, n_2, \cdots, n_K\} \leftarrow \mathbf{0}^{K \times 1}$.

Iteration:
    for $k = 1 : K$

1. $\Lambda_{test} = \Lambda_{\text{old}} \cup \Lambda_c(k)$;

2. $n_k = \sum\limits_{q=1}^{Q} \left\| \mathbf{y}^{(q)} - \left(\mathbf{\Phi}_{\Lambda_{test}}^{(q)} \right) \left(\mathbf{\Phi}_{\Lambda_{test}}^{(q)} \right)^{\dagger} \mathbf{y}^{(q)} \right\|_2^2$.

    end for

Output:
    $i = \Lambda_c(\widehat{k})$, where $\widehat{k} = \arg\min\limits_{k}\{n_k, k = 1, 2, \cdots, K\}$.

2.4.1   Algorithm description

Compared to the HMP algorithm, the additional operation in the LAHMP algorithm is the employment of the look-ahead strategy. The look-ahead strategy is described in Table 2.5, which selects an optimal basis-signal/index from the candidate set by evaluating its effect on the future recovery error in the sense of minimizing the residual error after the support set expansion. This algorithm is referred to as look-ahead atom selection (LAAS). In Table 2.5, the input variables include the measurements $\{\mathbf{y}^{(q)}, \mathbf{\Phi}^{(q)}, q = 1, 2, \cdots, Q\}$, the previous estimate for the support set Λ_{old}, the sparsity level K, and the candidate support set Λ_c containing K indices. First, the basis-signal indexed by the kth element of Λ_c is merged into the previous support set, and a temporary set is formed as $\Lambda_{\text{test}} = \Lambda_{\text{old}} \cup \Lambda_c(k)$. Then by projecting the measurements onto the subspace spanned by the columns indexed by Λ_{test} for each receiving channel, the total residual error of all receiving channels, denoted as n_k, can be calculated. It is clear that n_k represents the effect of the basis-signal selection on the recovery error. After testing all the elements of Λ_c, the index corresponding to the smallest value in $\{n_k, k = 1, 2, \cdots, K\}$ is regarded as the optimal selection. When more than one indices correspond to the minimum recovery error, one among them is arbitrarily determined. A compact notation of LAAS is defined as

$$i = \text{LAAS}\left(\{\mathbf{y}^{(q)}, \mathbf{\Phi}^{(q)}, q = 1, 2, \cdots, Q\}, K, \Lambda_{\text{old}}, \Lambda_c \right) \tag{2.6}$$

By embedding LAAS into HMP, the LAHMP algorithm is summarized in Table 2.6. Similar to existing greedy algorithms based on the MMV model, LAHMP algorithm also aims to reconstruct the common support set. The initialization phase of LAHMP is the same with that of HMP. After the initialization, three main steps are contained in the iterative process:

1.   Index selection. The locally solved indices are obtained by utilizing the OMP algorithm at each channel, and the indices corresponding to the largest K

Table 2.6 The LAHMP algorithm

Input:

    The measurement vectors $\mathbf{y}^{(q)}$, the dictionary $\boldsymbol{\Phi}^{(q)}$, for $q = 1, 2, \ldots, Q$, and the sparsity K.

Initialization:

1. $\mathbf{x}'^{(q)}_{\text{omp}} = \text{OMP}(\mathbf{y}^{(q)}, \boldsymbol{\Phi}^{(q)}, K)$, where $\mathbf{x}'^{(q)}_{\text{omp}}$ is the output of the standard OMP algorithm at the qth receiving channel, for $q = 1, 2, \cdots, Q$.

2. $\Lambda_{\text{old}} = \text{max\_ind}\left(\sum_{q=1}^{Q}\left|\mathbf{x}'^{(q)}_{\text{omp}}\right|, K\right)$.

3. $\mathbf{r}^{(q)}_{\text{old}} = \mathbf{y}^{(q)} - \left(\boldsymbol{\Phi}^{(q)}_{\Lambda_{\text{old}}}\right)\left(\boldsymbol{\Phi}^{(q)}_{\Lambda_{\text{old}}}\right)^{\dagger}\mathbf{y}^{(q)}$, for $q = 1, 2, \cdots, Q$.

Iteration:

4. $\mathbf{x}''^{(q)}_{\text{omp}} = \text{OMP}\left(\mathbf{r}^{(q)}_{\text{old}}, \boldsymbol{\Phi}^{(q)}, K\right)$, for $q = 1, 2, \cdots, Q$.

5. $\Lambda_c = \text{max\_ind}\left(\sum_{q=1}^{Q}\left|\mathbf{x}''^{(q)}_{omp}\right|, K\right)$.

6. $i = \text{LAAS}\left(\{\mathbf{y}^{(q)}, \boldsymbol{\Phi}^{(q)}, q = 1, 2, \cdots, Q\}, K, \Lambda_{\text{old}}, \Lambda_c\right)$.

7. $\Lambda_{\text{temp}} = \Lambda_{\text{old}} \cup \{i\}$.

8. $\Lambda_{\text{new}} = \text{max\_ind}\left(\sum_{q=1}^{Q}\left|(\boldsymbol{\Phi}^{(q)}_{\Lambda_{\text{temp}}})^{\dagger}\mathbf{y}^{(q)}\right|, K\right)$.

9. $\mathbf{r}^{(q)}_{\text{new}} = \mathbf{y}^{(q)} - (\boldsymbol{\Phi}^{(q)}_{\Lambda_{\text{new}}})(\boldsymbol{\Phi}^{(q)}_{\Lambda_{\text{new}}})^{\dagger}\mathbf{y}^{(q)}$, for $q = 1, 2, \cdots, Q$

10.   if $\sum_{q=1}^{Q}\left\|\mathbf{r}^{(q)}_{\text{old}}\right\|_2^2 > \sum_{q=1}^{Q}\left\|\mathbf{r}^{(q)}_{\text{new}}\right\|_2^2$, $\mathbf{r}^{(q)}_{\text{old}} = \mathbf{r}^{(q)}_{\text{new}}$, $\Lambda_{\text{old}} = \Lambda_{\text{new}}$, and return to step 4;

    otherwise, let $\widehat{\mathbf{x}}_{\Lambda_{\text{old}}}(q) = (\boldsymbol{\Phi}^{(q)}_{\Lambda_{\text{old}}})^{\dagger}\mathbf{y}^{(q)}$ and $\widehat{\mathbf{x}}^{(q)}_{\{1,2,\cdots,N\}-\Lambda_{\text{old}}} = \mathbf{0}$, and quit the iteration.

Output:

    Λ_{old} and $\widehat{\mathbf{x}}(q)$, for $q = 1, 2, \ldots, Q$.

---

entries in the sum fusion result of all the local OMP outputs are put into the candidate support set Λ_c.

2. Support set expansion. The LAAS algorithm is utilized to evaluate each element in Λ_c. The output of LAAS, that is, the index i corresponding to the smallest future recovery error, is merged into the previous estimate of the support set to form the temporarily expanded support set Λ_{temp}.

3. Pruning. This is the same with the reevaluation operation in HMP. The new global estimation of common support set Λ_{new} is obtained by finding the K indices corresponding to the largest coefficients in the sum fusion result of the projection coefficients of all the receiving channels, and the local residual vectors are accordingly updated by using Λ_{new}.

The iterations over all the receiving channels are terminated when the global recovery error no longer decreases. As for the setting of the sparsity level K in LAHMP, the same procedure used in HMP can be carried out, as discussed in Section 2.3.2.

2.4.2 The effect of the look-ahead operation

In step 2 of HMP provided in Table 2.3, the previous support set estimate Λ_{old} is expanded by adding K indices newly selected by the standard OMP algorithm. This expansion does not consider the effect of newly added basis-signals on the recovery error in the future and, therefore, cannot guarantee continuous reduction of the total recovery error. In contrast, the look-ahead operation in LAHMP selects an optimal basis-signal per iteration by evaluating its effect on the future reconstruction quality. This provides a way to continuously reduce the total recovery error during the iterative process.

Let us look at step 3 in HMP and step 8 in LAHMP, both of which solve a least:

squares approximation, that is, $\left[\left(\mathbf{\Phi}_{\Lambda_{\text{temp}}}^{(q)} \right)^H \left(\mathbf{\Phi}_{\Lambda_{\text{temp}}}^{(q)} \right) \right]^{-1} (\mathbf{\Phi}_{\Lambda_{\text{temp}}}^{(q)})^H \mathbf{y}^{(q)}$. Here the stability of the least squares approximation before and after embedding LAAS is analyzed. For a least squares problem:

$$\boldsymbol{\alpha} = \mathbf{\Psi}\boldsymbol{\beta} \tag{2.7}$$

where $\boldsymbol{\alpha} \in \mathbb{C}^{a \times 1}$, $\mathbf{\Psi} \in \mathbb{C}^{a \times b}$, $\boldsymbol{\beta} \in \mathbb{C}^{b \times 1}$, and $a > b$, the stability of the solution depends on how well conditioned $\mathbf{\Psi}$ is, which is determined by the condition number $\kappa(\mathbf{\Psi})$ [21]. That is

$$\frac{\|\Delta \mathbf{x}\|}{\|\mathbf{x}\|} < \frac{\kappa(\mathbf{\Psi})}{1 - \kappa(\mathbf{\Psi})\frac{\|\Delta \mathbf{\Psi}\|}{\|\mathbf{\Psi}\|}} \left(\frac{\|\Delta \mathbf{\Psi}\|}{\|\mathbf{\Psi}\|} + \frac{\|\Delta \mathbf{y}\|}{\|\mathbf{y}\|} \right) \tag{2.8}$$

where $\Delta(\cdot)$ denotes a perturbed term. A smaller value of $\kappa(\mathbf{\Psi})$ is helpful to guarantee a stable least squares solution [21]. Let $\{\mu_1, \mu_2, \cdots, \mu_R\}$ denote the nonzero singular values of $\mathbf{\Psi}$ arranged in nonincreasing order, where $R = \text{rank}(\mathbf{\Psi})$. The condition number $\kappa(\mathbf{\Psi})$ can be expressed as [21]:

$$\kappa(\mathbf{\Psi}) = \frac{\mu_1}{\mu_R} \tag{2.9}$$

Considering the least squares problems in HMP and LAHMP, now we compare the condition numbers of $\mathbf{\Phi}_{\Lambda_{\text{temp}}}^{(q)}$ before and after using the look-ahead operation.

First, a theorem from [22] is reviewed.

Theorem 2.1 *[22] Given a matrix* $\mathbf{\Psi} \in \mathbb{C}^{a \times b}$, *let* $\widetilde{\mathbf{\Psi}} \in \mathbb{C}^{a \times v}$ *be the matrix generated by removing any* $b - v$ *columns from* $\mathbf{\Psi}$, *where* $1 \leq v < b$. *Let* $\{\mu_1, \mu_2, \cdots, \mu_R\}$ *and* $\{\widetilde{\mu}_1, \widetilde{\mu}_2, \cdots, \widetilde{\mu}_{\widetilde{R}}\}$ *denote the singular values of* $\mathbf{\Psi}$ *and* $\widetilde{\mathbf{\Psi}}$, *respectively, both arranged in nonincreasing order, where* $\widetilde{R} = \text{rank}(\widetilde{\mathbf{\Psi}})$. *If* $a \geq b$, *for any integer* d *with* $1 \leq d \leq \widetilde{R}$, *the following conclusion holds:*

$$\mu_d \geq \widetilde{\mu}_d \geq \mu_{d+R-\widetilde{R}} \tag{2.10}$$

By utilizing the definition of the condition number in (2.9), it can be deduced that

$$\kappa(\mathbf{\Psi}) = \frac{\mu_1}{\mu_R} = \frac{\mu_1}{\mu_{\widetilde{R}+R-\widetilde{R}}} \geq \frac{\mu_1}{\widetilde{\mu}_{\widetilde{R}}} \geq \frac{\widetilde{\mu}_1}{\widetilde{\mu}_{\widetilde{R}}} = \kappa(\widetilde{\mathbf{\Psi}}) \tag{2.11}$$

□

At each iteration of HMP, where the look-ahead operation is not embedded, $\Phi_{\Lambda_{\text{temp}}}^{(q)}$ contains no more than $2K$ columns. In contrast, in LAHMP, no more than $K+1$ columns are involved in $\Phi_{\Lambda_{\text{temp}}}^{(q)}$ after using the look-ahead operation. From the above Theorem 2.1, it follows that the embedding of the look-ahead operation lowers the condition number of $\Phi_{\Lambda_{\text{temp}}}^{(q)}$. Thus, the step 8 of LAHMP can produce more stable projection coefficients to form the refined support set estimate Λ_{new}, which will be used at the next iteration. This is the reason why the look-ahead operation is beneficial to the next iteration. Accordingly, it is expected that LAHMP outperforms HMP in terms of the accurate selection of basis-signals corresponding to the true dominant scatterers in the scene to be imaged.

2.4.3 Computational complexity of LAHMP

In this section, we discuss the computational complexity of the LAHMP algorithm. Since one new basis-signal is added into the previous support set estimate at every iteration of LAHMP, the number of iterations required for LAHMP is about $O(K)$. As discussed in Section 2.3.3, HMP also requires $O(K)$ iterations. However, at each iteration of HMP, more than one basis-signal (K basis-signals) are merged into the previous support set estimate. Inevitably, in comparison with HMP, more iterations are required for LAHMP to select all correct basis-signals. The new operation in the LAHMP is the application of LAAS. In LAAS, averaging the norm of the residual vectors is performed K times and each calculation costs $O(K^2QML)$ by using the modified Gram–Schmidt algorithm [14]. Hence, the additional computational complexity owing to LAAS is about $O(K^4QML)$ through the iterative process of LAHMP. In step (3) of HMP and step (8) of LAHMP, the least squares estimations require $(2K)^2QML$ and $(K+1)^2QML$ floating point operations, respectively. That is to say that embedding LAAS into HMP reduces the computational complexity of the least squares estimation per iteration. Thus, when LAHMP is terminated, the reduced computational complexity caused by the embedding of LAAS is $O(K^3QML)$, which is negligible compared to the additional computational complexity. Since the complexity of HMP is $O(K^2NQML)$, it is clear that the total computational complexity of LAHMP is $O(K^2(K^2+N)QML)$. As analyzed in Section 2.3.3, the computational complexities of SOMP and SSP are about $O(KMLNQ)$. We can see that the computational complexity of LAHMP is several times higher than that of other greedy algorithms. Compared to HMP, the increase in the computational complexity of LAHMP is affordable when $K^2 \ll N$. It is worth mentioning that the computational complexity of LAHMP is still lower than that of convex optimization algorithms.

2.4.4 Experimental results

In this section, four greedy algorithms, that is, LAHMP, HMP, SOMP, and SSP, are compared by using real through-wall radar data. The data were collected by the Radar Imaging Lab of the Center for Advanced Communications at Villanova University [23]. A stepped frequency radar is used to acquire multipolarization data. The radar bandwidth is 1 GHz, the center frequency is 2.5 GHz, and the frequency step size of 5 MHz. Here the measurements from three polarimetric channels, S11 (HH), S12 (VV), and S22 (HV) are used to evaluate the above greedy algorithms. At each polarimetric channel, the data comprises 201 frequency points and 69 antenna positions. Two horn antennas were mounted side-by-side on a field probe scanner, one oriented for vertical polarization and the other for horizontal polarization. The transmit power of signal was set to 5 dBm. The synthetic aperture length is 1.51 m in cross-range and the imaging scene includes one 6-inch trihedral, three 3-inch trihedrals, three

(a)

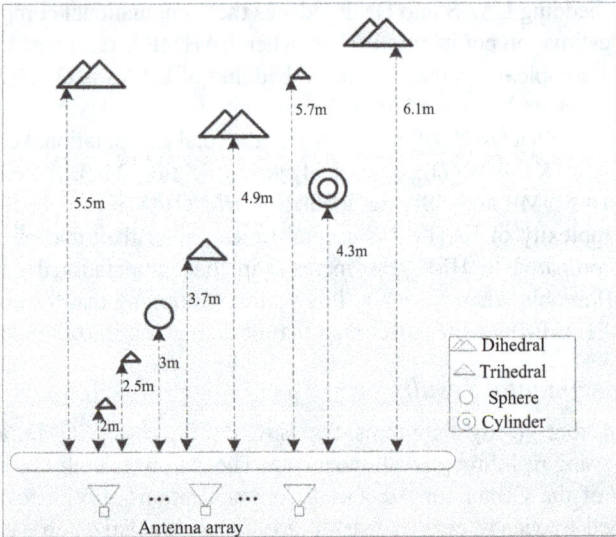

(b)

Figure 2.5   The scene to be imaged: (a) the image depicting the nine targets and (b) the ground-truth image

12-inch dihedrals, a 12-inch diameter sphere, and a 3-inch diameter cylinder, all placed at different spatial positions as shown in Figure 2.5. The through-wall data were collected through a 127 mm thick nonhomogeneous plywood and gypsum board wall, which is positioned 1.27 cm in down-range from the front of the antennas. The data in free-space scenario were collected in the same setup but without the wall. The observed down-range-cross-range plane to be imaged is discretized as 121×81 pixels so that the resolution of the scene is 0.05 m $\times$ 0.05 m, which is comparable with the

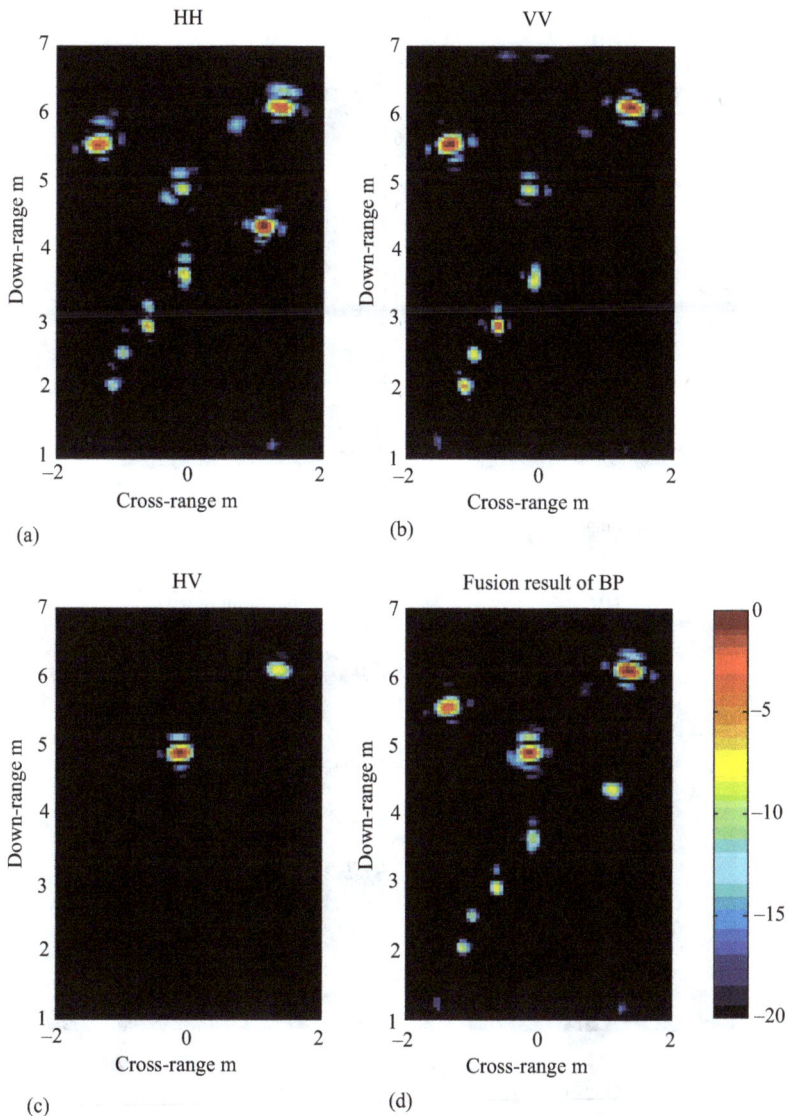

Figure 2.6 Imaging results produced by the BP algorithm in free-space scenario: (a) HH, (b) VV, (c) HV, and (d) composite image by sum fusion

size of the smallest 3-inch trihedral in the scene. In all the following imaging results, the imaging plane is at the height of 1.06 m above the ground.

The imaging results of the traditional BP algorithm based on the full measurement data in free-space and through-wall scenarios are shown in Figure 2.6 and Figure 2.7, respectively, where the color scale is in dB relative to the peak magnitude. From Figure 2.6 and Figure 2.7, one can observe the common sparsity pattern and the diversity of the scattering behavior at different polarization

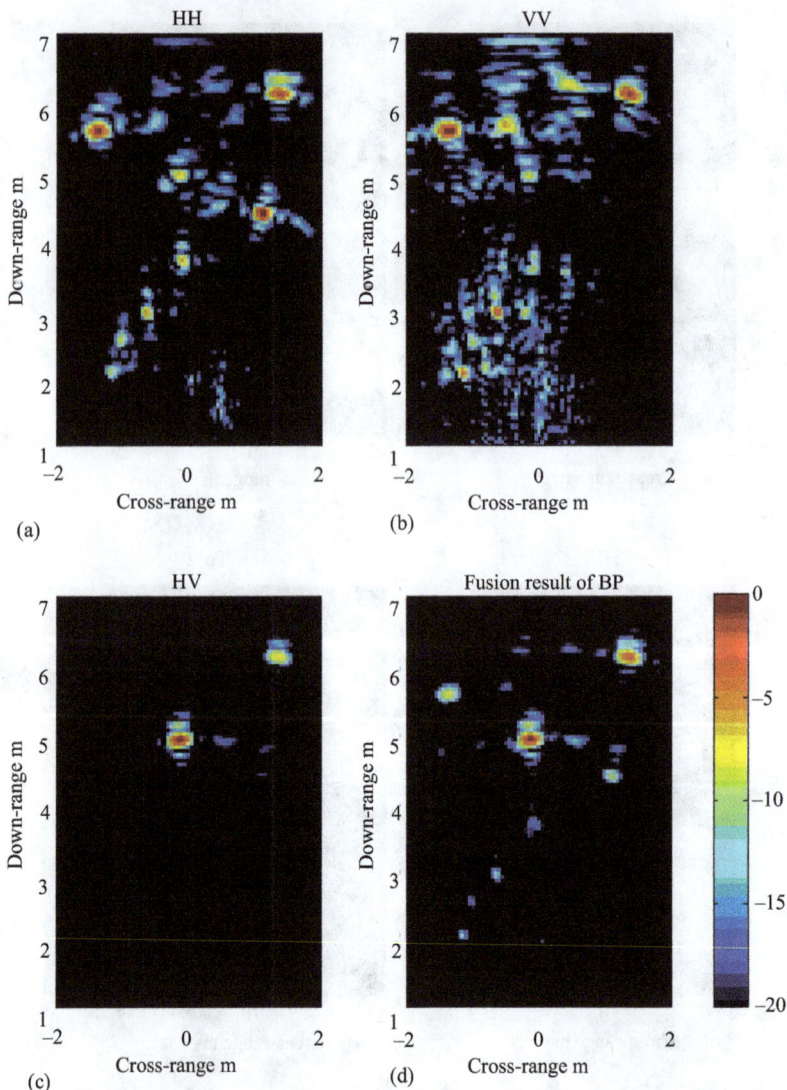

Figure 2.7   Imaging results produced by the BP algorithm in through-wall scenario:
(a) HH, (b) VV, (c) HV, and (d) composite image by sum fusion

channels. For example, the cylinder located at down-range 4.3 m in HH channel is stronger than that at other channels, the targets in HV channels are weak under −20 dB except the dihedrals located at down-range 4.9 m and 6.1 m. Furthermore, the change of clutter and the distortion of the image due to the presence of the wall can also be observed in Figure 2.6 and Figure 2.7. It is obvious that BP suffers from low resolution and high sidelobe levels.

In what follows, four greedy algorithms, that is, SOMP, SSP, HMP, and LAHMP, are evaluated by using part of data in both free-space and through-wall scenarios. Here 100 frequency points and 69 antenna positions are selected for all the receiving channels. The sparsity level K is set to be 70 for both free-space and

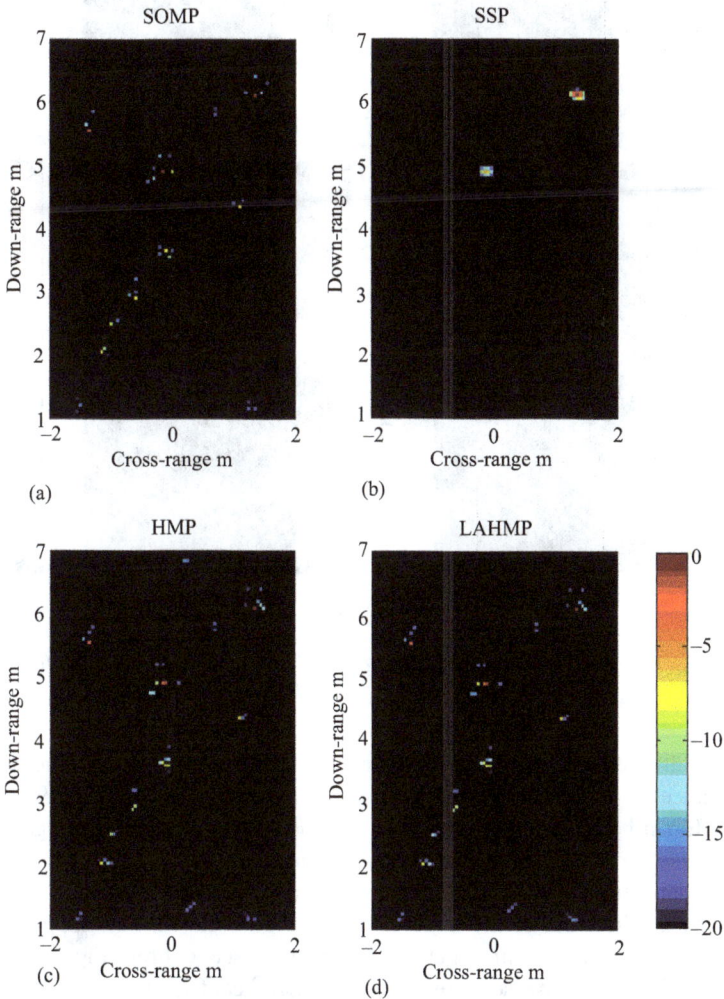

Figure 2.8   Composite images in free-space scenario produced by (a) SOMP, (b) SSP, (c) HMP, and (d) LAHMP

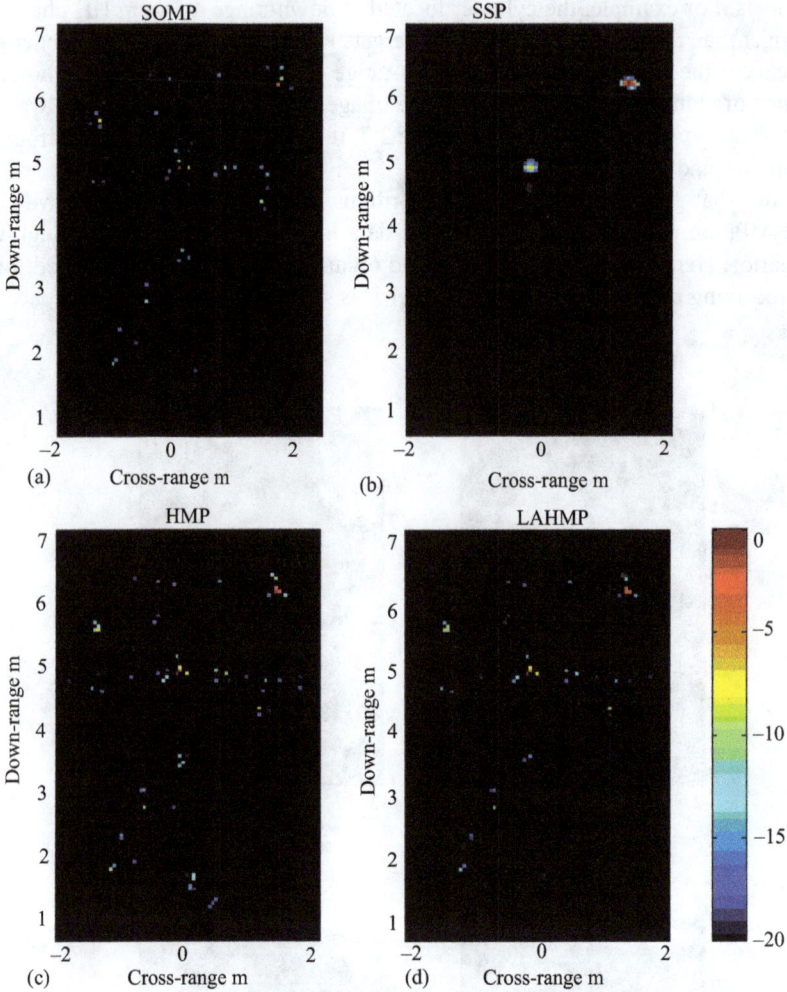

Figure 2.9   Composite images in through-wall scenario produced by (a) SOMP,
(b) SSP, (c) HMP, and (d) LAHMP

through-wall scenarios. This sparsity level is sufficient to image the scene of
interest since the number of dominant pixels in Figure 2.8 and Figure 2.9 is smaller
than 70. It can be observed from Figure 2.8 and Figure 2.9 that the targets can be
discriminated from clutter and noise by these greedy algorithms with less measurement data. In Figure 2.8(a), SOMP finds all point-like scatterers but suffers
from many artifacts out of the target areas. In Figure 2.8(b), SSP suppresses artifacts well but fails to find all targets in the observed scene. From Figure 2.8(c) and
Figure 2.8(d), it can be seen that the dominant coefficients reconstructed by
LAHMP are more concentrated than that produced by HMP. In Figure 2.9, even if
the presence of the wall leads to a large increase in propagation complexity,

LAHMP also outperforms other algorithms in the following two aspects: (1) the dominant coefficients are more concentrated around the positions of the true scatterers; and (2) the clutter and artifacts are more effectively suppressed. All of these greedy algorithms fail to detect the trihedral located at down-range 5.7 m due to its weak scattering behavior. To quantitatively compare the performance of these algorithms, the TCR is defined as

$$\text{TCR} = \frac{(1/N_{\text{B}})\sum_{(x,y)\in\text{B}}|I(x,y)|^2}{(1/N_{\text{C}})\sum_{(x,y)\in\text{C}}|I(x,y)|^2} \tag{2.12}$$

where $I(x,y)$ is the magnitude of the pixel (x,y) in the composite image, B and C denote the target and clutter regions, N_{B} and N_{C} are the numbers of pixels in the regions B and C, respectively. The target and nontarget areas are shown in Figure 2.10, which are selected according to the ground truth [24]. The definitions of TCR in (2.5) and (2.12) are slightly different but have similar meaning. That is, a large value of TCR indicates that the targets are exactly located and the clutter and artifacts outside the target areas are effectively suppressed. In Table 2.7, the TCR values produced by the four greedy algorithms are compared with different percentage of measurements. Because SSP only recovers two targets and cannot reflect the true information about the observed scene, its TCR values are not included in Table 2.7. As shown in Table 2.7, the quality of the reconstructed image

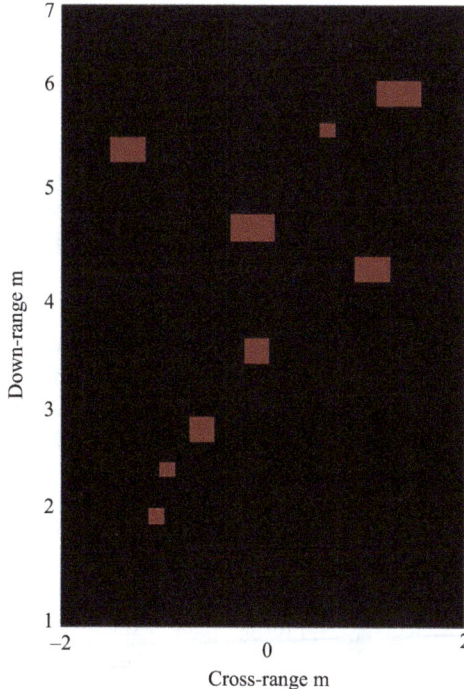

Figure 2.10   The target areas (red areas) and nontarget areas (black areas)

Table 2.7   TCR values of four algorithms

Scenario	Algorithm	Percentage of measurements		
		50%	37.5%	25%
Free-space	LAHMP	261.64	240.14	233.02
	HMP	242.42	238.64	221.39
	SOMP	235.36	231.67	229.98
	BP	87.39	68.65	46.15
Through-wall	LAHMP	155.82	144.14	114.98
	HMP	137.64	133.43	99.40
	SOMP	108.07	95.54	92.41
	BP	41.23	34.61	25.71

Table 2.8   Running time and iterations of HMP and LAHMP

Algorithm	$K = 10$		$K = 40$		$K = 70$	
	Average running time (s)	Average number of iterations	Average running time (s)	Average number of iterations	Average running time (s)	Average number of iterations
HMP	3.72	2.0	47.08	3.0	119.35	3.4
LAHMP	4.56	2.3	121.39	7.0	501.99	8.0

improves as the number of measurements increases. The TCR values of SOMP and HMP are lower due to the existence of large number of artifacts. A consistently higher TCR is provided by LAHMP since it achieves more accurate localization of targets and better suppression of artifacts. It is worth mentioning that the presence of wall lowers the TCR values due to the complexity of the through-wall propagation, which is consistent with the comparison between Figure 2.8 and Figure 2.9.

The higher image quality generated by LAHMP is at the cost of the increased computational complexity. In what follows, HMP and LAHMP are compared in terms of the running time and the number of iterations. We perform 50 trials on through-wall data, in which 50 frequency points and 30 antenna positions are randomly selected from every receiving channel for each trial. The average running time and the average number of iterations of HMP and LAHMP are listed in Table 2.8. It is observed that the running time and the number of iterations of LAHMP are larger than that of HMP, and furthermore, the increase of the computational complexity of LAHMP becomes more serious as the value of K increases.

The initial step of the original look-ahead algorithm in [20] is that the basis-signal candidate is merged into the previous support set to form the test support set and the recovery error is accordingly obtained. After that, the orthogonality projections are applied to iteratively add new basis-signals into the test support set until the recovery error no longer decreases. As summarized in Table 2.6, the look-ahead operation used in LAHMP can be regarded as a "one-step" look-ahead strategy. Without loss of generality, we can also set up a "p-step" look-ahead strategy, which evaluates the effect of the basis-signal candidate on the recovery error after p new basis-signals are merged into the previous support set. As seen in Figure 2.11, even if the original look-ahead strategy [20] is embedded in HMP,

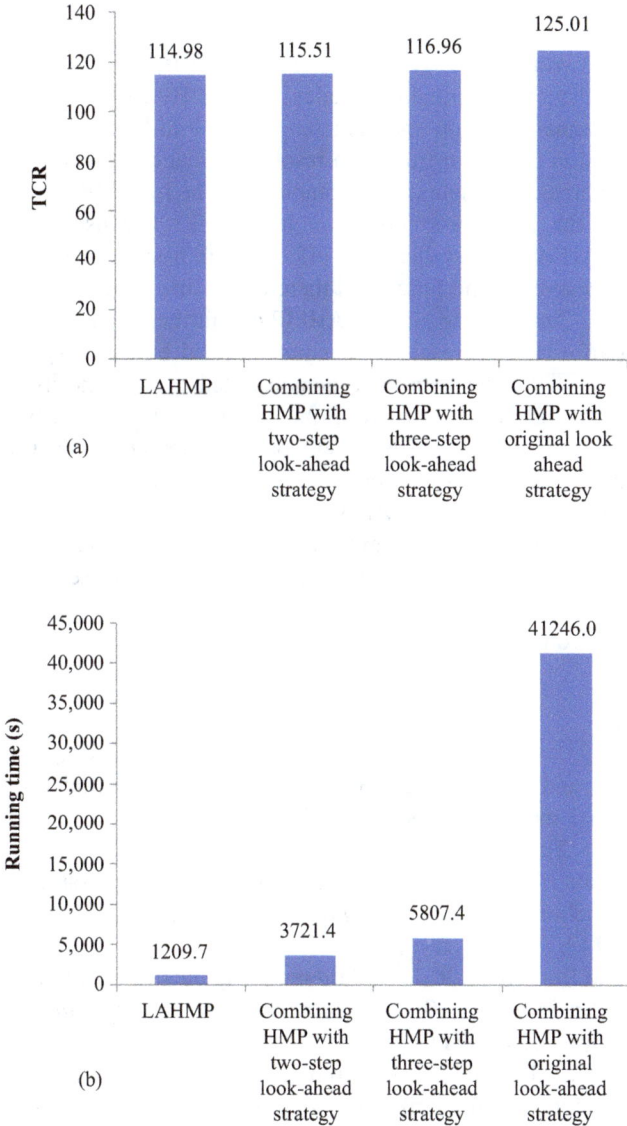

Figure 2.11   Comparisons of LAHMP and the algorithms which combine HMP with different versions of look-ahead strategies: (a) TCR and (b) running time. Here 25% measurements of the through-wall scenario are selected

TCR is only 8.7% higher than that of LAHMP, but the running time is about 34 times as much as that of LAHMP. Therefore, the simple "one-step" look-ahead strategy adopted in LAHMP leads to a good tradeoff between performance and complexity.

2.5   Conclusion

In this chapter, two advanced greedy algorithms, that is, HMP and LAHMP, were presented for enhancing the quality of radar imaging. By combining the strength of OMP in basis-signal selection and the strength of SP in basis-signal reevaluation, the HMP algorithm can reconstruct high-resolution images with no visible artifacts, at the cost of increased computational complexity. The LAHMP algorithm is based on embedding the look-ahead operation into HMP. The use of a look-ahead operation in LAHMP can evaluate the effect of the basis-signal selection on the recovery performance in the future and therefore ensure better reliability of basis-signal selection. Compared to HMP, LAHMP can further enhance the radar image quality at an affordable increase of the computational burden. Experiments based on real through-wall radar data have demonstrated the superiority of these two advanced greedy algorithms. It is worth mentioning that (1) these two algorithms can be applied to not only TWRI but also more radar imaging scenarios, where the dictionary may appear in various forms, and (2) they are suitable for both single-channel and multichannel radar systems. The possible extensions include combination of the look-ahead strategy and other greedy algorithms and more fusion approaches of the multichannel data during the basis signal selection. It is also worth investigating how to guarantee successful image formation for weak targets in presence of strong scatterers with the greedy algorithms.

References

[1]   Tropp J. and Gilbert A.C. "Signal recovery from partial information via orthogonal matching pursuit." *IEEE Transactions on Information Theory.* 2007; 53(12): 4655–4666.

[2]   Dai W. and Milenkovic O. "Subspace pursuit for compressive sensing signal reconstruction." *IEEE Transactions on Information Theory.* 2009; 55(5): 2230–2249.

[3]   Li G. and Burkholder R. J. "Hybrid matching pursuit for distributed through-wall radar imaging." *IEEE Transactions on Antennas and Propagation.* 2015; 63(4): 1701–1711.

[4]   Wang X., Li G., Wan Q., and Burkholder R. J. "Look-ahead hybrid matching pursuit for multipolarization through-wall radar imaging." *IEEE Transactions on Geoscience and Remote Sensing.* 2017; 55(7): 4072 – 4081.

[5]   Amin M. G. (Ed.). *Through-the-wall radar imaging.* Boca Raton, FL: CRC Press, 2010.

[6]   Chang P. C., Burkholder R. J., Marhefka R. J., Bayram Y., and Volakis J. L. "High-frequency EM characterization of through-wall building imaging." *IEEE Transactions on Geoscience and Remote Sensing.* 2009; 47(5): 1375–1387.

[7]   Chang P. C., Burkholder R. J., and Volakis J. L. "Adaptive CLEAN with target refocusing for through-wall image improvement." *IEEE Transactions on Antennas and Propagation.* 2010; 58(1): 155–162.

[8]   Tropp J. A., Gilbert A. C., and Strauss, M. J., "Algorithms for simultaneous sparse approximation. Part I: Greedy pursuit." *Signal Processing.* 2006; 86 (3): 572–588.

[9]   Duarte M. F., Sarvotham S., Baron D., Wakin M. B., and Baraniuk R. G. "Distributed compressed sensing of jointly sparse signals." *Proceedings of Conference Record of the Thirty-Ninth Asilomar Conference on Signals, Systems and Computers.* Pacific Grove, CA, USA: IEEE, 2005, pp. 1537–1541.

[10]   Feng J.-M., and Lee C.-H. "Generalized subspace pursuit for signal recovery from multiple-measurement vectors." *Proceedings of IEEE Wireless Communication Networks Conference (WCNC).* Shanghai: IEEE; 2013, pp. 2874–2878.

[11]   Maechler P., Felber N., and Kaeslin H. "Compressive sensing for wifi based passive bistatic radar." *Proceedings of 20th European Signal Processing Conference (EUSIPCO'12).* Bucharest, Romania: IEEE, 2012, pp. 1444–1448.

[12]   Needell D. and Tropp J.A. "CoSaMP: Iterative signal recovery from incomplete and inaccurate samples." *Applied and Computational Harmonic Analysis.* 2009; 26(3): 301–321.

[13]   Dai W. and Milenkovic O. "Subspace pursuit for compressive sensing signal reconstruction." *IEEE Transactions on Information Theory.* 2009; 55(5): 2230–2249.

[14]   Å. Björck. *Numerical methods for least squares problems.* Philadelphia, PA: SIAM; 1996.

[15]   Browne K. E., Burkholder R. J., and Volakis J. L. "Through-wall opportunistic sensing system utilizing a low-cost flat-panel array." *IEEE Transactions on Antennas Propagation.* 2011; 59(3): 859–868.

[16]   Browne K. E., Burkholder R. J., and Volakis J. L. "Fast optimization of through-wall radar images via the method of Lagrange multipliers." *IEEE Transactions on Antennas Propagation.* 2013; 61(1): 320–328.

[17]   Burkholder R. J. and Browne E. "Coherence factor enhancement of through-wall radar images." *IEEE Antennas Wireless Propagation Letters.* 2010; 9: 842–845.

[18]   Wang G. and Amin M. G. "Imaging through unknown walls using different standoff distances." *IEEE Transactions on Signal Processing.* 2006; 54(10): 4015–4025.

[19]   Tivive F. H. C., Bouzerdoum A., and Amin M. G. "An SVD-based approach for mitigating wall reflections in through-the-wall radar imaging."

Proceedings of IEEE Radar Conference. Kansas City, MO, USA, May 2011, pp. 519–524.

[20] Chatterjee S., Sundman D., Vehkapera M., and Skoglund M. "Projection-based and look-ahead strategies for basis-signal selection." *IEEE Transactions on Signal Processing.* 2012; 60(2): 634–647.

[21] Horn R. A. and Johnson C. R. "Norms for vectors and matrices." in *Matrix analysis.* 2nd ed. Cambridge: Cambridge University Press, 1994, 383–385.

[22] Deepa K. G., Sooraj K. A., and Hari K. V. S. "Modified greedy pursuits for improving sparse recovery." *Proceedings of Twentieth National Conference on Communications (NCC).* Kanpur, India: IEEE, 2014, 1–5.

[23] Dilsavor R., Ailes W., Rush P., *et al.* "Experiments on wideband through the wall imaging." *Proceedings of SPIE Symposium on Defense and Security, Algorithms for Synthetic Aperture Radar Imagery XII Conference.* Orlando, Florida, USA: SPIE, vol. 5808. 2005, 196–206.

[24] Seng C. H., Bouzerdoum A., Amin M. G., and Phung S. L. "Probabilistic fuzzy image fusion approach for radar through wall sensing." *IEEE Transactions on Image Processing.* 2013; 22(12): 4938–4951.

Chapter 3
Two-level block sparsity model for multichannel radar signals

3.1 Introduction

In multichannel radar systems, e.g., polarimetric through-wall radar imaging (TWRI) system [1,2], airborne space-time adaptive processing (STAP) system [3,4], multistatic radar system, and distributed radar system, the multichannel data often have the following two structures:

1. At a single channel, the dominant scatterers appear in clusters and occupy more than one radar resolution cell. This structure is referred to as the clustered sparsity or block sparsity [5].
2. By adopting the same method of discretizing the parameter domain, the locations of the dominant coefficients are common to all the receiving channels. This structure is called the common sparsity or joint sparsity and also viewed as a special case of the block sparsity from the global viewpoint [5].

The combination of the above two sparse structures is referred as the two-level block sparsity model [6]. Under the constraint of the two-level block sparsity, the problem of multichannel radar signal processing can be converted into the problem of recovering a group of clustered sparse signals that have the common sparsity pattern. Compared to only promoting the clustered sparsity or the common sparsity, the two-level block sparsity model can improve the reconstruction of the dominant coefficients and the suppression of the artifacts in the sparse solutions. In this chapter, we formulate the two-level block sparsity model in multichannel radar systems and apply it to enhance the performances of TWRI [6] and STAP [7]. Simulations and experimental results on real radar data demonstrate the superiority of the two-level block sparsity model.

3.2 Formulation of the two-level block sparsity model

In this section, we first review the models of the clustered sparsity and the joint sparsity and then formulate the two-level block sparsity model in multichannel radar systems.

3.2.1   Clustered sparsity of single-channel data

The clustered sparsity means that there are a few dominant coefficients in the parameter domain and they appear in a clustered manner [8–10]. For example, an extended
target occupies more than one resolution cell. At the qth receiving channel, the coefficients vector in the parameter domain is denoted as $\boldsymbol{\theta}^{(q)}$, the indices of the dominant
coefficients are recorded by the support set $\Lambda^{(q)}$, and the complementary set of $\Lambda^{(q)}$ is
denoted by $\bar{\Lambda}^{(q)}$. A support area $\mathbf{s}$ is denoted by a vector composed of 1 and -1, where
$\mathbf{s}_{\Lambda^{(q)}} = \mathbf{1}$ and $\mathbf{s}_{\bar{\Lambda}^{(q)}} = -\mathbf{1}$. A widely used tool to formulate the clustered property of the
coefficients is the Markov random field (MRF), which captures the spatial dependence
between a coefficient and its neighbors by defining a conditional probability distribution. For example, the support area $\mathbf{s}$ in an image can be described by the second-
order auto-logistic MRF model, in which each pixel is assumed to be correlated to its
eight surrounding neighbors. The influence of the neighbors on the ith entry of $\mathbf{s}$ can be
described by the conditional probability density function (PDF) [11]:

$$P(s_i|\mathbf{s}_{\mathrm{N}_i}) = \frac{1}{Z(\mathbf{s}_{\mathrm{N}_i})} \exp\left(\sum_{i' \in \mathrm{N}_i} s_i s_{i'}\right) \tag{3.1}$$

where N_i is the index set of the neighbors of the ith entry of $\mathbf{s}$, and $Z(\mathbf{s}_{\mathrm{N}_i})$ is a
normalizing factor.

3.2.2   Joint sparsity of the multichannel data

Taking the same method of discretizing the parameter domain at all the Q receiving
channels ensures that $\left\{\boldsymbol{\theta}^{(1)}, \boldsymbol{\theta}^{(2)}, \cdots, \boldsymbol{\theta}^{(Q)}\right\}$ share the joint sparsity pattern. In other
words, the positions of dominant coefficients in $\boldsymbol{\theta}^{(q)}$ are the same for different
receiving channels. The values of dominant coefficients in $\boldsymbol{\theta}^{(q)}$ may vary with the
channel index q, as a target may have different scattering characteristics due to the
polarization diversity, the frequency diversity, or the aspect angle diversity of multiple
receiving channels. Denote the common support set of $\left\{\boldsymbol{\theta}^{(1)}, \boldsymbol{\theta}^{(2)}, \cdots, \boldsymbol{\theta}^{(Q)}\right\}$ as Λ,
that is, $\Lambda = \Lambda^{(q)}$ for $q = 1, 2, \cdots, Q$. Accordingly, the support area $\mathbf{s}$ is also common
for all the receiving channels, with $\mathbf{s}_\Lambda = \mathbf{1}$ and $\mathbf{s}_{\bar{\Lambda}} = -\mathbf{1}$. The sparsity level of each
one of $\left\{\boldsymbol{\theta}^{(1)}, \boldsymbol{\theta}^{(2)}, \cdots, \boldsymbol{\theta}^{(Q)}\right\}$ is denoted as K.

3.2.3   Two-level block sparsity

Figure 3.1 presents an example of the two-level block sparsity. Figure 3.2 describes
the relationship among $\left\{\theta_i^{(1)}, \theta_i^{(2)}, \cdots, \theta_i^{(Q)}\right\}$, s_i, and $\mathbf{s}_{\mathrm{N}_i}$. It can be observed that

1. the ith entry of $\mathbf{s}$, s_i, is dependent on its neighbors $\mathbf{s}_{\mathrm{N}_i}$ due to the clustered
   sparsity constraint; and
2. the value of s_i determines whether $\left\{\theta_i^{(1)}, \theta_i^{(2)}, \cdots, \theta_i^{(Q)}\right\}$ are zero according to
   the joint sparsity pattern.

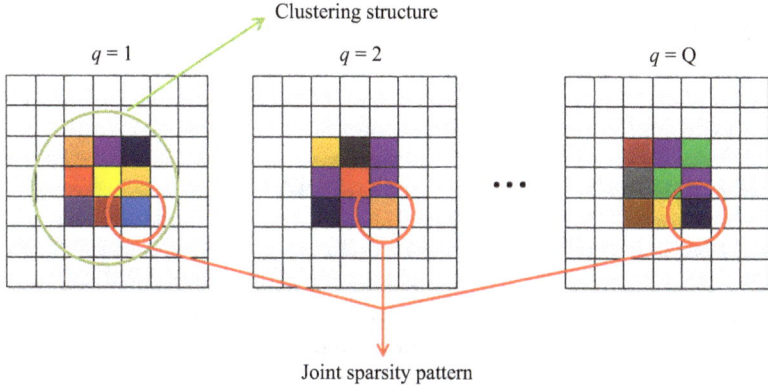

Figure 3.1   *An example of the two-level block sparsity. The colorized lattices denote the dominant coefficients in the parameter domain. Different colors illustrate the diversity of scattering behavior of the same target area at different receiving channels*

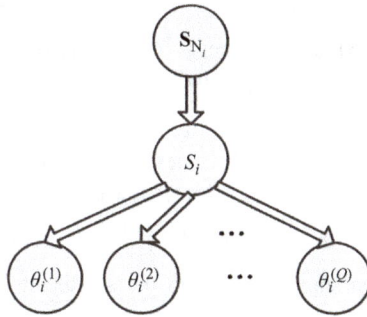

Figure 3.2   *The relationship among $\left\{\theta_i^{(1)}, \theta_i^{(2)}, \cdots, \theta_i^{(Q)}\right\}$, s_i, and $\mathbf{s}_{N_i}$*

Assume that $\left\{\theta_i^{(1)}, \theta_i^{(2)}, \cdots, \theta_i^{(Q)}\right\}$ are independent given the value of s_i. This assumption usually holds in those multichannel radar systems acquiring data from different polarimetric channels, different frequency bands, or different aspect angles [8,12]. Thus we have:

$$p\left(\theta_i^{(1)}, \theta_i^{(2)}, \cdots, \theta_i^{(Q)}\big|s_i\right) = \prod_{q=1}^{Q} p\left(\theta_i^{(q)}\big|s_i\right) \tag{3.2}$$

The goal of the sparse signal recovery is to determine Λ (or $\mathbf{s}$) and the nonzero coefficients. The values of $\mathbf{s}$ and $\left\{\boldsymbol{\theta}^{(1)}, \boldsymbol{\theta}^{(2)}, \cdots, \boldsymbol{\theta}^{(Q)}\right\}$ influence each other. Given $\mathbf{s}$, the nonzero coefficients in $\left\{\boldsymbol{\theta}^{(1)}, \boldsymbol{\theta}^{(2)}, \cdots, \boldsymbol{\theta}^{(Q)}\right\}$ can be easily reconstructed by the least squares estimator. In what follows, we explain how the two-level block sparsity contributes to the determination of s_i, given $\left\{\theta_i^{(1)}, \theta_i^{(2)}, \cdots, \theta_i^{(Q)}\right\}$ and $\mathbf{s}_{N_i}$. The conditional distribution of s_i can be expressed as

$$
\begin{aligned}
P&\left(s_i \middle| \mathbf{s}_{N_i}, \theta_i^{(1)}, \cdots, \theta_i^{(Q)}\right) \\
&\propto P(\mathbf{s}_{N_i}) \cdot P(s_i|\mathbf{s}_{N_i}) \cdot p\left(\theta_i^{(1)}, \cdots, \theta_i^{(Q)} \middle| \mathbf{s}_{N_i}, s_i\right) \\
&= P(\mathbf{s}_{N_i}) \cdot P(s_i|\mathbf{s}_{N_i}) \cdot p\left(\theta_i^{(1)}, \cdots, \theta_i^{(Q)} \middle| s_i\right)
\end{aligned}
\tag{3.3}
$$

The last equation in (3.3) holds because $\left\{\theta_i^{(1)}, \theta_i^{(2)}, \cdots, \theta_i^{(Q)}\right\}$ is independent of $\mathbf{s}_{N_i}$ given the value of s_i. Submitting (3.2) into (3.3) yields:

$$
P\left(s_i \middle| \mathbf{s}_{N_i}, \theta_i^{(1)}, \cdots, \theta_i^{(Q)}\right) \propto P(s_i|\mathbf{s}_{N_i}) \cdot \prod_{q=1}^{Q} p\left(\theta_i^{(q)}|s_i\right) \triangleq J(s_i)
\tag{3.4}
$$

Thus s_i can be determined by finding the maximal value of $J(s_i)$:

$$
\widehat{s}_i = \arg\max_{s_i \in \{-1,1\}} J(s_i)
\tag{3.5}
$$

Note that solving (3.5) is equivalent to promoting the two-level block sparsity that consists of the following two sparsity aware structures:

1. The clustered sparsity. The probability $P(s_i|\mathbf{s}_{N_i})$ is a measure of the influence of the neighbors $\mathbf{s}_{N_i}$ on s_i, which is described by (3.1). Maximizing $p(s_i|\mathbf{s}_{N_i})$ tends to preserve the underlying clustered structure.

2. The joint sparsity. The product $\prod_{q=1}^{Q} p\left(\theta_i^{(q)}|s_i\right)$ represents the relationship between the common support area and the scattering coefficients at different receiving channels. According to the definition of $\mathbf{s}$, when $\theta_i^{(q)}$ is a dominant coefficient for all the channels, $\prod_{q=1}^{Q} p\left(\theta_i^{(q)}|s_i\right)$ reaches a maximum if and only if $s_i = 1$. On the other hand, when $\theta_i^{(q)} \approx 0$ for all the receiving channels, $\prod_{q=1}^{Q} p\left(\theta_i^{(q)}|s_i\right)$ is expected to be large if and only if $s_i = -1$. Hence maximizing $\prod_{q=1}^{Q} p\left(\theta_i^{(q)}|s_i\right)$ tends to enforce the joint sparsity pattern.

Since $s_i \in \{1, -1\}$, solving (3.5) can be converted to the problem of hypothesis testing. Consider the following two hypotheses H_0 and H_1:

$$
\begin{cases} H_0 : s_i = -1 \\ H_1 : s_i = 1 \end{cases}
\tag{3.6}
$$

Then the value of s_i can be determined by comparing the values of $J(s_i = -1)$ and $J(s_i = 1)$. Define the following two functions:

$$F_0 = \log(J(s_i = -1)) \qquad (3.7)$$

and

$$F_1 = \log(J(s_i = 1)) \qquad (3.8)$$

Consequently, the solution of (3.5) can be given by

$$\begin{cases} \widehat{s}_i = 1, & \text{when } \Delta > 0 \\ \widehat{s}_i = -1, & \text{when } \Delta \leq 0 \end{cases} \qquad (3.9)$$

where

$$\begin{aligned} \Delta &= F_1 - F_0 \\ &= 2 \sum_{i' \in N_i} s_{i'} + \sum_{q=1}^{Q} \log \left[\frac{p\left(\theta_i^{(q)} | s_i = 1\right)}{p\left(\theta_i^{(q)} | s_i = -1\right)} \right] \end{aligned} \qquad (3.10)$$

The second equation in (3.10) is deduced from (3.1) and (3.4).

The analytical expressions of $p\left(\theta_i^{(q)} | s_i = -1\right)$ and $p\left(\theta_i^{(q)} | s_i = 1\right)$ are required to calculate Δ in (3.10). Inspired by [8,13], here the geometrical approximations of $p\left(\theta_i^{(q)} | s_i = -1\right)$ and $p\left(\theta_i^{(q)} | s_i = 1\right)$ are taken into account. Since $\mathbf{s_\Lambda} = \mathbf{1}$ and $\mathbf{s_{\bar{\Lambda}}} = -\mathbf{1}$, it can be concluded that $p\left(\theta_i^{(q)} | s_i\right) \approx 0$ in the following cases:

1. $s_i = -1$ and $\left|\theta_i^{(q)}\right| \gg 0$;

2. $s_i = 1$ and $\theta_i^{(q)} \approx 0$

An example of the geometrical approximations of $p\left(\theta_i^{(q)} | s_i = -1\right)$ and $p\left(\theta_i^{(q)} | s_i = 1\right)$ is given in Figure 3.3, where α is the maximum magnitude of $\left|\theta_i^{(q)}\right|$, τ is a threshold for separating large and small coefficients, and $\{\varepsilon_1, \varepsilon_2, \varepsilon_3, \varepsilon_4\}$ are the density level parameters in the geometrical approximations of the PDFs. As can be seen in Figure 3.3(a) and (b), $\theta_i^{(q)}$ tends to be close to zero when $s_i = -1$ and tends to be a dominant value when $s_i = 1$. Thus, the geometrical approximation of $p\left(\theta_i^{(q)} | s_i = 1\right) / p\left(\theta_i^{(q)} | s_i = -1\right)$ in (3.10) can be directly expressed as

$$\frac{p\left(\theta_i^{(q)} | s_i = 1\right)}{p\left(\theta_i^{(q)} | s_i = -1\right)} = \begin{cases} \dfrac{\varepsilon_3}{\varepsilon_2}, & \text{for } \left|\theta_i^{(q)}\right| \geq \tau \\[2mm] \dfrac{\varepsilon_4}{\varepsilon_1}, & \text{for } \left|\theta_i^{(q)}\right| < \tau \end{cases} \qquad (3.11)$$

(a)

(b)

(c)

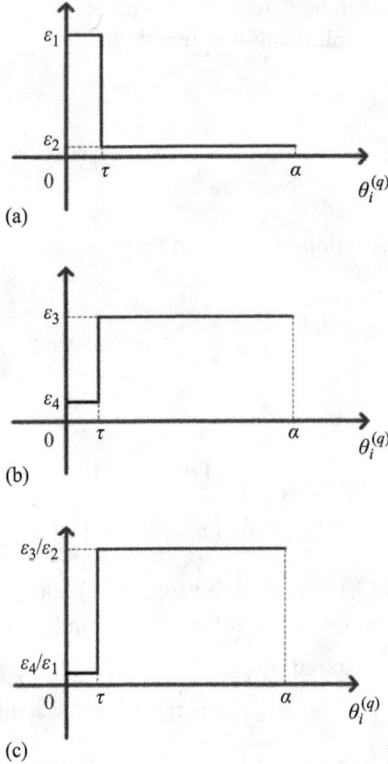

Figure 3.3   Geometrical approximations: (a) $p\left(\theta_i^{(q)}|s_i = -1\right)$, (b) $p\left(\theta_i^{(q)}|s_i = 1\right)$, and (c) $p\left(\theta_i^{(q)}|s_i = 1\right)/p\left(\theta_i^{(q)}|s_i = -1\right)$

which is shown in Figure 3.3(c). Define an indicator function $I(x)$ as

$$I(x) = \begin{cases} 1 & |x| \geq \tau \\ 0 & |x| < \tau \end{cases} \tag{3.12}$$

Then (3.10) can be rewritten as

$$\Delta = 2\sum_{i' \in N_i} s_{i'} + \sum_{q=1}^{Q} \log\left[\frac{\varepsilon_3}{\varepsilon_2} I\left(\theta_i^{(q)}\right) + \frac{\varepsilon_4}{\varepsilon_1}\left(1 - I\left(\theta_i^{(q)}\right)\right)\right] \tag{3.13}$$

In summary, s_i can be solved by calculating Δ according to (3.13) and then substituting it to (3.9).

3.3   TWRI based on two-level block sparsity

In this section, we describe how to apply the two-level block sparsity model to multichannel radar imaging. A greedy algorithm referred to as the two-level block matching pursuit (TLBMP) [6] is presented for enhancing the quality of TWRI by combining the two-level block sparsity model and the greedy pursuit framework.

3.3.1   Signal model and algorithm description

Assume that the two-dimensional scene at a specified height behind the wall is discretized as $N \triangleq P_x \times P_y$ resolution cells, and the distance from the origin to the ith resolution cell is represented by $\mathbf{r}_i$. There are Q receiving channels in the through-wall radar system. At the qth polarimetric channel, M equivalent antenna phase centers are located at $\left\{ \mathbf{r}_1^{(q)}, \mathbf{r}_2^{(q)}, \cdots, \mathbf{r}_M^{(q)} \right\}$ at L frequency points $\{f_1, f_2, \cdots, f_L\}$. The radar echo received from the qth polarimetric channel at the lth frequency point and the mth antenna phase center can be expressed as

$$y_{m,l}^{(q)} = \sum_{i=1}^{N} \theta_i^{(q)} \exp\left[-j \frac{4\pi}{c} f_l \left(\|\mathbf{r}_i - \mathbf{r}_m^{(q)}\|_2 \right) \right] + w_{m,l}^{(q)} \tag{3.14}$$

where $\theta_i^{(q)}$ is the complex reflectivity of the ith resolution cell at the qth polarmetric channel, $w_{m,l}^{(q)}$ is the additive noise, and c is the propagation speed. The target reflectivity $\theta_i^{(q)}$ is assumed to be constant across all the L frequency points. If the ith resolution cell is occupied by a target, $\theta_i^{(q)}$ is a nonzero value; otherwise $\theta_i^{(q)} \approx 0$.

The signal in (3.14) can be rewritten as

$$\mathbf{y}^{(q)} = \mathbf{\Phi}^{(q)} \mathbf{\theta}^{(q)} + \mathbf{w}^{(q)} \tag{3.15}$$

where $\mathbf{y}^{(q)} \in \mathbb{C}^{ML \times 1}$ is the measurement vector at the qth polarimetric channel, $\mathbf{\theta}^{(q)} \in \mathbb{C}^{P_x P_y \times 1}$ is the discrete description of the reflectivity density function of the observed scene, and $\mathbf{w}^{(q)}$ denotes the additive noise. The element of the dictionary matrix $\mathbf{\Phi}^{(q)} \in \mathbb{C}^{ML \times N}$ is given by

$$\mathbf{\Phi}^{(q)}(l + (m-1)L, i) = \exp\left[-j \frac{4\pi}{c} f_l \left(\|\mathbf{r}_i - \mathbf{r}_m^{(q)}\|_2 \right) \right] \tag{3.16}$$

where $l = 1, 2, \cdots, L$, $m = 1, 2, \cdots, M$, $i = 1, 2, \cdots, N$, and $q = 1, 2, \cdots, Q$. Each column in the dictionary matrix $\mathbf{\Phi}^{(q)}$ is referred to as an atom or a basis-signal. The imaging result produced by the traditional back-projection (BP) algorithm based on matched filtering can be expressed as

$$\widehat{\mathbf{\theta}}(q) = \left(\mathbf{\Phi}^{(q)} \right)^H \mathbf{y}^{(q)} \tag{3.17}$$

This is basically equivalent to a coherent accumulation over all the measurements, without consideration of any prior knowledge of the image to be formed.

Before introducing the TLBMP algorithm, we first present a majority-vote algorithm in Table 3.1 for estimating the common support set. The input variables of this algorithm include the local support sets estimated at different receiving channels, and its output is the estimate of the joint sparsity pattern by

Table 3.1   The majority-vote algorithm

Input:
$$\left\{\Lambda^{(q)}|q=1,2,\cdots,Q\right\}.$$
Initialization:
$$\mathbf{n}=\mathbf{0}^{P_xP_y\times 1}$$
Operations:
For each $q=1,2,\cdots,Q$
$$\mathbf{n}=\mathrm{add}\left(\mathbf{n},\Lambda^{(q)}\right)$$
End for
$$\Lambda_{\mathrm{vote}}=\mathrm{max\_ind}(\mathbf{n},K)$$
Output:
$$\Lambda_{\mathrm{vote}}$$

---

majority-vote fusion. The majority-vote algorithm essentially relies on the fact that the elements in the common support set have the highest score in terms of the occurrence. A proof on the validity of majority-vote has been addressed in [14]. For expression convenience, we define the following two algorithmic notations:

$$\mathrm{max\_ind}(\mathbf{x},K)=\{\text{indices of the }K\text{ largest magnitude entries in }\mathbf{x}\}\quad(3.18)$$

and

$$\mathrm{add}\,(\mathbf{x},\Lambda)=\{x_i=x_i+1,\forall i\in\Lambda\}\quad(3.19)$$

The TLBMP algorithm is detailed in Table 3.2. At the start of each iteration, the temporary coefficient vectors, denoted by $\breve{\theta}^{(q)}$ for $q=1,2,\cdots,Q$, are first estimated by taking the sum of the previous estimates $\widehat{\theta}^{(q,\mathrm{old})}$ and $\left(\Phi^{(q)}\right)^H\mathbf{r}^{(q,\mathrm{old})}$. In step 2, the local support sets $\left\{\Lambda^{(q)}|q=1,2,\cdots,Q\right\}$ are obtained by seeking K indices corresponding to the largest values of $\breve{\theta}^{(q)}$ for $q=1,2,\cdots,Q$. It is worth mentioning that the combination of steps 1 and 2 is a variant of the iterative hard thresholding [15,16]. In step 3, the temporary common support set Λ_{vote} is estimated by majority voting of $\left\{\Lambda^{(q)}|q=1,2,\cdots,Q\right\}$ followed by setting $\mathbf{s}_{\Lambda_{\mathrm{vote}}}=\mathbf{1}$ and $\mathbf{s}_{\{1,2,\cdots,P_xP_y\}-\Lambda_{\mathrm{vote}}}=-\mathbf{1}$. In step 4, the ith entry of $\widehat{\mathbf{s}}$ is updated according to (3.9) and (3.13) for each $i\in\left\{1,2,\cdots,P_xP_y\right\}$ by using its neighbors $\mathbf{s}_{N_i}$ and the temporary coefficient estimates $\left\{\breve{\theta}_i^{(1)},\breve{\theta}_i^{(2)},\cdots,\breve{\theta}_i^{(Q)}\right\}$. As analyzed in Section 3.2.3, this enhances the two-level block sparsity, that is, the clustered sparsity and the joint sparsity. In step 5, the common support set Λ_{temp} is updated by selecting the set of indices which correspond to the locations of "1" in $\widehat{\mathbf{s}}$. In step 6, the signal estimate

$\widehat{\theta}^{(q,\text{new})}$ is updated by using the least squares estimation with the selected basis-signals recorded in $\Phi^{(q)}_{\Lambda_{\text{temp}}}$. In order to ensure that the sparsity level of the solution is equal to K, the smallest $P_xP_y - K$ coefficients of $\widehat{\theta}^{(q,\text{new})}$ are set to be zero. In step 7, the residual vectors are updated. The iterations over all the channels are terminated when the global recovery error no longer decreases.

3.3.2 Model parameter selection

Next, we discuss how to select the model parameters used in the TLBMP algorithm. As can be seen from Table 3.2 and Figure 3.3, the parameters to be properly selected include K, τ, $\varepsilon_4/\varepsilon_1$, and $\varepsilon_3/\varepsilon_2$. The values of these parameters may be empirically determined as described below.

Setting the value of K can be carried out as introduced in Section 2.3.2 of Chapter 2. As suggested in [17], $K \approx ML/(2\log N)$ is a good estimate for the successful recovery of most sparse signals.

The threshold parameter τ separates large and small coefficients. By sorting the magnitudes of the coefficients of $\breve{\theta}^{(q)}$ in descending order, the $2K$-th magnitude is selected as the value of τ. This gives preference to the largest $2K$ coefficients of $\breve{\theta}^{(q)}$, that is, the largest $2K$ coefficients achieve the state $s_i = 1$. The pruning operation contained in step 6 of TLBMP guarantees the desired sparsity K.

Table 3.2 The TLBMP algorithm

Input:
$\{y^{(q)}, \Phi^{(q)}|q = 1, 2, \cdots, Q\}$, K, $\{\varepsilon_1, \varepsilon_2, \varepsilon_3, \varepsilon_4\}$.
Initialization: $r^{(q,old)} = y^{(q)}$, $\widehat{\theta}^{(q,old)} = 0^{P_xP_y \times 1}$, for $q = 1, 2, \cdots, Q$, $s = 0^{P_xP_y \times 1}$.
Iterations:
1. $\breve{\theta}^{(q)} = (\Phi^{(q)})^H r^{(q,old)} + \widehat{\theta}^{(q,old)}$, for $q = 1, 2, \cdots, Q$.
2. $\Lambda^{(q)} = \text{max\_ind}(\breve{\theta}^{(q)}, K)$, for $q = 1, 2, \cdots, Q$.
3. $\Lambda_{\text{vote}} = \text{majority} - \text{vote}(\{\Lambda^{(q)}|q = 1, 2, \cdots, Q\})$, and set $s_{\Lambda_{\text{vote}}} = 1$, $s_{\{1,2,\cdots,P_xP_y\}-\Lambda_{\text{vote}}} = -1$.
4. For each $i \in \{1, 2, \cdots, P_xP_y\}$, compute Δ according to (3.13) by utilizing $\{\breve{\theta}_i^{(1)}, \breve{\theta}_i^{(2)}, \cdots, \breve{\theta}_i^{(Q)}\}$ and s_{N_i}. If $\Delta > 0$, let $\widehat{s}_i = 1$; else let $\widehat{s}_i = -1$.
5. $\Lambda_{\text{temp}} = \{$the set of indices which correspond to the locations of "1" in $\widehat{s}\}$.
6. $\widehat{\theta}^{(q,\text{new})}_{\Lambda_{\text{temp}}} = (\Phi^{(q)}_{\Lambda_{\text{temp}}})^\dagger y^{(q)}$ and the smallest N-K coefficients of $\widehat{\theta}^{(q,\text{new})}$ are set to be zero, for $q = 1, 2, \cdots, Q$.
7. $r^{(q,\text{new})} = y^{(q)} - \Phi^{(q)}\widehat{\theta}^{(q,\text{new})}$, for $q = 1, 2, \cdots, Q$.
8. if $\sum_{q=1}^{Q} \|r^{(q,old)}\|_2^2 > \sum_{q=1}^{Q} \|r^{(q,\text{new})}\|_2^2$, let $r^{(q,old)} = r^{(q,\text{new})}$ and $\widehat{\theta}^{(q,old)} = \widehat{\theta}^{(q,\text{new})}$, for $q = 1, 2, \cdots, Q$, and return to step 1; otherwise, stop the iteration.
Output: $\{\widehat{\theta}^{(q,old)}|q = 1, 2, \cdots, Q\}$.

According to (3.13), only the values of $\varepsilon_4/\varepsilon_1$ and $\varepsilon_3/\varepsilon_2$ are required instead of the values of $\{\varepsilon_1, \varepsilon_2, \varepsilon_3, \varepsilon_4\}$. Considering the natural constraint that $\varepsilon_4/\varepsilon_1 < \varepsilon_3/\varepsilon_2$, a simple approach for setting the values of $\varepsilon_4/\varepsilon_1$ and $\varepsilon_3/\varepsilon_2$ is

$$\frac{\varepsilon_4}{\varepsilon_1} = \delta, \quad \frac{\varepsilon_3}{\varepsilon_2} = \frac{1}{\delta}, \quad 0 < \delta < 1 \tag{3.20}$$

Substituting (3.20) into (3.13), we have:

$$\Delta = 2 \sum_{i' \in N_i} \mathbf{s}_{i'} + \sum_{q=1}^{Q} \log \left[\frac{1}{\delta} I\left(\theta_i^{(q)}\right) + \delta \left(1 - I\left(\theta_i^{(q)}\right)\right) \right] \tag{3.21}$$

According to (3.20) and the fact that the integral of each PDF in Figure 3.3(a) and (b) equals to 1, it can be deduced that

$$\frac{\varepsilon_1}{\varepsilon_2} = \frac{\alpha - \tau}{\delta \tau}, \frac{\varepsilon_3}{\varepsilon_4} = \frac{\tau}{\delta(\alpha - \tau)} \tag{3.22}$$

Since $\varepsilon_1 > \varepsilon_2$ and $\varepsilon_3 > \varepsilon_4$ as shown in Figure 3.3, we have

$$\delta < \min\left\{\frac{\alpha - \tau}{\tau}, \frac{\tau}{\alpha - \tau}\right\} \tag{3.23}$$

The above example of parameter selection corresponds to the geometrical approximations of PDFs given in Figure 3.3. These approximations are geometrically simple at the cost that a number of model parameters are needed to be properly selected. If other approximations of PDFs are taken into account, the number of model parameters and the strategy of parameter selection will change accordingly. Note that the empirical selection of the approximation model and its parameters may be application dependent.

3.3.3   Computational complexity

For simplicity of expression, we let $M^{(q)} = M$ and $L^{(q)} = L$, for $q = 1, 2, \cdots, Q$. In step 1 of TLBMP, the calculation of a temporary signal estimate costs $O(NQML)$. In step 3, the complexity of majority-vote is $O(KQ)$. In steps 4 and 5, the computation involved in estimating Λ_{temp} is $O(N)$. The least squares estimation in step 6 requires $O(K^2QML)$ floating point operations by using the modified Gram–Schmidt algorithm [18]. In step 7, updating of the residual costs the complexity of $O(KQML)$. Under the assumption that $2K < ML < N$, the computational complexity of TLBMP is $T_{\text{TLBMP}} \cdot O((N + K^2)QML)$, where T_{TLBMP} is the number of iterations. Since TLBMP adopts the popular greedy iterative framework, the required number of iterations of TLBMP is expected to be $T_{\text{TLBMP}} = O(K)$. Thus, the total computational complexity of TLBMP is given by $O(K(N + K^2)QML)$. As a comparison, the computational complexities of the hybrid matching pursuit (HMP) algorithm [9] and the simultaneous orthogonal matching pursuit (SOMP) algorithm [19] are $O(K^2NQML)$ and $O(KNQML)$, respectively. Thus, it can be concluded that the computational complexity of TLBMP is lower than that of HMP and slightly higher than that of SOMP.

Note that the sparse signal recover problem under the constraint of two-level block sparsity can be also solved in the Bayesian optimization framework, as done in the clustered multitask Bayesian compressive sensing (CMT-BCS) algorithm [8]. The computational complexity of the CMT-BCS algorithm is $O\left(T_{CMT-BCS}NQ(ML)^3\right)$, where $T_{CMT-BCS}$ is the number of required iterations. It is obvious that the complexity of CMT-BCS is much higher than that of all the above greedy algorithms.

3.3.4 Experimental results

In this section, TLBMP is compared with some sparsity-aware methods, including, CMT-BCS, SOMP, and HMP, in terms of the quality of radar imaging, based on the simulated and real radar data. The values of hyper-parameters of CMT-BCS are selected as the same as in [8].

In all the reconstructed images, the image intensity is plotted in the range of $[-30, 0]$ dB, with the maximum intensity value of each image normalized to 0 dB. The performance of the reconstructed images is evaluated by the target-to-clutter ratio (TCR), which is defined as

$$\text{TCR} = \frac{(1/P_{\Omega_1})\sum_{(x,y)\in\Omega_1}|G(x,y)|^2}{(1/P_{\Omega_2})\sum_{(x,y)\in\Omega_2}|G(x,y)|^2} \tag{3.24}$$

where $G(x,y)$ is the magnitude of the resolution cell (x,y) in the composite image based on the imaging results at all the receiving channels, Ω_1 and Ω_2 denote the target area and clutter area, respectively, P_{Ω_1} and P_{Ω_2} denote the numbers of resolution cells in Ω_1 and Ω_2, respectively. The target and clutter areas are selected according to the truth target locations, as reported in [20]. A large value of TCR indicates that the dominant coefficients are exactly clustered in the target area and the artifacts outside the target area are effectively suppressed.

3.3.4.1 Simulations

The simulated TWRI system consists of 61 antenna centers with an array aperture of 0.6 m. The antenna array is placed against the wall. The stepped-frequency signal covers 1 GHz bandwidth, ranging from 2 to 3 GHz, with 201 frequency points. The observed scene is situated behind a homogeneous wall of thickness 0.15 m and dielectric constant 7.6632, which is typical for block concrete. The multipath returns are all assumed to be 10 dB weaker than the direct path and only single-bounce multipath is considered. Here, three polarization channels, that is, HH, VV, and HV are used for data collection. Measurements in free-space scenario were collected in the same setup but without the wall. The white Gaussian noise with 25 dB signal-to-noise ratio (SNR) is added to the simulated measurements. Figure 3.4 provides the layout of the simulated scene.

Figure 3.5 illustrates the sum fusion result of the images produced by BP [21] with 12.5% of the measurements (51 frequencies and 30 antenna positions are

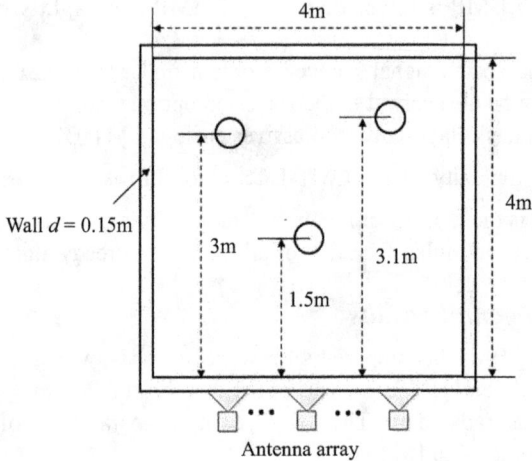

Figure 3.4   Layout of the simulated scene

selected) in free-space and through-wall scenarios. It can be observed that the formed images contain serious sidelobes around the targets. In comparison with Figure 3.5(a), Figure 3.5(b) shows a displacement of the targets, mainly in range and marginally in azimuth, owing to the presence of the wall. In addition, there exist some ghost targets in Figure 3.5 (b) due to the indoor multipath environment.

The imaging results of the sparsity-aware methods in free-space and through-wall scenarios are presented in Figure 3.6 and Figure 3.7, respectively. The sparsity level for all the greedy algorithms is set to $K = 30$. As shown in Figure 3.6, CMT-BCS, HMP, and TLBMP provide relatively concentrated target area, while SOMP suffers from some isolated pixels in the target area.

In Figure 3.7, CMT-BCS, SOMP, and HMP fail to remove the spurious pixels outside the target area. The image resulting from applying the TLBMP algorithm is better than that of other methods in terms of clustering the dominant coefficients and removing the artifacts. The TCRs of the reconstructed images corresponding to BP and sparsity-aware methods are listed in Table 3.3, where TLBMP yields a higher TCR than that achieved by other methods, in both free-space and through-wall scenarios.

The running time results of different algorithms are compared in Table 3.4. The BP algorithm has the minimum running time because it contains only simple correlation operation. The CMT-BCS is more computationally expensive in comparison to the greedy methods. Among the greedy algorithms, TLBMP is more computationally efficient than SOMP and HMP. Moreover, the numbers of iterations of these three greedy algorithms are compared in Table 3.5. The number of iterations of SOMP is fixed to the sparsity level K [19], while the numbers of iterations of both HMP and TLBMP are much lower. The reason is that SOMP selects only one reliable basis-signal per iteration, while K basis-signals are selected at each iteration of both HMP and TLBMP. From Table 3.4 and Table 3.5, it can be deduced that the computational efficiency of TLBMP comes from the

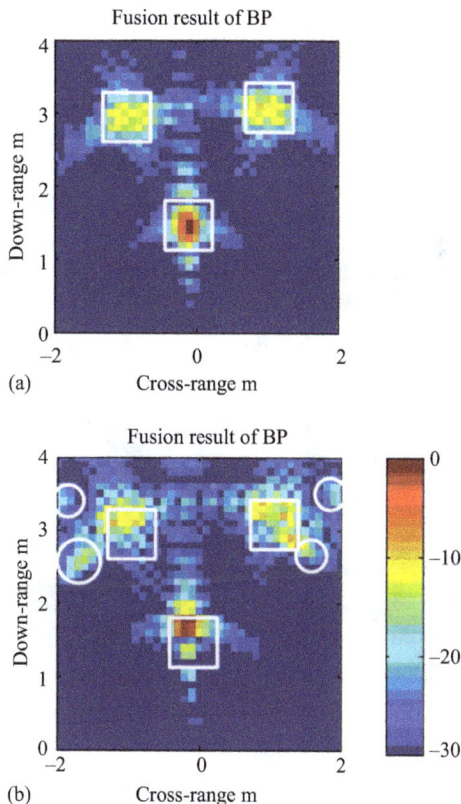

Figure 3.5   Composite image obtained by sum fusion of the BP images with 12.5% of the simulated data: (a) free-space scenario, and (b) through-wall scenario. The targets are noted by white rectangles, while the multipath ghosts are noted by white circles

accurate selection of the basis-signals at each iteration and the decrease of the required number of iterations.

3.3.4.2   Experiments on real radar data

The sparsity-aware algorithms mentioned above are further compared by using real polarimetric TWRI data. The measured data are the same as that used in Section 2.3.4 of Chapter 2. The sparsity level K is set to 50. In Figure 3.8, the images reconstructed by these sparsity-aware methods in free-space are provided based on 12.5% of the full measurement data. From Figure 3.8 (a), it is clear that CMT-BCS tends to highlight the four strongest targets but ignore other weak targets. In Figure 3.8(b) and (c), both SOMP and HMP correctly locate the targets but suffer from some artifacts outside the target area. The imaging result of the TLBMP algorithm is given in Figure 3.8(d), where the dominant resolution cells are well concentrated with clear background.

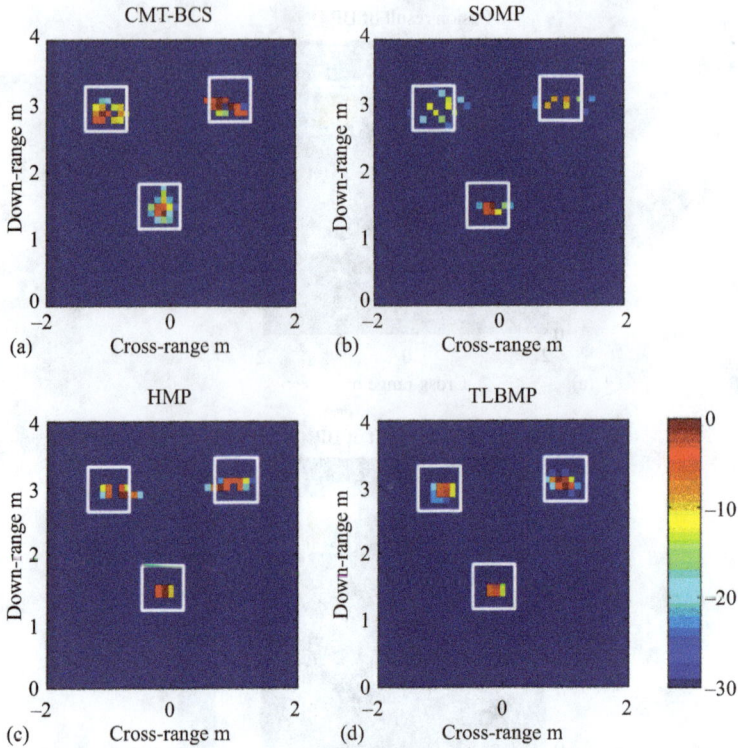

Figure 3.6   Composite image using 12.5% of the simulated data in free-space: (a) CMT-BCS, (b) SOMP, (c) HMP, and (d) TLBMP

The above experiment is repeated with the same parameter setting in the through-wall scenario. The corresponding imaging results of these four sparsity aware algorithms are presented in Figure 3.9. As can be seen in Figure 3.9(a), the strong targets are well reconstructed by CMT-BCS, but some weak targets are still missing. In contrast, the weak targets are clearly reconstructed by the greedy algorithms thanks to the guarantee of the orthogonality between the selected basis-signals and the residual at each iteration. In Figure 3.9(b)–(c), the SOMP and HMP methods, which promote the joint sparsity but do not consider the clustered sparsity, are capable of locating the targets but fail to attenuate artifacts outside the target area. It is clear from Figure 3.9(d) that TLBMP not only locates the targets correctly but also significantly suppresses the artifacts outside the target area.

The quantitative comparison of the radar images generated by the above algorithms is given in Table 3.6. The highest TCR values in free-space and through-wall scenarios are produced by TLBMP. This also implies that TLBMP can achieve more accurate localization of the targets and more effective suppression of the artifacts, which is consistent with the visualized comparisons in Figure 3.8 and

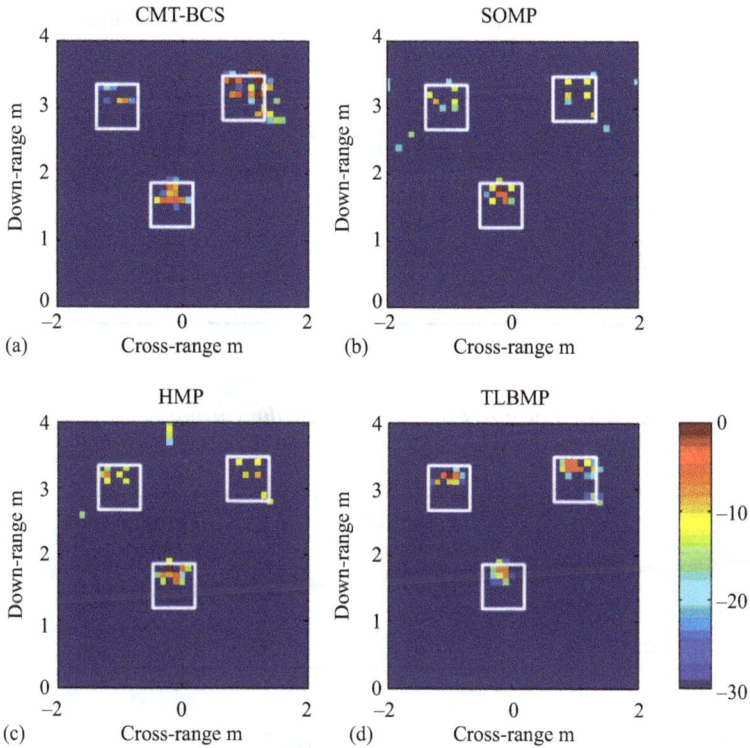

Figure 3.7   Composite images using 12.5% of the simulated data in through-wall scenario: (a) CMT-BCS, (b) SOMP, (c) HMP, and (d) TLBMP

Table 3.3   TCRs of reconstructed images using simulated data (in dB)

	Free-space scenario	Through-wall scenario
BP	19.25	16.57
CMT-BCS	25.56	21.40
SOMP	32.13	24.74
HMP	27.67	22.84
TLBMP	56.51	31.14

Figure 3.9. Note that, for each algorithm, the TCR value produced in through-wall scenario is smaller than that obtained in free-space, which is attributed to the complex propagation in the presence of the wall. A trihedral at 5.7 m down-range is difficult to detect for some algorithms due to its weak scattering reflectivity. The running time and the numbers of iterations of these algorithms are compared in

Table 3.4   Computational time of radar imaging methods
using simulated data (in second)

	Free-space scenario	Through-wall scenario
BP	0.1555	0.1809
CMT-BCS	2039.9	2179.1
SOMP	4.6626	4.6676
HMP	13.950	14.252
TLBMP	1.4389	1.7114

Table 3.5   Iterations of greedy algorithms using
simulated data

	Free-space scenario	Through-wall scenario
SOMP	30	30
HMP	2	3
TLBMP	3	4

Table 3.7 and Table 3.8, respectively, which again demonstrates the superiority of TLBMP in terms of computational complexity.

3.4   STAP based on two-level block sparsity

In this section, we discuss how to apply the two-level block sparsity model to STAP for ground moving target detection with the airborne antenna array radar systems. An algorithm referred to as two-level block STAP (TBS-STAP) is presented for improving the performance of clutter suppression and target detection.

3.4.1   Signal model and algorithm description

The common geometry of an airborne antenna array radar is shown in Figure 3.10, where the aircraft flies along x axis with a velocity v, H denotes the height of the aircraft, the triangle S^k stands for the kth clutter patch in the observed scene, and ψ is the crab angel between the flight direction and the baseline of the array. Here the side-looking scenario is considered, that is, $\psi = 0$. Suppose that a uniform linear array consisting of N antenna elements is equipped and M pulses are received in a coherent processing interval (CPI).

At each range cell, the radar echo can be modeled as the superposition of the signals reflected from K clutter patches. The corresponding elevation angle and the azimuth angle of the kth clutter patch are denoted as θ_c^k and φ_c^k, respectively,

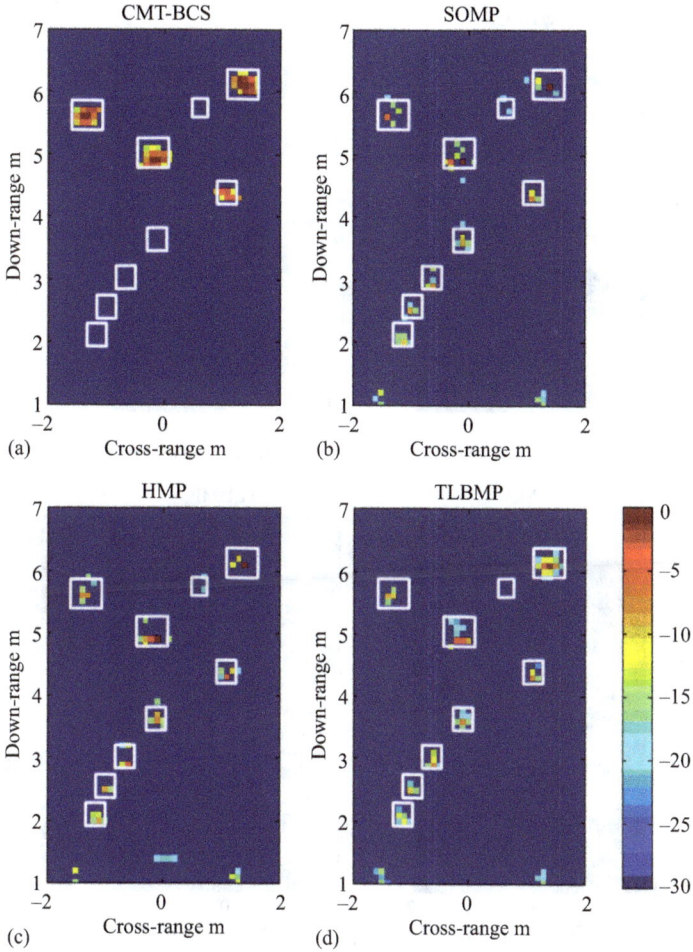

Figure 3.8   Composite images using 12.5% of the real data in free-space scenario: (a) CMT-BCS, (b) SOMP, (c) HMP, and (d) TLBMP

$k = 1, 2, \cdots, K$. The normalized Doppler frequency $f_{d,c}^k$ and spatial frequency $f_{s,c}^k$ of the kth clutter patch can be expressed as [4]

$$f_{d,c}^k = \frac{2v}{\lambda f_r} \cos \varphi_c^k \cos \theta_c^k$$

$$f_{s,c}^k = \frac{d}{\lambda} \cos \varphi_c^k \cos \theta_c^k \qquad (3.25)$$

for $k = 1, 2, ..., N_c$, where d and λ are the antenna spacing and the wavelength of radar, respectively, the subscript c indicates the clutter, and f_r denotes the pulse repetition frequency (PRF).

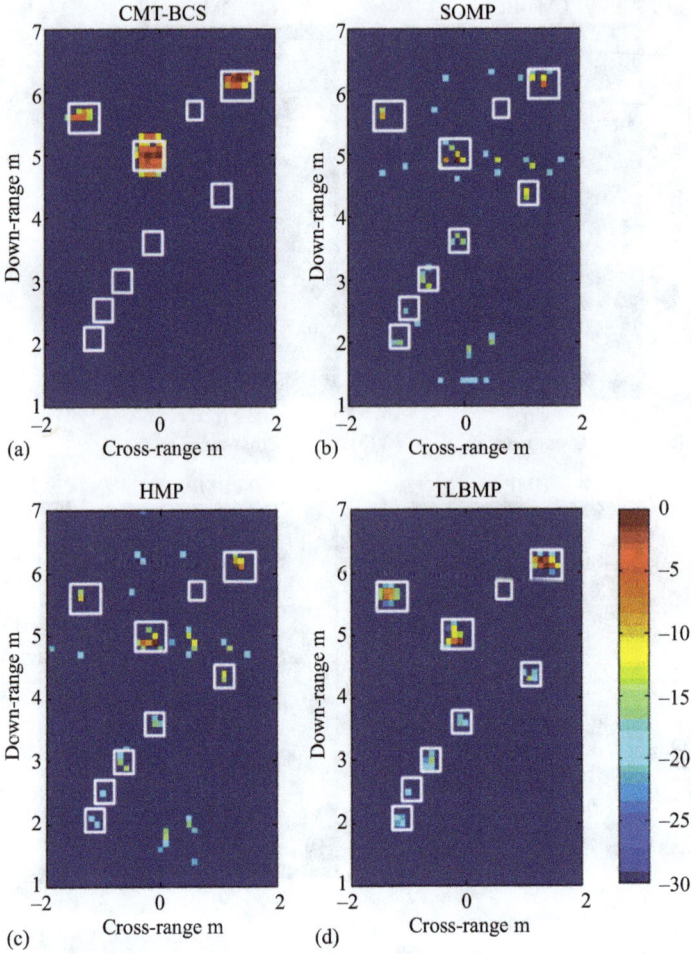

Figure 3.9   Composite images using 12.5% of the real data in through-wall scenario: (a) CMT-BCS, (b) SOMP, (c) HMP, and (d) TLBMP

Table 3.6   TCR of reconstructed images using real data (in dB)

	Free-space scenario	Through-wall scenario
BP	16.41	13.80
CMT-BCS	21.64	15.49
SOMP	18.56	17.66
HMP	20.68	17.55
TLBMP	23.85	21.72

Table 3.7   Computational time of reconstructed
images using real data (in second)

	Free-space scenario	Through-wall scenario
BP	0.1738	0.1732
CMT-BCS	2684.8	3123.3
SOMP	12.709	14.496
HMP	56.146	75.505
TLBMP	5.0642	7.5815

Table 3.8   Iterations of greedy algorithms using
real data

	Free-space scenario	Through-wall scenario
SOMP	50	50
HMP	3	4
TLBMP	7	7

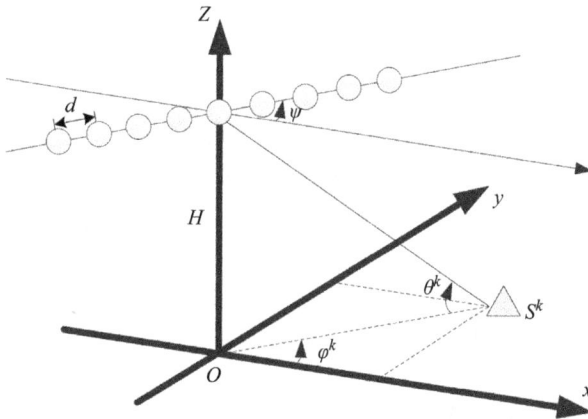

Figure 3.10   The geometry of the airborne antenna array radar

The space-time steering vector of the kth clutter patch at the qth range cell can
be expressed as

$$\mathbf{v}\left(f_{d,c}^{k}, f_{s,c}^{k}\right) = \mathbf{v}_{d}\left(f_{d,c}^{k}\right) \otimes \mathbf{v}_{s}\left(f_{s,c}^{k}\right) \tag{3.26}$$

where $\mathbf{v}_d\left(f_{d,c}^k\right) = \left[1, e^{j2\pi f_{d,c}^k}, \ldots, e^{j2\pi(M-1)f_{d,c}^k}\right]^T$ is the temporal steering vector, and $\mathbf{v}_s\left(f_{s,c}^k\right) = \left[1, e^{j2\pi f_{s,c}^k}, \ldots, e^{j2\pi(N-1)f_{s,c}^k}\right]^T$ is the spatial steering vector. Then the radar echo at the qth range cell can be expressed as [4]

$$\mathbf{x}^{(q)} = \sum_{k=1}^{K} \theta_k^{(q)} \mathbf{v}\left(f_{d,c}^k, f_{s,c}^k\right) + \mathbf{w}^{(q)} \tag{3.27}$$

where $\mathbf{x}^{(q)} \in \mathbb{C}^{NM \times 1}$, $\theta_k^{(q)}$ denotes the complex reflectivity of the kth clutter patch, $\mathbf{w}^{(q)} \in \mathbb{C}^{NM \times 1}$ is the additive Gaussian noise, $q = 1, 2, \cdots, Q$, and Q is the number of range cells involved in STAP. The radar echoes collected at all the Q range cells are called training snapshots. Note that K and $\mathbf{v}\left(f_{d,c}^k, f_{s,c}^k\right)$ are independent of the range cell index q, which holds in homogeneous environments.

By discretizing the angle-Doppler domain into $N_s \times N_d$ grids, where N_s and N_d are the numbers of spatial samples and Doppler samples, respectively, (3.27) can be rewritten as

$$\mathbf{x}^{(q)} = \mathbf{\Phi}\boldsymbol{\theta}^{(q)} + \mathbf{w}^{(q)} \tag{3.28}$$

where $\boldsymbol{\theta}^{(q)} \in \mathbb{C}^{N_d N_s \times 1}$ denotes the clutter profile in the angle-Doppler domain at the qth range cell, $q = 1, 2, \cdots, Q$, $\mathbf{\Phi} \in \mathbb{C}^{NM \times N_d N_s}$ with $NM < N_d N_s$ is an overcomplete dictionary matrix, which is composed of all the space-time steering vectors corresponding to all the positions of the angle-Doppler domain:

$$\mathbf{\Phi} = \left[\mathbf{v}\left(f_d^1, f_s^1\right), \ldots, \mathbf{v}\left(f_d^1, f_s^{N_s}\right), \ldots, \mathbf{v}\left(f_d^{N_d}, f_s^1\right), \ldots \mathbf{v}\left(f_d^{N_d}, f_s^{N_s}\right)\right] \tag{3.29}$$

One can see that the expressions in (3.28) and (3.15) are almost the same. Based on the fundamentals of STAP [3,4], it can be deduced that the two-level block sparsity model is suitable for STAP. On one hand, the dominant coefficients of $\boldsymbol{\theta}^{(q)}$ appear in a clustered shape, since the clutter received by an airborne antenna array normally are spread in both Doppler and spatial domains due to the movement of the platform. On the other hand, $\left\{\boldsymbol{\theta}^{(1)}, \boldsymbol{\theta}^{(2)}, \cdots, \boldsymbol{\theta}^{(Q)}\right\}$, the angle-Doppler-domain profiles at different range cells, share the joint sparsity pattern in homogenous environments. This allows us to solve $\left\{\boldsymbol{\theta}^{(1)}, \boldsymbol{\theta}^{(2)}, \cdots, \boldsymbol{\theta}^{(Q)}\right\}$ by using the TLBMP algorithm presented in Table 3.2.

Once the clutter profile estimates $\left\{\widehat{\boldsymbol{\theta}}^{(1)}, \widehat{\boldsymbol{\theta}}^{(2)}, \cdots, \widehat{\boldsymbol{\theta}}^{(Q)}\right\}$ are solved by the TLBMP algorithm, the clutter covariance matrix (CCM) can be accordingly estimated by [22,23]

$$\widehat{\mathbf{R}} = \frac{1}{Q}\sum_{q=1}^{Q}\sum_{i=1}^{N_d N_s} \left|\widehat{\theta}_i^{(q)}\right|^2 \mathbf{v}\left(f_d^i, f_s^i\right)\mathbf{v}\left(f_d^i, f_s^i\right)^H + \sigma^2\mathbf{I} \tag{3.30}$$

where σ^2 is a small loading factor related to the noise level, and $\mathbf{I}$ is the identity matrix. Note that (3.30) holds under the assumption that the nonzero coefficients in $\boldsymbol{\theta}^{(q)}$ are independent and identically distributed (iid). Then the optimal STAP filter can be obtained by solving the following optimization problem:

$$\min_{\mathbf{w}} \ \mathbf{w}^H \widehat{\mathbf{R}} \mathbf{w} \text{ s.t. } \mathbf{w}^H \mathbf{v}_t\left(f_{d,t}, f_{s,t}\right) = 1 \tag{3.31}$$

where $\mathbf{v}_t\left(f_{d,t}, f_{s,t}\right)$ is the steering vector of the target and the subscript t denotes the target to be detected. The solution of (3.31) can be given by

$$\widehat{\mathbf{w}} = \frac{\widehat{\mathbf{R}}^{-1} \mathbf{v}_t\left(f_{d,t}, f_{s,t}\right)}{\mathbf{v}_t\left(f_{d,t}, f_{s,t}\right)^H \widehat{\mathbf{R}}^{-1} \mathbf{v}_t\left(f_{d,t}, f_{s,t}\right)} \tag{3.32}$$

3.4.2   Experimental results

In this section, the TBS-STAP algorithm is compared with two sparsity-based methods, that is, the sparse recovery with multiple measurement vectors (SRMMV) [24] and the group least absolute shrinkage and selection operator (GLasso) [25], and the classical diagonally loaded sample matrix inversion (LSMI) method [26], by using the simulated data and the real radar data. Note that, SRMMV and Glasso enhance the joint sparsity and the clustered sparsity, respectively, while the TBS-STAP algorithm promotes both of these two kinds of signal sparsity.

3.4.2.1   Simulations

The improvement factor (IF) is widely employed as the metric to evaluate the performance of STAP, which is defined by [3]

$$\text{IF} = \frac{\left|\widehat{\mathbf{w}}^H \mathbf{v}_t\left(f_d, t, f_{s,t}\right)\right|^2 / \widehat{\mathbf{w}}^H \mathbf{R} \widehat{\mathbf{w}}}{\mathbf{v}_t\left(f_d, t, f_{s,t}\right)^H \mathbf{v}_t\left(f_{d,t}, f_{s,t}\right) / \text{tr}(\mathbf{R})} \tag{3.33}$$

where $\mathbf{R}$ denotes the true covariance matrix of the clutter-plus-noise. The higher IF value means better detection performance of clutter suppression.

The parameters of the simulated airborne radar system are listed in Table 3.9. The angle-Doppler domain is discretized into 64 Doppler bins and 64 angle bins, that is, $N_d = N_s = 64$. In the TBS-STAP algorithm, the sparsity level K is set to 64. In the GLasso algorithm, the group size of the clutter profile is set to 2.

The IF value versus the Doppler frequency of the target is plotted in Figure 3.11. For covariance matrix estimation, the training snapshots are collected from $Q = 8$ and $Q = 20$ range cells in Figure 3.11(a) and (b), respectively. Both of these two numbers of training snapshots are insufficient for the traditional LSMI algorithm, resulting in small IF values of LSMI in Figure 3.11. As can be seen from Figure 3.11, around the zero Doppler, the IF value produced by TBS-STAP is larger than that of SRMMV and GLasso. This benefits from the two-level block

Table 3.9   *Parameters of the simulated airborne radar system*

Antenna array	Side-looking uniform linear array
Antenna spacing	Half-wavelength
Bandwidth	10 MHz
PRF	3,000 Hz
Platform velocity	150 m/s
Platform height	10,000 m
Number of antennas	14
Number of pulses per CPI	16
Clutter-to-noise ratio	40 dB

sparsity model utilized during the CCM estimation, that is, both of the clustered sparsity of the angle-Doppler domain at a single range cell and the common sparsity pattern across all the training snapshots collected at multiple range cells are promoted simultaneously. It is expected from Figure 3.11 that the TBS-STAP algorithm has better detection performance for slowly moving targets.

In addition, the true clutter profile and the estimation results at the cell under test using different algorithms are shown in Figure 3.12. The intensity of the clutter profile is set in the range of $[-35, 0]$ dB, with the maximum intensity value normalized to 0 dB. Figure 3.12(a) provides the true clutter profile. The results estimated by TBS-STAP, SRMMV, and GLasso based on the training snapshots collected at 20 range cells are provided in Figure 3.12(b)–(d), respectively. It is obvious that the clutter profile estimated by TBS-STAP is closer to the true value. This explains why TBS-STAP produces better estimation of CCM and accordingly outputs higher IF values.

3.4.2.2   Experiments on real radar data

In this section, the STAP algorithms mentioned above are compared by using the real radar data. The Mountain-top radar data were acquired by a stationary antenna array which was designed to emulate an airborne radar [27]. The measurements collected by 14 antennas at 16 pulses are available in a CPI. A target is moving at the 147th range cell with a normalized Doppler frequency of 0.25. The training snapshots are collected from those range cells around the range cell under test, excluding four guard cells.

The range profiles produced by different algorithms based on 8 and 20 training snapshots are plotted in Figure 3.13 and Figure 3.14, respectively. The maximum value of each curve is normalized to 0 dB. The target position and the gap between the largest and the second largest values of every range profile are specifically indicated. The larger is the gap, the easier is the moving target detection. It is clear that moving target detection is easier to be carried out on the range profile produced by TBS-STAP. One may set a detection threshold in Figure 3.13 and Figure 3.14 according to the desired probability of false alarm, and it can be intuitively seen that TBS-STAP will achieve the highest

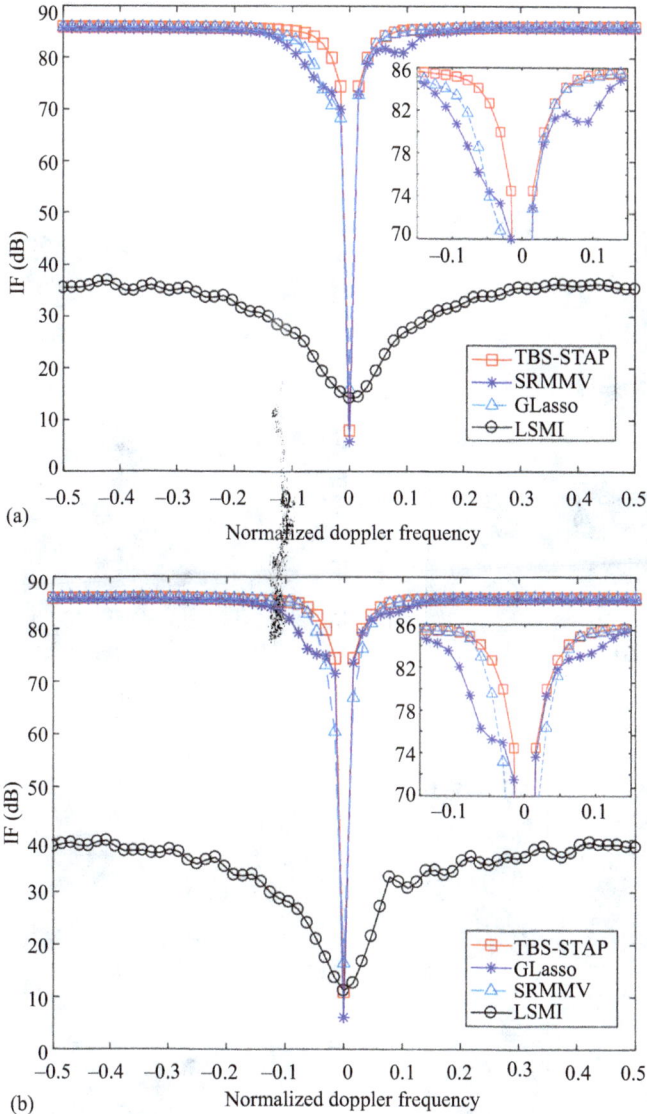

Figure 3.11   The IF curves with the training snapshots collected at (a) 8 range cells and (b) 20 range cells

probability of detection among these algorithms. In Table 3.10, the gaps between the largest and the second largest values of the range profiles outputted by these algorithms are summarized when $Q = 8$, 14, and 20. All the above experimental results demonstrate the superiority of TBS-STAP in clutter

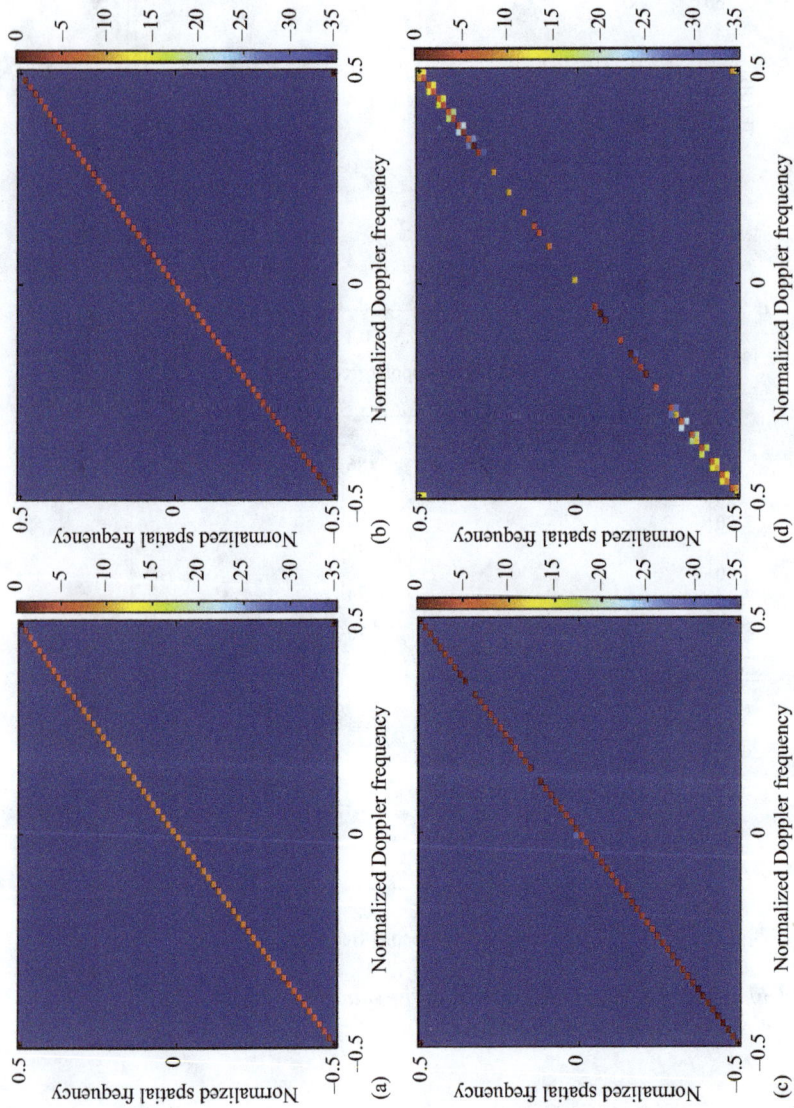

Figure 3.12 True clutter profile and the estimation results: (a) true clutter profile; (b) the result estimated by TBS-STAP; (c) the result estimated by SRMMV; and (d) the result estimated by GLasso

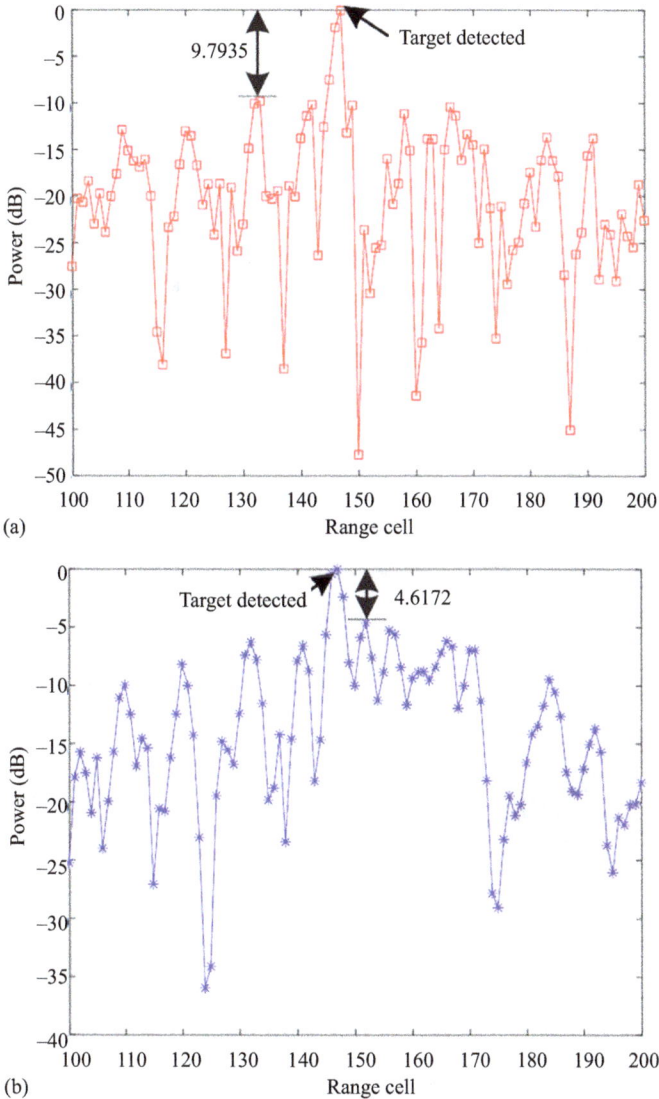

Figure 3.13 Range profiles produced by (a) TBS-STAP, (b) GLasso, (c) SRMMV, and (d) LSMI, based on the training snapshots collected at 8 range cells

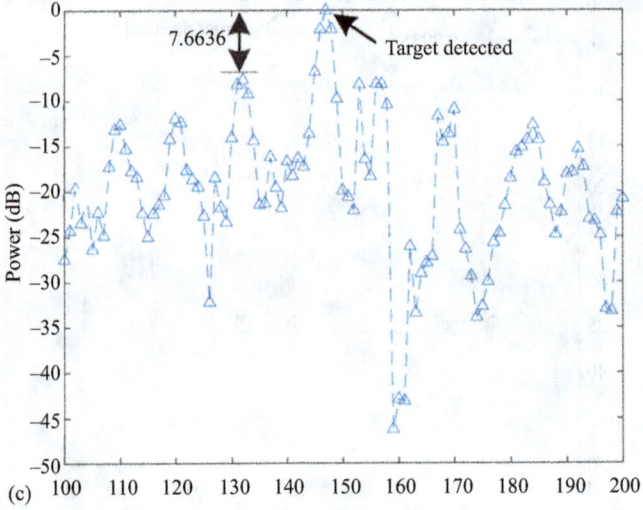

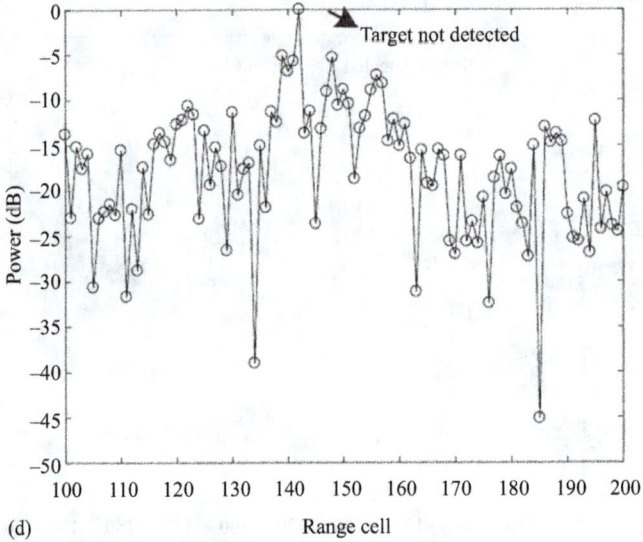

Figure 3.13   (Continued)

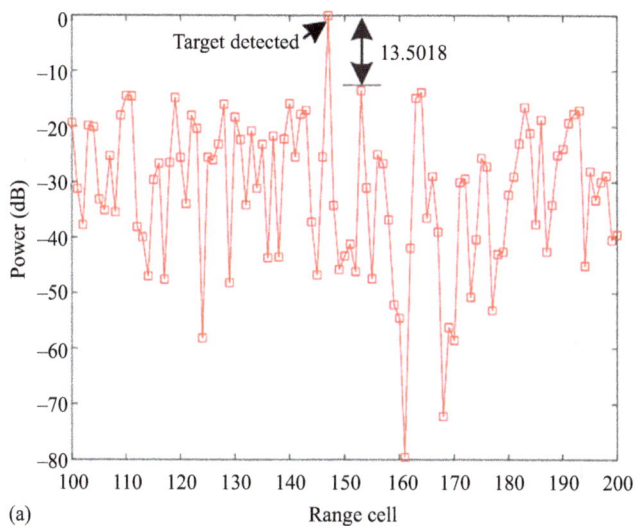

(a)

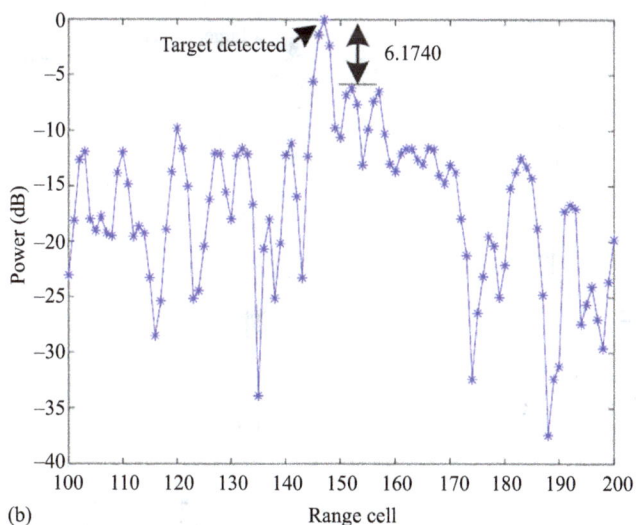

(b)

Figure 3.14   *Range profiles produced by (a) TBS-STAP, (b) GLasso, (c) SRMMV, and (d) LSMI, based on the training snapshots collected at 20 range cells*

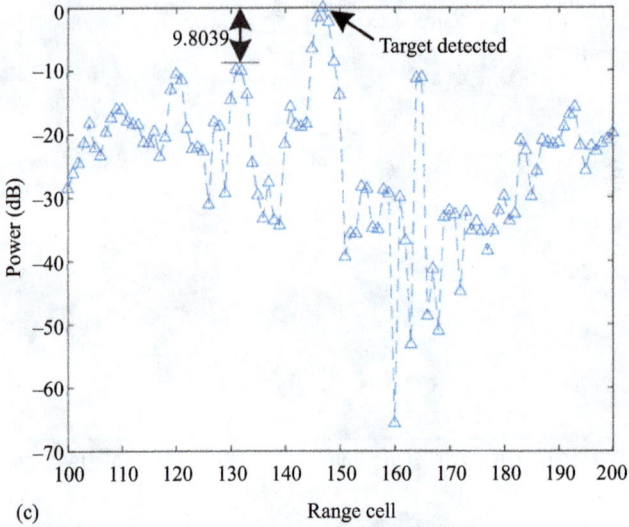

(c)

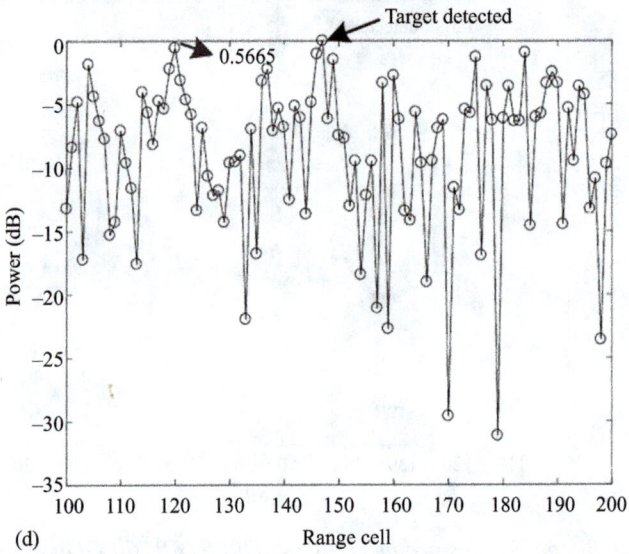

(d)

Figure 3.14   (Continued)

Table 3.10 The gaps between the largest and the second largest values of the range profiles produced by different algorithms

Algorithms	With 8 training snapshots	With 14 training snapshots	With 20 training snapshots
TBS-STAP	9.7935 dB	12.3502 dB	13.5018 dB
SRMMV	7.6636 dB	9.2092 dB	9.8039 dB
GLasso	4.6172 dB	5.5285 dB	6.1740 dB
LSMI	−9.0815 dB	−0.6325 dB	0.5665 dB

suppression and moving target detection, especially in the case of small number of training snapshots.

3.5 Conclusion

In this chapter, we presented an advanced sparse signal model referred to as two-level block sparsity model and introduced its applications in multichannel radar signal processing such as TWRI and STAP. By enforcing both the clustered sparsity of each single-channel signal and the joint sparsity pattern of the signals across all the channels, the two-level block sparsity model can help in clustering the dominant components and suppressing the artifacts. In the case of TWRI, the two-level block sparsity model was directly applied to radar image formation in free-space and through-wall scenarios. In the case of STAP, the two-level block sparsity model was utilized to first reconstruct the angle-Doppler domain and then estimate CCM. The experimental results on simulations and real radar data have demonstrated the positive effect of the two-level block sparsity model on improving the quality of TWRI and enhancing the detection performance of STAP. The applications of the two-level block sparsity model can also be extended to other multichannel radar systems, such as multiple-input multiple-output (MIMO) radar, multistatic radar, and distributed radar. To further develop the two-level block sparsity model in various applications, it is worth studying more accurate formulation of the prior knowledge about the clustered structure and more simplified approaches for selection of model parameters.

References

[1] Yang J., Bouzerdoum A., Tivive F. H. C., and Amin M. G. "Multiple-measurement vector model and its application to through-wall radar imaging." *Proceedings of the IEEE International Conference on Acoustics, Speech and Signal Processing (ICASSP)*. Prague, Czech Republic: IEEE; 2011, pp. 2672–2675.

[2] Bouzerdoum A., Yang J., and Tivive F. H. C. "Compressive sensing for multi-polarization through wall radar imaging." In M. G. Amin (ed.). *Compressive sensing for urban radar*. Boca Raton, FL: CRC Press; 2014.

[3]   Klemm R. *Principles of space-time adaptive processing*. London, UK: IEEE Press; 2002.

[4]   Ward J. "Space-time adaptive processing for airborne radar." *MIT Technical Report 1015*. MIT Lincoln Laboratory, 1994.

[5]   Eldar Y. C., Kuppinger P., and Bolcskei H. "Block-sparse signals: Uncertainty relations and efficient recovery." *IEEE Transactions on Signal Processing*. 2010; 58(6): 3042–3054.

[6]   Wang X., Li G., Liu Y., and Amin M. G. "Two-level block matching pursuit for polarimetric through-wall radar imaging." *IEEE Transactions on Geoscience and Remote Sensing*. 2018; 56(3): 1533–1545.

[7]   Jiang Z., Li G., Wang X., Zhang X.-P., and He Y. "STAP based on two-level block sparsity." *Proceedings of IEEE Radar Conference*. Boston, MA, USA: IEEE; 2019.

[8]   Wu Q., Zhang Y. D., Ahmad F., and Amin M. G. "Compressive-sensing-based high-resolution polarimetric through-the-wall radar imaging exploiting target characteristics." *IEEE Antennas and Wireless Propagation Letters*. 2015; 14: 1043–1047.

[9]   Li G. and Burkholder R. J. "Hybrid matching pursuit for distributed through-wall radar imaging." *IEEE Transactions on Antennas and Propagation*. 2015; 63(4): 1701–1711.

[10]   Leigsnering M., Ahmad F., Amin M. G., and Zoubir A. "Multipath exploitation in through-the-wall radar imaging using sparse reconstruction." *IEEE Transactions on Aerospace Electronic Systems*. 2014; 50(2): 920–939.

[11]   Li S. Z. *Markov random field modeling in image analysis*, 2nd ed. New York, NY: Springer-Verlag; 2001.

[12]   Bouzerdoum A., Tivive F. H. C., and Tang V. H. "Multi-polarization through-the-wall radar imaging using joint Bayesian compressed sensing." *Proceedings of 2014 19th IEEE International Conference on Digital Signal Processing*, 2014, pp. 783–788.

[13]   Zhou Z., Wagner A., Mobahi H., Wright J., and Ma Y. "Face recognition with contiguous occlusion using Markov random fields." *Proceeding of IEEE International Conference on Computer Vision*. 2009, pp. 1050–1057.

[14]   Sundman D., Chatterjee S., and Skoglund M. "Analysis of democratic voting principles used in distributed greedy algorithms." Available at: https://arxiv.org/abs/1407.4491 [accessed 16 July 2014].

[15]   Blumensath T. and Davies M. E. "Iterative hard thresholding for compressed sensing." *Applied and Computational Harmonic Analysis*. 2009; 27(3): 265–274.

[16]   Cartis C. and Thompson A. "A new and improved quantitative recovery analysis for iterative hard thresholding algorithms in compressed sensing." *IEEE Transactions on Information Theory*. 2015; 61(4): 2019–2042.

[17]   Needell D. and Tropp J. A. "CoSaMP: Iterative signal recovery from incomplete and inaccurate samples." *Applied and Computational Harmonic Analysis*. 2009; 26(3): 301–321.

[18]   Björck Å. *Numerical methods for least squares problems*. Philadelphia, PA: SIAM; 1996. pp. 60–62.

[19] Tropp J. A., Gilbert A. C., and Strauss M. J. "Algorithms for simultaneous sparse approximation. Part I: Greedy pursuit." *Signal Processing.* 2006; 86 (3): 572–588.

[20] Seng C. H., Bouzerdoum A., Amin M. G., and Phung S. L. "Probabilistic fuzzy image fusion approach for radar through wall sensing." *IEEE Transactions on Image Processing.* 2013; 22(12): 4938–4951.

[21] Wang G. and Amin M. G. "Imaging through unknown walls using different standoff distances." *IEEE Transactions on Image Processing.* 2006; 54(10): 4015–4025.

[22] Sen S. "Low-rank matrix decomposition and spatio-temporal sparse recovery for STAP radar." *IEEE Journal of Selected Topics in Signal Processing.* 2015; 9(8): 1510–1523.

[23] Yang Z., Lamare R. C., and Li X. "L1-regularized STAP algorithms with a generalized sidelobe canceler architecture for airborne radar." *IEEE Transactions on Signal Processing.* 2012; 60(2): 674–686.

[24] Sun K., Zhang H., Li G., Meng H., and Wang X. "Airborne radar STAP using sparse recovery of clutter spectrum." arXiv preprint arXiv:1008.4185, 2010.

[25] Yuan M. and Lin Y. "Model selection and estimation in regression with grouped variables." *Journal of the Royal Statistical Society: Series B (Statistical Methodology).* 2006; 68(1): 49–67.

[26] Carlson B. D. "Covariance matrix estimation errors and diagonal loading in adaptive arrays." *IEEE Transactions on Aerospace and Electronic Systems.* 1988; 24(4): 397–401.

[27] Titi G. W., and Marshall D. F. "The ARPA/NAVY mountaintop program: Adaptive signal processing for airborne early warning radar." *Proceedings of International Conference on Acoustics, Speech, and Signal Processing (ICASSP).* Atlanta, GA, USA: IEEE; 1996, pp. 1165–1168.

Chapter 4
Parametric sparse representation for radar imaging with model uncertainty

4.1 Introduction

The traditional methods of radar imaging relies on the model certainty. For example, the pulse compression needs the exact knowledge of the waveform of the transmitted signal for producing the high-resolution range profile. Another example is that the commonly used algorithms of synthetic aperture radar (SAR) imaging such as the range-Doppler algorithm [1], the chirp scaling algorithm [2], and the $\omega-$K algorithm [3] require the exact geometric relationship between the radar platform and the observed scene. In sparsity-driven algorithms for radar imaging, model certainty means that the basis-signal dictionary is predesigned and fixed during the imaging process. However, the model uncertainty may occur during the radar imaging process, for example, the unknown motion of the targets and the undesired trajectory error of the radar platform, which may result in the error of raw radar data and seriously degrade the radar imaging quality. There are a number of traditional algorithms for compensating the model uncertainty, without exploiting the sparsity of the signals as a prior knowledge. In order to eliminate the influence of the model uncertainty on sparsity-driven radar imaging, in this chapter, we present a method referred to as parametric sparse representation (PSR). The key idea of PSR is that the model uncertainty is formulated as a group of parameters inside the dictionary so that the dictionary updating and the sparse signal recovery can be simultaneously accomplished. The updating of the parametric dictionary can be viewed as an adaptive adjustment for correcting the error of the radar data caused by the model uncertainty. When the model uncertainty is eliminated by training a proper dictionary, the sparse solution is expected to be able to well represent the reflectivity density function of the observed scene. We also demonstrate the applications of the PSR method to some radar imaging scenarios with the model uncertainty, including SAR refocusing of moving targets [4], SAR motion compensation [5] and inverse synthetic aperture radar (ISAR) imaging of maneuvering targets [6]. Simulations and experimental results on real radar data show that the PSR method is capable of significantly improving the quality of radar imaging in presence of the model uncertainty.

4.2   Parametric dictionary

Consider the following model of radar imaging [7]:

$$\mathbf{y} = \mathbf{\Phi}\mathbf{x} + \mathbf{w}, \tag{4.1}$$

where $\mathbf{y}$ is the radar measurement vector, $\mathbf{\Phi}$ is the basis-signal dictionary, $\mathbf{x}$ is the sparse imaging result, and $\mathbf{w}$ is the additive noise. The goal of radar imaging is to reconstruct $\mathbf{x}$ by using the dictionary $\mathbf{\Phi}$ and the measurement vector $\mathbf{y}$. The traditional method based on matched filtering can be expressed as

$$\widehat{\mathbf{x}} = \mathbf{\Phi}^H \mathbf{y} \tag{4.2}$$

The sparsity-driven methods for radar imaging can be expressed as [7]

$$\widehat{\mathbf{x}} = f(\mathbf{\Phi}, \mathbf{y}) \tag{4.3}$$

where $f(\cdot)$ is a function with sparsity constraint. Note that (4.3) is dependent on the value of $\mathbf{\Phi}$, which is predesigned before the imaging process based on the radar system parameters such as the wavelength, the waveform, the platform motion, and the imaging geometry. When some uncertainties such as the unknown motion of the targets and the undesired trajectory error of the radar platform exist, the radar imaging quality may be seriously degraded due to the difference between the true value of $\mathbf{\Phi}$ and that used in the imaging procedure.

To solve this problem, in this chapter we present the PSR method for sparsity-driven radar imaging. The key idea of PSR is to design a parametric dictionary, which is related to the model uncertainty. By modifying (4.1), the model of PSR can be expressed as [4–6,8,9]

$$\mathbf{y} = \mathbf{\Phi}(\omega)\mathbf{x} + \mathbf{w} \tag{4.4}$$

where ω denotes the vector of unknown parameters that represents the model uncertainty. The physical meaning of ω may vary with different applications. There are some examples of the model uncertainty: in the scenario of SAR/ISAR imaging of moving targets, ω may represent the unknown motion status of a moving target; in the scenario of airborne SAR motion compensation, ω may represent the non-ideal trajectory parameters of the radar platform; in the case of through-wall radar imaging, ω may represent the characteristics of the unknown propagation due to the presence of the wall. Accordingly, we need to solve not only $\mathbf{x}$ but also ω from (4.4), which increases the computational cost of PSR in comparison with the traditional sparse signal recovery. As shown later in Sections 4.3, 4.4, and 4.5, the solution of $\{\mathbf{x}, \omega\}$ can be obtained by alternating iterations or parameter searching.

The advantage of the parametric dictionary lies in that the presence of the adjustable parameter vector ω in (4.4) allows us to dynamically update the dictionary as the sparse solution is iteratively refined. Thus, the PSR method can be viewed as a special case of dictionary learning [10]. Different from the general approaches of dictionary learning, the dictionary updating in PSR is equivalent to the estimation of ω that has definite physical meaning. This may provide more information on the observed targets in addition to the radar image.

4.3 Application to SAR refocusing of moving targets

In this section, the PSR method is applied to SAR refocusing of moving targets.

4.3.1 Problem formulation

Since the SAR imaging algorithms are originally designed for stationary targets, the main challenge of SAR imaging of moving targets is to compensate the phase error caused by the noncooperative motion of the targets. It is well known that the target movement results in not only the azimuth displacement but also the defocus of the moving targets in a regular SAR image [11]. The complex image of the region of interest (ROI) that contains the defocused moving target can be cropped from the regular SAR image. By doing so, the amount of data to be processed is significantly reduced and a large part of the stationary clutter is effectively removed. The goal of SAR refocusing is to obtain the focused image of the moving target by using the ROI data.

The geometry of a side-looking SAR is shown in Figure 4.1. The horizontal axis denotes the azimuth direction and the vertical axis denotes the slant-range direction. Assume that the SAR platform flies straight at a speed of V along the x-axis, the velocities of a moving target in azimuth and range are v_x and v_r, respectively.

The slow time is denoted as $t = nT$, where n is an integer and T is the pulse repetition interval. Thus, the radar antenna phase center (APC) position at t is $(Vt, 0)$, and the instantaneous position of the moving target is $(x_0 + v_x t, r_0 + v_r t)$, where (x_0, r_0) is the initial position at $t = 0$. The instantaneous distance between

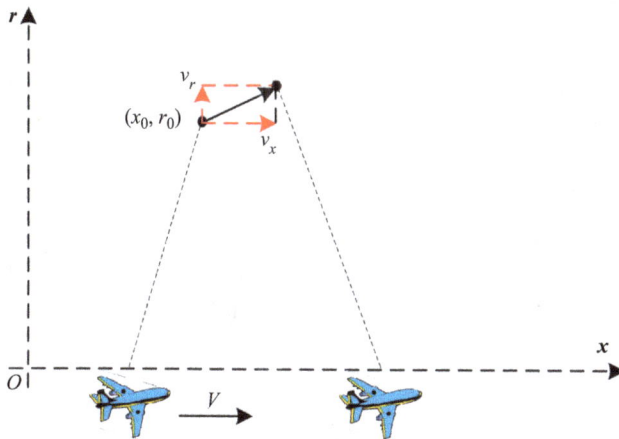

Figure 4.1   Geometry of SAR imaging of a moving target

the moving target and the radar can be expressed as

$$R(t) = \sqrt{(Vt - x_0 - v_x t)^2 + (r_0 + v_r t)^2} \qquad (4.5)$$

Suppose that the radar transmits a linear frequency-modulated (LFM) signal:

$$s(\tau) = \text{rect}\left(\frac{\tau}{T_p}\right) \cdot \exp\left(j2\pi f_c \tau + j\pi\gamma\tau^2\right) \qquad (4.6)$$

where τ is the fast time, T_p is the pulse width, $\text{rect}(\cdot)$ denotes the rectangular function, f_c is the carrier frequency, and γ is the chirp rate of the transmitted signal. The baseband echo of the moving target can be expressed as

$$
\begin{aligned}
s_r(\tau, t) = \sigma \cdot &\text{rect}\left(\frac{\tau - 2R(t)/c}{T_p}\right) \cdot \text{rect}\left(\frac{t}{T_a}\right) \\
&\cdot \exp\left(-j4\pi\frac{f_c R(t)}{c} + j\pi\gamma\left(\tau - \frac{2R(t)}{c}\right)^2\right)
\end{aligned}
\qquad (4.7)
$$

where T_a is the synthetic aperture time and the target scattering coefficient σ is assumed to be constant during the observation time interval. By taking the two-dimensional (2D) Fourier transform, the signal of the moving target in 2D frequency domain can be expressed as

$$
\begin{aligned}
S_r(f_r, f_a) = &\sigma \cdot W_r(f_r) \cdot W_a(f_a) \\
&\cdot \exp\left\{j \cdot \left[-2\pi f_a \frac{\delta}{v_e^2} - \pi\frac{f_r^2}{\gamma} - \frac{4\pi[x_0 v_r + r_0(V - v_x)]}{c v_e}\sqrt{(f_c + f_r)^2 - \frac{c^2 f_a^2}{4 v_e^2}}\right]\right\}
\end{aligned}
\qquad (4.8)
$$

where f_r and f_a are the range and azimuth frequencies, respectively, $W_r(f_r)$ and $W_a(f_r)$ are the envelope functions in range and azimuth frequency domains, respectively, $\delta = x_0(V - v_x) - r_0 v_r$, and $v_e = \sqrt{(V - v_x)^2 + v_r^2}$.

Assume that the $\omega-$K algorithm [3] is used to image the stationary scene. In other words, the result of the $\omega-$K algorithm is the input data of the procedure of moving target refocusing. After performing the Stolt interpolation of the $\omega-$K algorithm, (4.8) becomes [12]

$$
\begin{aligned}
S_{r\_\text{Stolt}}(f_r, f_a) = &\sigma \cdot W_r(f_r) \cdot W_a(f_a) \\
&\cdot \exp\left\{j \cdot \left[-2\pi f_a \frac{\delta}{v_e^2} - \frac{4\pi[x_0 v_r + r_0(V - v_x)]}{c v_e}\sqrt{(f_c + f_r)^2 + \frac{c^2 f_a^2}{4}\left(\frac{1}{V^2} - \frac{1}{v_e^2}\right)}\right]\right\}
\end{aligned}
\qquad (4.9)
$$

For stationary targets with $v_r = v_x = 0$, the phase term in (4.9) becomes $\exp\left\{j \cdot \left[-2\pi f_a \frac{x_0}{V} - \frac{4\pi r_0}{c}(f_c + f_r)\right]\right\}$, so a focused image can be obtained by taking the 2D inverse Fourier transform in (4.9). For moving targets, the phase error still exists in (4.9). This means that the image of the moving target in the output of the $\omega - K$ algorithm is blurred. The phase error caused by the target motion can be approximately compensated by using the following filter [12]:

$$H_a(f_r, f_a) = \exp\left\{ j \cdot \left[\frac{4\pi R_{\text{ref}}}{c} \sqrt{(f_c + f_r)^2 + \frac{c^2 f_a^2}{4}\left(\frac{1}{V^2} - \alpha\right)} - \frac{4\pi R_{\text{ref}}}{c}(f_c + f_r) \right] \right\}$$

(4.10)

where R_{ref} is the reference range, and

$$\alpha \overset{\Delta}{=} \frac{1}{v_e^2} = \frac{1}{(V - v_x)^2 + v_r^2}$$

(4.11)

is the phase compensation parameter related to the motion parameters of the target. When the target is stationary, $H_a(f_r, f_a) = 1$ in (4.9), which means that no phase compensation is required for the data reflected from the stationary targets.

4.3.2 Algorithm description

In this section, the PSR method for SAR refocusing of moving targets is formulated [4]. In the image of the entire scene generated by the standard $\omega - K$ algorithm, the subimage containing the moving target is cropped and denoted by s_{r_ROI}. The refocused image of the moving target can be obtained from s_{r_ROI} through the following refocusing transform $\Gamma_\alpha(\cdot)$:

$$\Theta = \Gamma_\alpha(s_{r\_\text{ROI}}) = \Psi_r^{-1} \cdot \left[\left(\Psi_r \cdot s_{r\_\text{ROI}} \cdot \Psi_a \right) \circ H_a \right] \cdot \Psi_a^{-1}$$

(4.12)

where Θ is the refocusing result of the moving target, Ψ and Ψ^{-1} are the Fourier matrix and the inverse Fourier matrix, respectively, the subscripts r and a denote the range direction and azimuth direction, respectively, and the elements of H_a are given in (4.10). As shown in (4.12), through a series of matrix operations, the refocusing transform $\Gamma_\alpha(\cdot)$ can achieve the conversion from the defocused ROI data to the refocusing result of moving target. It is clear that the refocusing quality of Θ depends on the phase compensation parameter α. Different value of the phase compensation parameter α produces different filter function H_a and, therefore, different refocusing result Θ. Thus, Θ can be regarded as a function of α and denoted by $\Theta(\alpha)$. With a wrong value of α, a blurred image $\Theta(\alpha)$ is most likely obtained. With an accurate estimate of α, a well-focused image $\Theta(\alpha)$ will be obtained. From the above observation, moving target refocusing can be carried out in an iterative fashion, that is, the sparse imaging result Θ and the phase compensation parameter α are iteratively updated.

The PSR method for SAR refocusing of moving targets can be implemented by alternating iterations, as summarized in Figure 4.2. The iterative process is terminated until the difference between the estimated values of α at two consecutive iterations is smaller than a convergence threshold η.

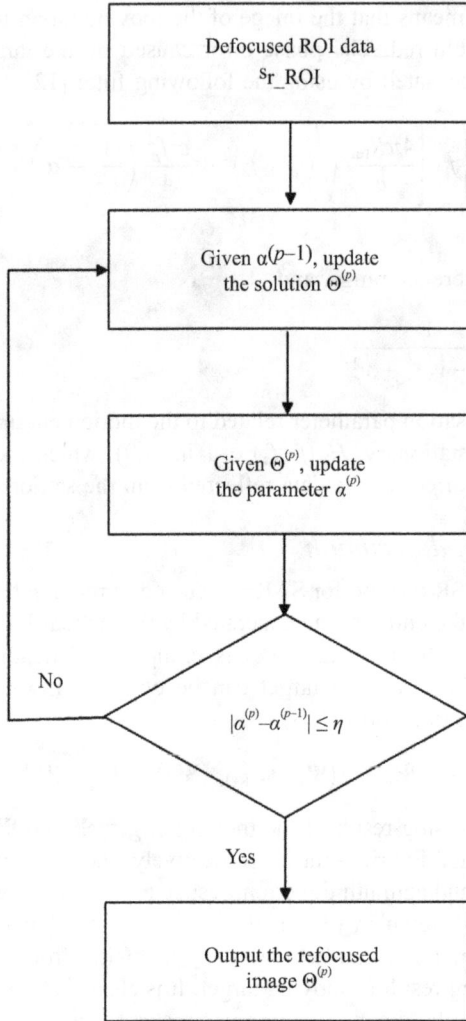

Figure 4.2   Flowchart of the PSR method for SAR refocusing of moving targets

4.3.2.1   Update the sparse solution

The initialized value of α is given by $\alpha^{(0)} = 1/V^2$, where the superscript denotes the iteration index. From (4.11), one can see that the value of $\alpha^{(0)}$ is set in accordance to the stationary target case. Given the value of $\alpha^{(p-1)}$, at the pth iteration, the imaging result $\Theta^{(p)}$ can be updated as follows. The inverse transform of (4.12) can be written as

$$\mathbf{s}_{r\_ROI} = \Gamma^{-1}_{\alpha^{(p-1)}}\left(\Theta^{(p)}\right) = \Psi_r^{-1} \cdot \left[\left(\Psi_r \cdot \Theta^{(p)} \cdot \Psi_a\right) \circ \mathbf{H}^*\left(\alpha^{(p-1)}\right)\right] \cdot \Psi_a^{-1}$$

$$(4.13)$$

where $\Gamma_{\alpha^{(p-1)}}^{-1}(\cdot)$ is the inverse operation of $\Gamma_\alpha(\cdot)$ with the parameter $\alpha^{(p-1)}$. The image of the moving target can be considered to be sparse, so $\Theta^{(p)}$ can be solved by

$$\widehat{\Theta}^{(p)} = \arg\min_{\Theta} \|\Theta\|_0, \quad \text{s.t.} \ \|s_{r\_ROI} - \Gamma_{\alpha^{(p-1)}}^{-1}(\Theta)\|_2^2 \leq \varepsilon \tag{4.14}$$

where $\varepsilon > 0$ is the reconstruction error tolerance. Given the value of $\alpha^{(p-1)}$, (4.14) is a standard problem of sparse signal recovery. Here the iterative hard thresholding (IHT) algorithm [13] is used to solve (4.14). The detailed steps of IHT are also provided in Chapter 1.

4.3.2.2   Update the estimate of the phase compensation parameter

Given the sparse solution $\widehat{\Theta}^{(p)}$, the phase compensation parameter can be updated by minimizing the recovery error, that is,

$$\alpha^{(p)} = \arg\min_{\alpha} \|s_{r\_ROI} - \Gamma_\alpha^{-1}(\widehat{\Theta}^{(p)})\|_2^2 \tag{4.15}$$

It is clear that (4.15) is a problem of single variable optimization, so it can be easily solved by searching approach such as bisection search or golden section search. Solving (4.15) can be also carried out by the Newton's method, that is, solving the following problem:

$$\Delta\alpha = \arg\min_{\Delta\alpha} \left\| \Psi_r \cdot s_{r\_ROI} \cdot \Psi_a - \left(\Psi_r \cdot \Theta^{(p)} \cdot \Psi_a\right) \right.$$
$$\left. \circ \left[H^*(\alpha^{(p-1)}) + \frac{dH^*(\alpha)}{d\alpha}\bigg|_{\alpha^{(p-1)}} \cdot \Delta\alpha \right] \right\|_2^2 \tag{4.16}$$

where $\Delta\alpha$ denotes increment to be estimated. Once $\Delta\alpha$ is solved from (4.16), the estimate of the phase compensation parameter can be updated by $\alpha^{(p)} = \alpha^{(p-1)} + \Delta\alpha$.

4.3.3   Experimental results

The experimental results based on simulated data and real SAR data collected by the GF-3 satellite are presented in this section to demonstrate the effectiveness of the PSR method for refocusing of moving targets.

4.3.3.1   Simulations

The parameters of the simulations are as follows. The radar carrier frequency is 10 GHz, the scene center range is 10 km, the velocity of the airborne radar platform is 150 m/s, the bandwidth of the radar signal is 300 MHz, and the pulse width is 2.2 μs. The simulated scene contains two stationary reference scatterers (S1–S2) and a rigid-body moving target consisting of four scatterers (M1–M4), as shown in Figure 4.3.

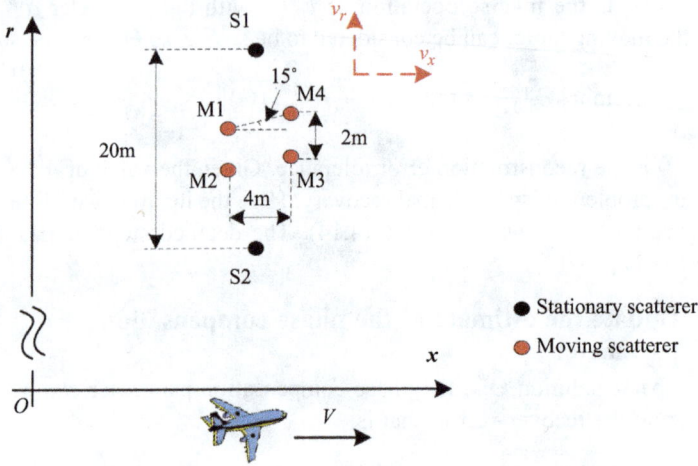

Figure 4.3   Simulated scene consisting of a rigid-body moving target and two stationary scatterers

In the first simulation, the target is assumed to move with a constant speed $v_x = 10$ m/s and $v_r = 5$ m/s. Figure 4.4 shows the imaging result of the entire scene obtained by the $\omega-$K algorithm, where the defocused ROI data containing the moving target is indicated by a red dashed box. The ROI subimage with the size of $30 \times 1,051$ is cropped from the regular SAR image and inputted into the PSR method. The initialized value of phase compensation parameter is set as $\alpha^{(0)} = 1/V^2 = 1/150^2$, and the convergence threshold in Figure 4.2 is set as $\eta = \alpha^{(0)}/10^4$. The convergence process of the phase compensation parameter estimation is shown in Figure 4.5. The defocused ROI data and the refocusing results obtained by the algorithm proposed in [12] and the PSR method are shown in Figure 4.6. It can be seen that both of these two methods can produce well refocused image of the moving target, while the PSR method also effectively suppresses the side-lobes.

In the second simulation, the target is assumed to move with the initial speed $v_x = 10$ m/s and $v_r = 5$ m/s and the acceleration $a_x = 1$ m/s^2 and $a_r = 1$ m/s^2, which deviates from the motion model used in the PSR method. As shown in Figure 4.7, the PSR method also converges well even in the presence of model mismatch. The refocusing results obtained by the method in [12] and the PSR method are compared in Figure 4.8. Especially, the range and azimuth profiles of a moving scatterer (M3) are plotted. One can find that the refocusing performance of the method in [12] is deteriorated with asymmetric side-lobes and wider main-lobe. In contrast, the PSR method has good tolerance to model mismatch.

4.3.3.2   Experiments on real radar data

The experimental results based on C-band SAR data collected by the GF-3 satellite are presented in this section to demonstrate the effectiveness of the PSR method. The regular SAR image of the sea surface is shown in Figure 4.9. It can be seen that

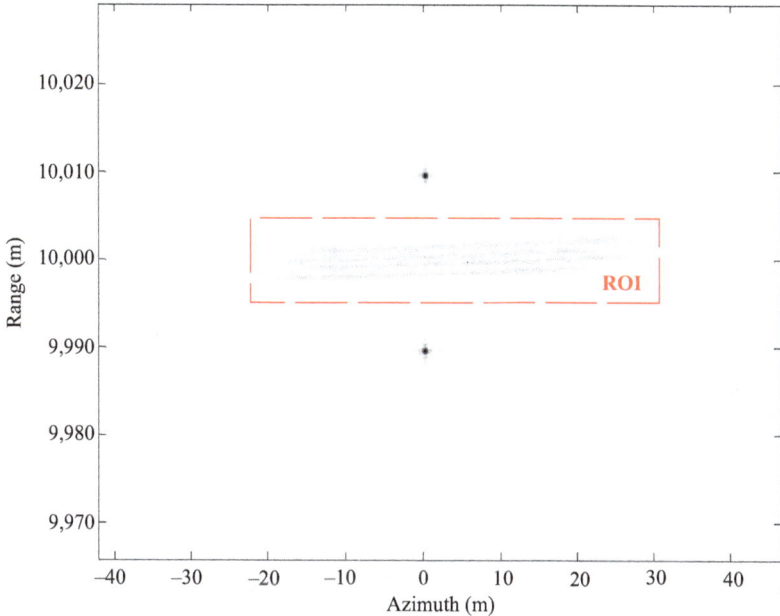

Figure 4.4   The imaging result of the entire scene produced by the $\omega - K$ algorithm

three ships are defocused due to their movements. Three ROIs named T1, T2, and T3 containing the moving ships are cropped from the original complex image, respectively, and then the refocusing of these ships is carried out by using the algorithm in [12] and the PSR method, respectively. Figure 4.10 shows the convergence processes of the PSR method for these moving ships. The refocusing results obtained by the method in [12] and the PSR method are compared in Figure 4.11, Figure 4.12, and Figure 4.13. It can be observed that the PSR method can successfully reconstruct the images of the moving ships and significantly suppress the side-lobes. To quantitatively compare different algorithms in terms of the image quality, the image entropy values of the refocused ROI images are listed in Table 4.1. The smaller value of the image entropy means better refocusing quality. As shown in Table 4.1, the PSR method can provide better image quality than the method in [12], thanks to the usage of the parametric dictionary.

4.4   Application to SAR motion compensation

In this section, the PSR method is applied to SAR motion compensation.

4.4.1   Problem formulation

In the scenario of airborne SAR imaging, the trajectory error of a radar platform results in the phase error of the raw data and seriously degrades the imaging quality

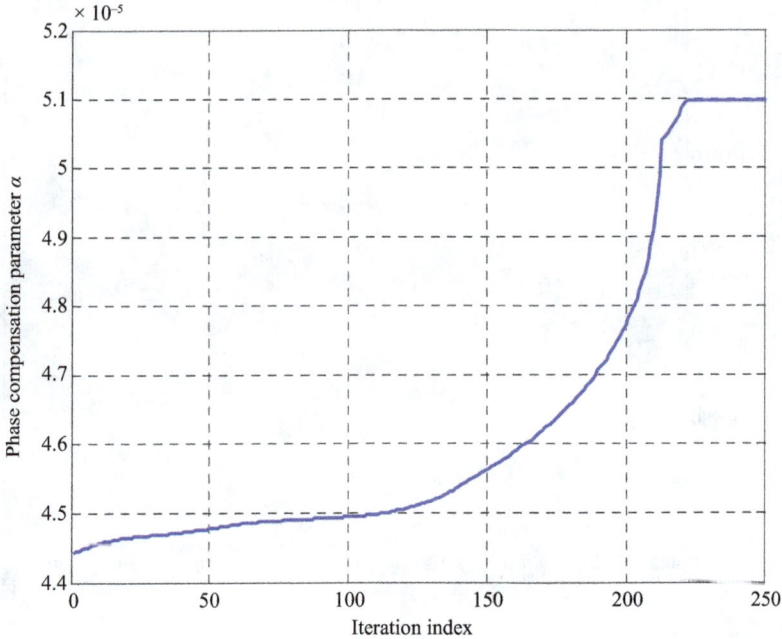

Figure 4.5   Convergence processes of the PSR method

[14,15]. The goal of SAR motion compensation is to correct the motion-induced phase error. The traditional algorithms of SAR motion compensation include the map-drift (MD) [16,17], the phase gradient autofocusing (PGA) [18], and the minimum entropy autofocus (MEA) [19]. In [20], a sparsity-driven method is proposed to deal with the SAR image formation from the data corrupted by the general phase error, which can be used for SAR motion compensation. This method is computationally expensive because a high-dimensional error vector is needed to be solved. Especially, for the 2D nonseparable phase error, the number of the variables to be solved in [20] is as many as the number of the SAR measurements. The PSR method attempts to formulate the effect of the phase error with a low-dimensional vector consisting of the motion parameters of the radar platform and accordingly to achieve the phase error correction with affordable computational complexity.

The side-looking geometry of a strip-map airborne SAR is shown in Figure 4.14, where the solid line indicates the actual flight path, the horizontal axis denotes the nominal path, and the vertical axis denotes the slant-range. In the ideal case, the antenna phase center (APC) of the radar is assumed to move along the nominal path with a constant velocity. In practice, however, the trajectory error usually exists in both horizontal direction and vertical direction. At the slow time $t = nT$, where T is the pulse repetition interval and n is the pulse index, and the actual APC position is denoted as $[X(t), Y(t)]$. The cross-range

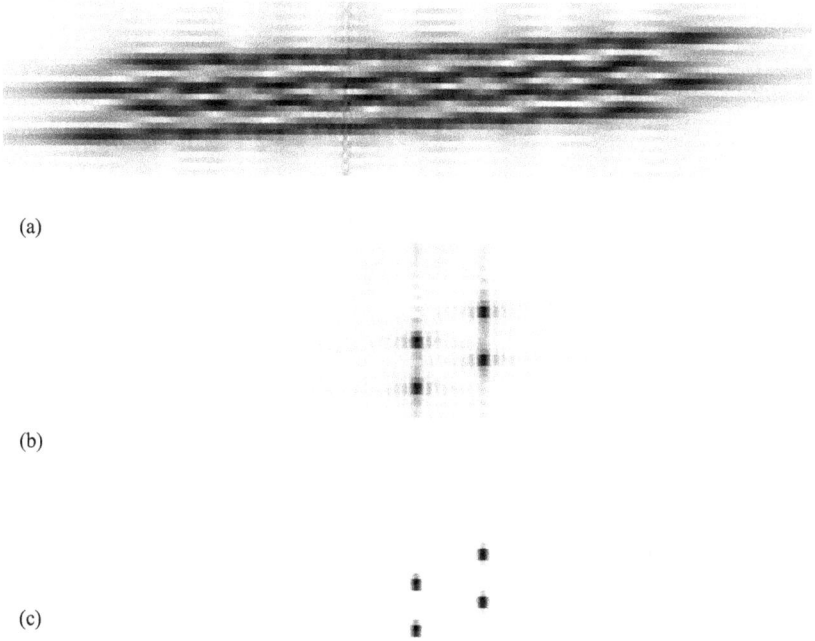

(a)

(b)

(c)

Figure 4.6   *(a) The defocused ROI data, (b) the refocused image obtained by the method in [12], and (c) the refocused image obtained by the PSR method*

position of the *i*th scatterer at the *m*th range bin R_m is denoted as X_i. Then the instantaneous distance between this scatterer and the radar can be expressed as

$$R_i(t) = \sqrt{[X(t) - X_i]^2 + [Y(t) - R_m]^2}$$
$$\approx R_m + \frac{[X(t) - X_i]^2}{2R_m} - Y(t) + \frac{Y^2(t)}{2R_m} \tag{4.17}$$

Since $Y(t) \ll R_m$, the last term in (4.17) can be ignored. Thus, the instantaneous Doppler frequency and the chirp rate of the baseband signal at the *m*th range cell can be calculated as

$$f_d(t) = \frac{1}{2\pi} \frac{d}{dt} \left(\frac{-4\pi R_i(t)}{\lambda} \right)$$
$$\approx -\frac{2[X(t) - X_i]v_x(t)}{\lambda R_m} + \frac{2}{\lambda} v_y(t) \tag{4.18}$$

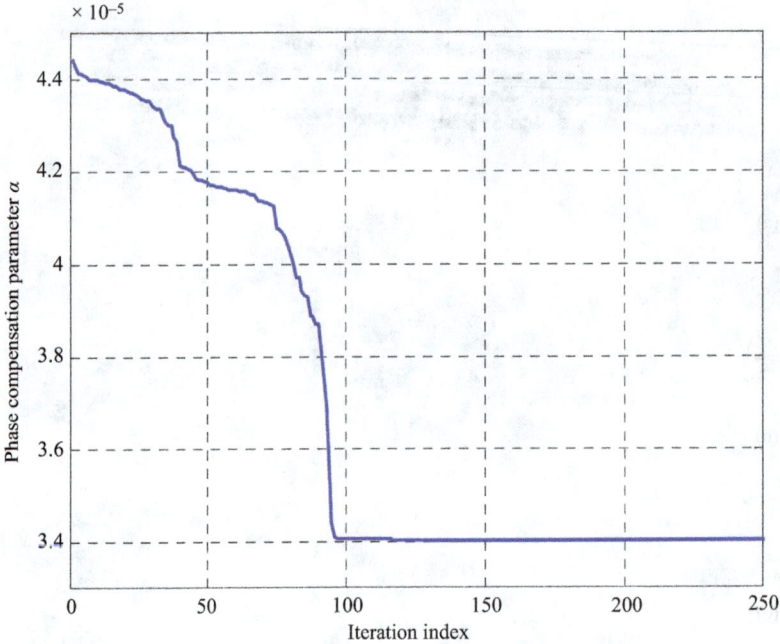

Figure 4.7   *Convergence processes of the PSR method in the presence of a model mismatch*

and

$$\gamma_m(t) = \frac{d}{dt} f_d(t)$$
$$= -\frac{2v_x^2(t)}{\lambda R_m} - \frac{2[X(t) - X_i]a_x(t)}{\lambda R_m} + \frac{2}{\lambda} a_y(t) \tag{4.19}$$

where λ is the radar wavelength, $v_x(t)$, $v_y(t)$, $a_x(t)$ and $a_y(t)$ denote the azimuth velocity, the radial velocity, the azimuth acceleration, and the radial acceleration of the APC, respectively. Due to the mechanical inertia of the radar platform, the azimuth velocity of the APC varies slowly [21]. This means that the contribution of the azimuth acceleration $a_x(t)$ to the chirp rate is ignorable in the case that the synthetic aperture is not too large. Thus (4.19) can be simplified as

$$\gamma_m(t) = -\frac{2v_x^2(t)}{\lambda R_m} + \frac{2}{\lambda} a_y(t) \tag{4.20}$$

Here the values of $v_x(t)$ and $a_y(t)$ are assumed to be range-invariant. In the case of wide-swath imaging, the wide range of swath can be divided into several narrow range strips so that this assumption holds. From (4.20), it is clear that the chirp rate $\gamma_m(t)$ is a common parameter for all the scatterers at the mth range bin.

(a) (b)

(c)

(d)

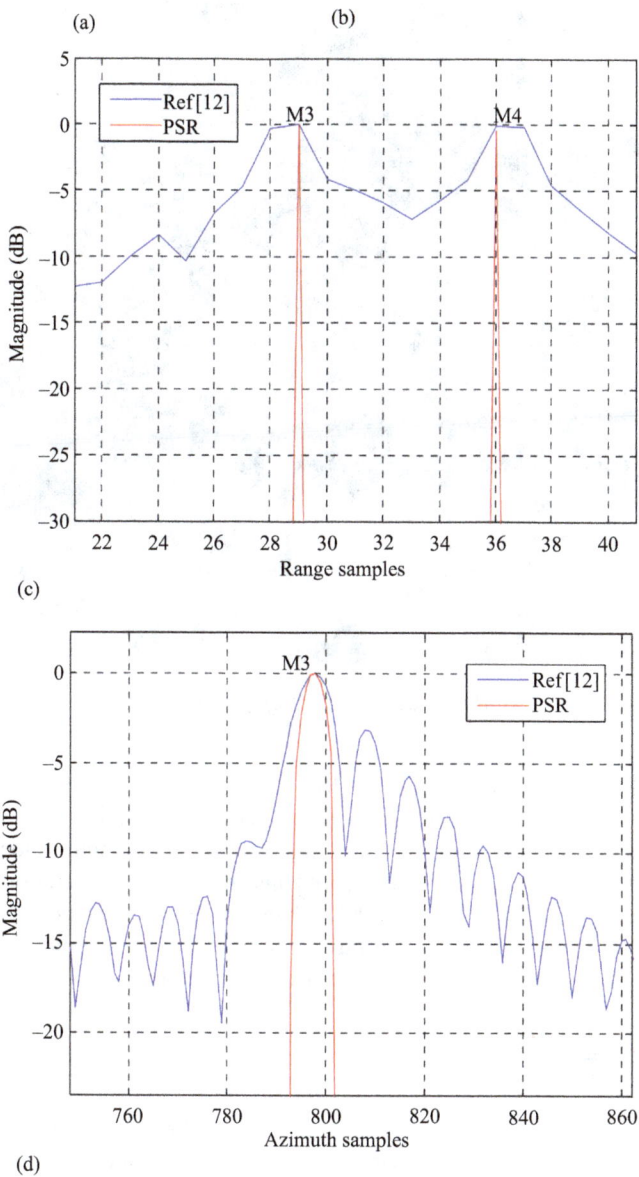

Figure 4.8 (a) The refocused image obtained by the method in [12], (b) the refocused image obtained by the PSR method, (c) the range profiles of a moving scatterer, and (d) the azimuth profiles of a moving scatterer

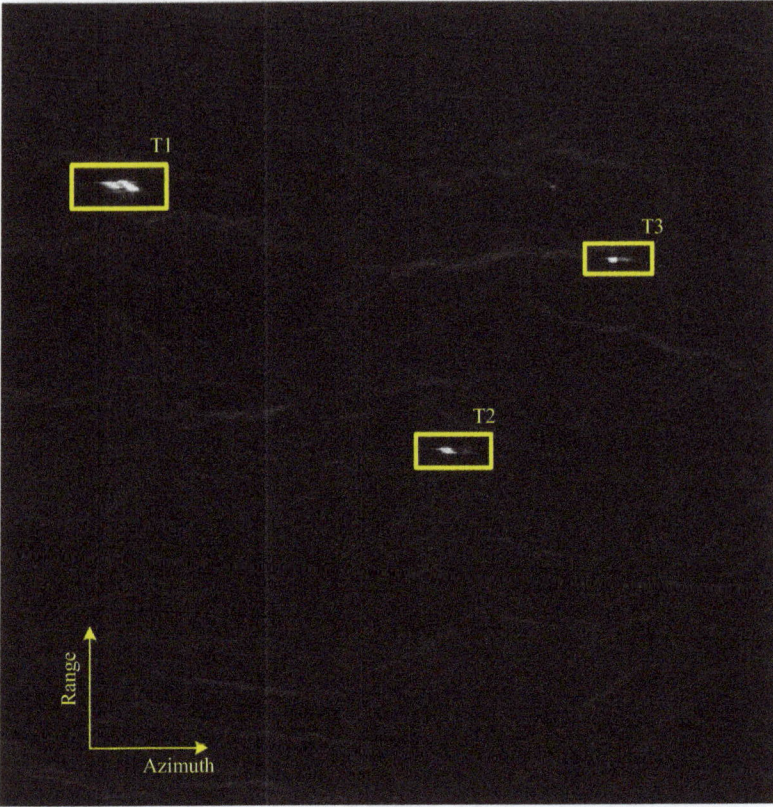

Figure 4.9   Regular imaging result of the GF-3 space-borne SAR

The full-aperture is decomposed into W subapertures, and each subaperture is composed of N sequential pulses. In each subaperture, the motion parameters of the radar platform are approximately constant. Denote the azimuth velocities in all the subapertures as $\{v_{x;w}, w = 1, 2, \ldots, W\}$ and the radial accelerations in all the subapertures as $\{a_{y;w}, w = 1, 2, \ldots, W\}$, which are the parameters to be estimated for motion compensation. After the pulse compression, the azimuth signal at the mth range bin within the wth subaperture can be written as

$$s_{m;w}(t_w) = \sum_{i=1}^{\mu_m} \sigma_i \exp\left(j2\pi\left(f_{i;m;w}t_w + \frac{1}{2}\gamma_{m;w}t_w^2\right)\right) + \xi_{m;w}(t_w) \tag{4.21}$$

where $t_w \in [(w-1)NT, wNT)$, μ_m denotes the number of scatterers at the mth range bin, σ_i denotes the reflectivity of the ith scatterer and it is assumed to be constant within the full aperture, $f_{i;m;w}$ and $\gamma_{m;w}$ denote the Doppler center frequency and the chirp rate of the ith scatterer, and $\xi_{m;w}(t_w)$ denotes the additive noise. When the value of N is small, (4.21) holds because the range migration can

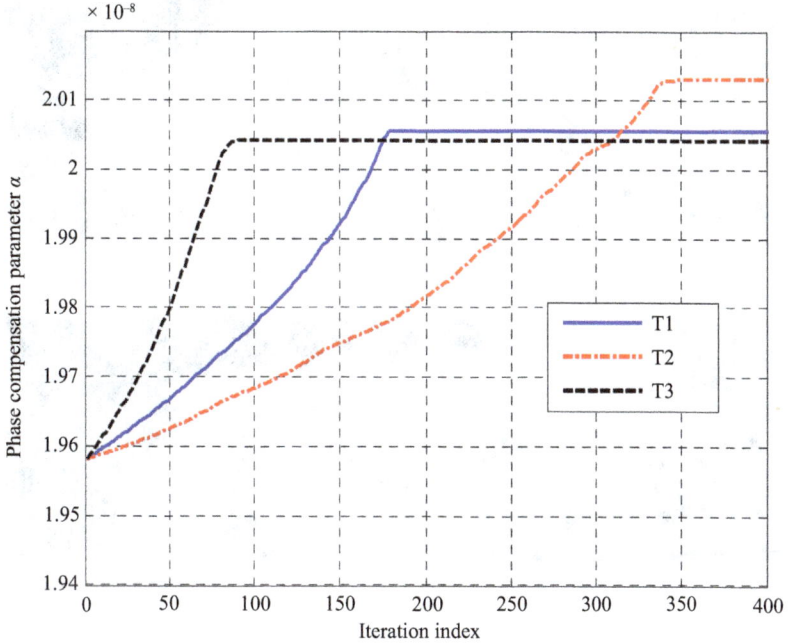

Figure 4.10 Convergence processes of the PSR method

be neglected within a short observation time. As reported in [14], in the side-looking scenario, the range migration can be ignored when:

$$\frac{\lambda^2 R_m}{32\rho_{a;w}^2} < \frac{\rho_r}{4} \tag{4.22}$$

where $\rho_{a;w}$ denotes the azimuth resolution obtained by the wth subaperture, and ρ_r denotes the range resolution. According to the fundamental knowledge of SAR imaging, we have [14]

$$\rho_{a;w} = \frac{\lambda R_m}{2v_{x;w}NT} \tag{4.23}$$

From (4.22) and (4.23), the condition that the range migration is ignorable can be expressed as

$$N < \frac{\sqrt{2R_{\min}\rho_r}}{v_{x;w}T} \tag{4.24}$$

where $R_{\min}$ is the nearest range between the radar and the observed region. In other words, the number of pulses in each subaperture should follow (4.24). In practical applications, the approximate boundary of N can be calculated based on the

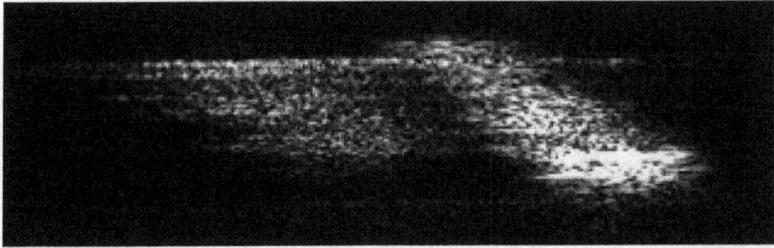

(a)

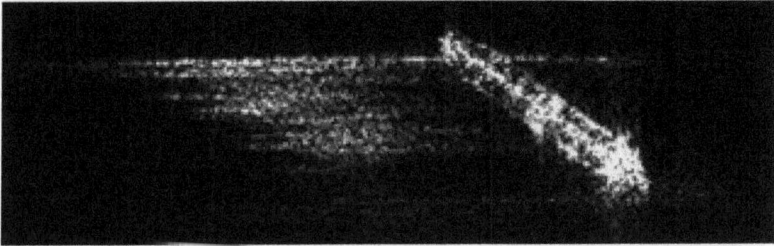

(b)

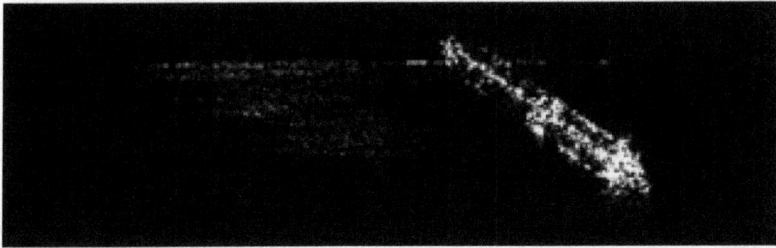

(c)

*Figure 4.11   (a) Defocused ROI data of ship T1, (b) refocused image obtained by
the method in [12], and (c) refocused image obtained by the
PSR method*

measurement of the platform velocity provided by the inertial navigation system. In
4.4.2, we will describe how to estimate the motion parameters of the APC in each
subaperture. Once this is done, by inputting the estimates of the motion parameters
of the APC in all the subapertures into the existing autofocusing schemes [21–23], a
well-focused SAR image can be obtained.

4.4.2   Algorithm description

In this section, we formulate how to estimate $\{v_{x;w}, a_{y;w} | w = 1, 2, \ldots, W\}$ via the
PSR method for SAR motion compensation [5]. Define the dictionary at the mth

(a)

(b)

(c)

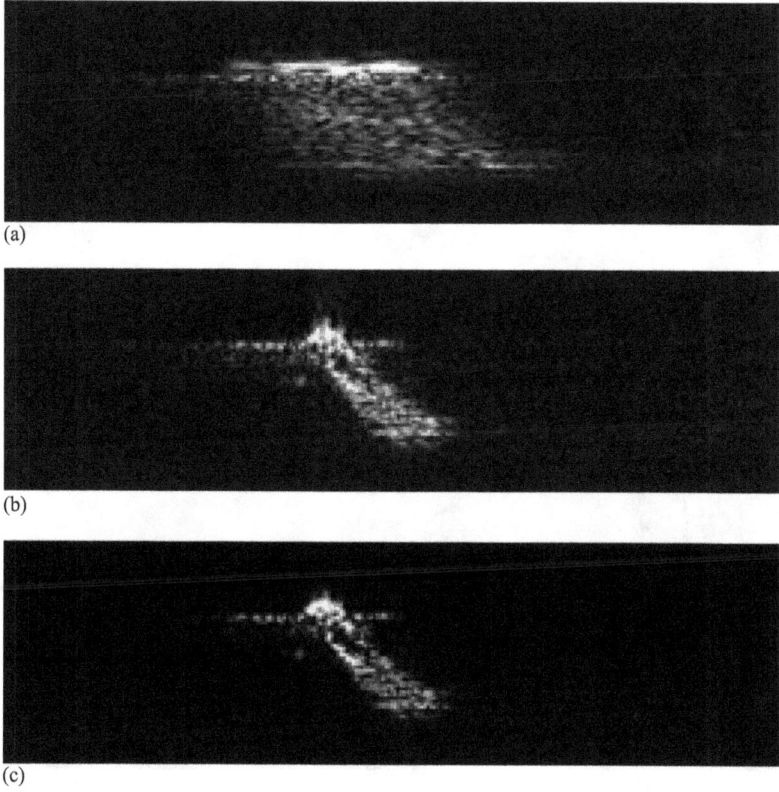

Figure 4.12 (a) The defocused ROI data of ship T2, (b) the refocused image obtained by the method in [12], and (c) the refocused image obtained by the PSR method

range bin and in the wth subaperture as

$$\boldsymbol{\Phi}_{m;w}\left(v_{x;w}, a_{y;w}\right) = \left[\boldsymbol{\phi}_{m,1}\left(v_{x;w}, a_{y;w}\right), \ldots, \boldsymbol{\phi}_{m,k}\left(v_{x;w}, a_{y;w}\right), \ldots, \boldsymbol{\phi}_{m,K}\left(v_{x;w}, a_{y;w}\right)\right]$$

(4.25)

where its kth column is expressed as

$$\boldsymbol{\phi}_{m,k}\left(v_{x;w}, a_{y;w}\right) = \left[e^{j2\pi f_k z_1 T + j\pi\left(-\frac{2v_{x;w}^2}{\lambda R_m} + \frac{2}{\lambda}a_{y;w}\right)(z_1 T)^2}, \ldots, e^{j2\pi f_k z_N T + j\pi\left(-\frac{2v_{x;w}^2}{\lambda R_m} + \frac{2}{\lambda}a_{y;w}\right)(z_N T)^2}\right]^T$$

(4.26)

where $z_n = (w-1)N + n - 1, n = 1, 2, \ldots, N$, and $f_k = (k-1)/KT$, $k = 1, 2, \ldots, K$, K denotes the number of the discretized Doppler frequency

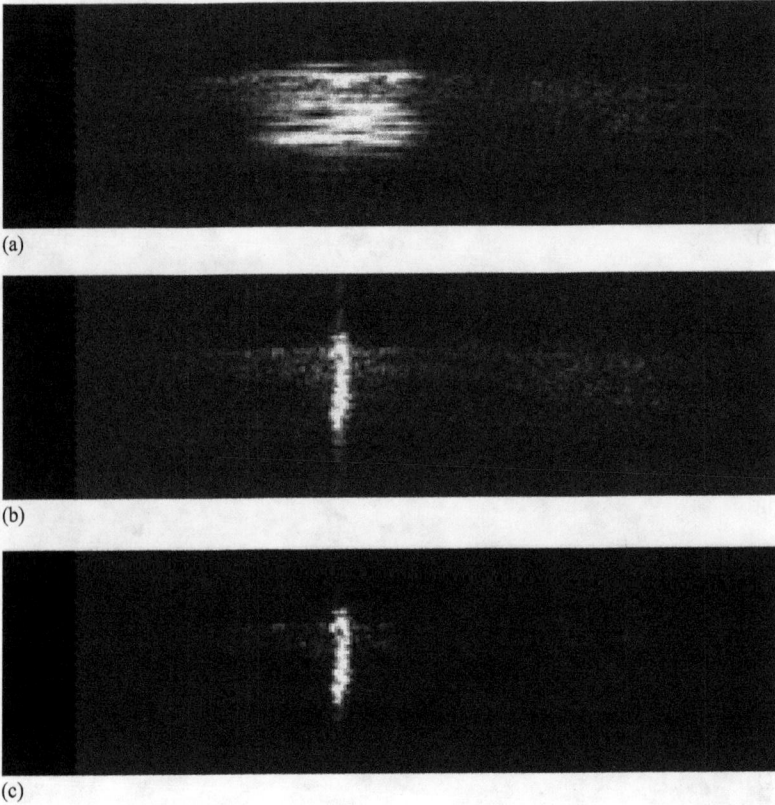

(a)

(b)

(c)

*Figure 4.13   (a) The defocused ROI data of ship T3, (b) the refocused image
obtained by the method in [12], and (c) the refocused image
obtained by the PSR method*

Table 4.1   The image entropy of the refocused ROI images

Targets	Methods	Image Entropy
Ship T1	Ref. [12]	6.4149
	PSR	3.5389
Ship T2	Ref. [12]	6.3011
	PSR	3.9863
Ship T3	Ref. [12]	5.0770
	PSR	4.0953

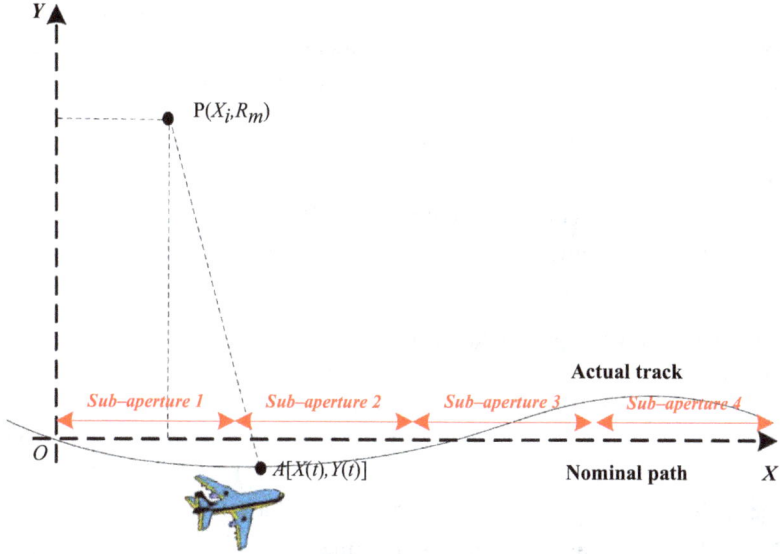

Figure 4.14   The geometry of airborne SAR

candidates. Then, (4.21) can be rewritten as

$$\mathbf{S}_{m;w} = \mathbf{\Phi}_{m;w}(v_{x;w}, a_{y;w})\mathbf{\theta}_{m;w} + \mathbf{\xi}_{m;w} \tag{4.27}$$

where $\mathbf{S}_{m;w} = [s_{m;w}(wNT - NT), \ldots, s_{m;w}(wNT - T)]^T$, $\mathbf{\xi}_{m;w} = [\xi_{m;w}(wNT - NT),$
$\ldots, \xi_{m;w}(wNT - T)]^T$, and the vector $\mathbf{\theta}_{m;w}$ consists of the reflectivities of all the
scatterers at the mth range bin within the wth subaperture. In the general case,
there is no guarantee for the sparsity of $\mathbf{\theta}_{m;w}$. As reported in [24], enforcing the sparsity
during SAR image formation of a nonsparse scene can successfully reconstruct the
dominant scatterers, at the cost of missing some weak scatterers. Here we also impose
the sparsity constraint on $\mathbf{\theta}_{m;w}$, because our goal is to accurately estimate the values of
$v_{x;w}$ and $a_{y;w}$ rather than the reconstruction of the entire image through the sparse signal
recovery. The SAR image formation will be implemented by inputting the estimates of
$\{v_{x;w}, a_{y;w} | w = 1, 2, \ldots, W\}$ into the autofocusing schemes [21–23], which are
suitable for both sparse and nonsparse scenes.

Let $\mathbf{\Theta}_w = [\mathbf{\theta}_{1;w}^T, \mathbf{\theta}_{2;w}^T, \ldots, \mathbf{\theta}_{M;w}^T]^T$, which is a column vector indicating the
scattering behavior of the dominant scatterers at all the M range cells in the wth
subaperture. It is clear that $\mathbf{\Theta}_w$ depends on the values of $v_{x;w}$ and $a_{y;w}$, that is,
different motion parameters produce different dictionary $\mathbf{\Phi}_{m;w}(v_{x;w}, a_{y;w})$ and,
therefore, different solution of $\mathbf{\Theta}_w$. Thus, $\mathbf{\Theta}_w$ can be considered as a function of $v_{x;w}$
and $a_{y;w}$, denoted by $\mathbf{\Theta}_w(v_{x;w}, a_{y;w})$. With correct values of $v_{x;w}$ and $a_{y;w}$, the dictionary matrix $\mathbf{\Phi}_{m;w}(v_{x;w}, a_{y;w})$ well matches the actual SAR echo and accordingly a
well-focused SAR image of the dominant scatterers will be obtained. With wrong
candidates of $v_{x;w}$ and $a_{y;w}$, the mismatch between the dictionary matrix

$\Phi_{m;w}(v_{x;w}, a_{y;w})$ and the actual SAR echo may be serious, and thus a blurred image is most likely obtained. From the above consideration, the estimation of the sparse solution Θ_w and the motion parameters $(v_{x;w}, a_{y;w})$ can be carried out by alternating iterations.

The flowchart of the PSR method for SAR autofocusing is summarized in Figure 4.15. The iterative process is terminated until $\left|\Delta v_{x;w}^{(p)}\right| < \eta_x$ and $\left|\Delta a_{y;w}^{(p)}\right| < \eta_y$, where η_x and η_y are the convergence thresholds. By inputting the estimated motion parameters into the traditional autofocussing algorithms [21–23], the focused SAR image can be obtained, no matter whether the observed scene is sparse or not.

4.4.2.1   Update the sparse solution

The initial value of $v_{x;w}^{(0)}$ is set according to the platform velocity provided by the inertial navigation system. The initial value of $a_{y;w}^{(0)}$ is set to zero. By using $(v_{x;w}^{(p-1)}, a_{y;w}^{(p-1)})$, the dictionary matrix at the mth range bin $\Phi_{m;w}(v_{x;w}^{(p-1)}, a_{y;w}^{(p-1)})$ can be constructed according to (4.25) and (4.26). Let $S_w = [S_{1;w}^T, S_{2;w}^T, \ldots, S_{M;w}^T]^T$ denote the radar signal at all the M range cells, and let $\Psi_w(v_{x;w}^{(p-1)}, a_{y;w}^{(p-1)})$ denote the jointly parametric dictionary matrix, which can be expressed as

$$
\Psi_w(v_{x;w}^{(p-1)}, a_{y;w}^{(p-1)}) =
\begin{bmatrix}
\Phi_{1;w}\left(v_{x;w}^{(p-1)}, a_{y;w}^{(p-1)}\right) & 0 & \cdots & 0 \\
0 & \Phi_{2;w}\left(v_{x;w}^{(p-1)}, a_{y;w}^{(p-1)}\right) & \cdots & 0 \\
\vdots & \vdots & \ddots & \vdots \\
0 & 0 & \cdots & \Phi_{M;w}\left(v_{x;w}^{(p-1)}, a_{y;w}^{(p-1)}\right)
\end{bmatrix}.
$$

(4.28)

From (4.27), the image of the dominant scatterers at all the M range cells can be reconstructed by solving the following problem:

$$
\Theta_w^{(p)} = \arg\min_{\Theta} \|\Theta\|_0, \text{ s.t.} \left\|S_w - \Psi_w\left(v_{x;w}^{(p-1)}, a_{y;w}^{(p-1)}\right)\Theta\right\|_2^2 \leq \varepsilon
$$

(4.29)

where $\varepsilon > 0$ is the reconstruction error tolerance. Given the values of $v_{x;w}^{(p-1)}$ and $a_{y;w}^{(p-1)}$, (4.29) is a standard problem of sparse signal recovery. Here the orthogonal matching pursuit (OMP) algorithm [25] is used to solve (4.29). The detailed steps of OMP are also provided in Chapter 1.

4.4.2.2   Update the motion parameter estimates

Given a sparse solution $\Theta_w^{(p)}$, the motion parameter estimates can be updated by minimizing the recovery error, that is,

$$
\left(v_{x;w}^{(p)}, a_{y;w}^{(p)}\right) = \arg\min_{v_{x;w}, a_{y;w}} \left\|S_w - \Psi_w\left(v_{x;w}, a_{y;w}\right) \cdot \Theta_w^{(p)}\right\|_2
$$

(4.30)

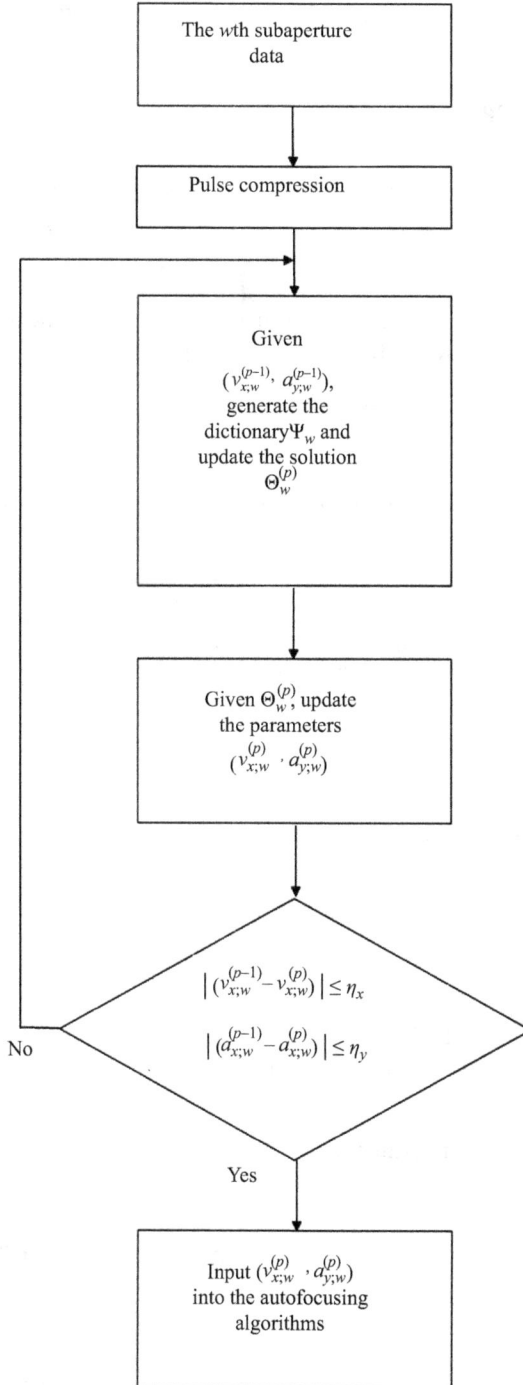

The wth subaperture data

Pulse compression

Given

$(v_{x;w}^{(p-1)}, a_{y;w}^{(p-1)})$, generate the dictionary Ψ_w and update the solution $\Theta_w^{(p)}$

Given $\Theta_w^{(p)}$, update the parameters $(v_{x;w}^{(p)}, a_{y;w}^{(p)})$

$\left| (v_{x;w}^{(p-1)} - v_{x;w}^{(p)}) \right| \le \eta_x$

$\left| (a_{x;w}^{(p-1)} - a_{x;w}^{(p)}) \right| \le \eta_y$

No

Yes

Input $(v_{x;w}^{(p)}, a_{y;w}^{(p)})$ into the autofocusing algorithms

Figure 4.15   Flowchart of the PSR method for SAR motion compensation

By taking the first-order Taylor expansion of $\boldsymbol{\Psi}_w\left(v_{x;w}, a_{y;w}\right)$ around $\left(v_{x;w}^{(p-1)}, a_{y;w}^{(p-1)}\right)$, we have

$$
\begin{aligned}
\boldsymbol{\Psi}_w\left(v_{x;w}, a_{y;w}\right) &\approx \boldsymbol{\Psi}_w\left(v_{x;w}^{(p-1)}, a_{y;w}^{(p-1)}\right) \\
&+\frac{\partial \boldsymbol{\Psi}_w\left(v_{x;w}, a_{y;w}\right)}{\partial v_{x;w}}\Bigg|_{\substack{v_{x;w}=v_{x;w}^{(p-1)} \\ a_{y;w}=a_{y;w}^{(p-1)}}} \cdot \Delta v_{x;w} + \frac{\partial \boldsymbol{\Psi}_w\left(v_{x;w}, a_{y;w}\right)}{\partial a_{y;w}}\Bigg|_{\substack{v_{x;w}=v_{x;w}^{(p-1)} \\ a_{y;w}=a_{y;w}^{(p-1)}}} \cdot \Delta a_{y;w}
\end{aligned}
$$

$$(4.31)$$

where $\Delta v_{x;w}$ and $\Delta a_{y;w}$ denote the azimuth velocity increment and the radial acceleration increment, respectively. From (4.26), we have

$$
\begin{aligned}
\frac{\partial \boldsymbol{\Phi}_{m;w}\left(v_{x;w}, a_{y;w}\right)}{\partial v_{x;w}} &= -\frac{j4\pi v_{x;w}}{\lambda R_m} \\
&\cdot \operatorname{diag}\left(\left(z_1 T\right)^2, \left(z_2 T\right)^2, \ldots, \left(z_N T\right)^2\right) \boldsymbol{\Phi}_{m;w}\left(v_{x;w}, a_{y;w}\right)
\end{aligned}
$$

$$(4.32)$$

and

$$
\frac{\partial \boldsymbol{\Phi}_{m;w}\left(v_{x;w}, a_{y;w}\right)}{\partial a_{y;w}} = \frac{j2\pi}{\lambda} \cdot \operatorname{diag}\left(\left(z_1 T\right)^2, \left(z_2 T\right)^2, \ldots, \left(z_N T\right)^2\right) \cdot \boldsymbol{\Phi}_{m;w}\left(v_{x;w}, a_{y;w}\right)
$$

$$(4.33)$$

Let

$$
\mathbf{A}_w^{(p-1)} = \frac{\partial \boldsymbol{\Psi}_w\left(v_{x;w}, a_{y;w}\right)}{\partial v_{x;w}}\Bigg|_{\substack{v_{x;w}=v_{x;w}^{(p-1)} \\ a_{y;w}=a_{y;w}^{(p-1)}}}
$$

$$(4.34)$$

and

$$
\mathbf{B}_w^{(p-1)} = \frac{\partial \boldsymbol{\Psi}_w\left(v_{x;w}, a_{y;w}\right)}{\partial a_{y;w}}\Bigg|_{\substack{v_{x;w}=v_{x;w}^{(p-1)} \\ a_{y;w}=a_{y;w}^{(p-1)}}}
$$

$$(4.35)$$

Then (4.30) can be rewritten as

$$
\begin{aligned}
&\left(\Delta v_{x;w}^{(p)}, \Delta a_{y;w}^{(p)}\right) = \\
&\underset{\Delta v_{x;w}, \Delta a_{y;w}}{\arg\min} \left\| \mathbf{S}_{vec;w} - \boldsymbol{\Psi}_w\left(v_{x;w}^{(p-1)}, a_{y;w}^{(p-1)}\right)\boldsymbol{\Theta}_w^{(p)} - \left[\mathbf{A}_w^{(p-1)}\boldsymbol{\Theta}_w^{(p)}, \mathbf{B}_w^{(p-1)}\boldsymbol{\Theta}_w^{(p)}\right] \cdot \begin{bmatrix}\Delta v_{x;w} \\ \Delta a_{y;w}\end{bmatrix} \right\|_2
\end{aligned}
$$

$$(4.36)$$

It is clear from (4.36) that $\left(\Delta v_{x;w}^{(p)}, \Delta a_{y;w}^{(p)}\right)$ can be solved by the least squares estimation. Then, the motion parameter estimates can be updated by

$$
\begin{aligned}
v_{x;w}^{(p)} &= v_{x;w}^{(p-1)} + \Delta v_{x;w}^{(p)} \\
a_{y;w}^{(p)} &= a_{y;w}^{(p-1)} + \Delta a_{y;w}^{(p)}
\end{aligned}
\tag{4.37}
$$

4.4.3   Experimental results

The experimental results based on simulated data and real airborne X-band SAR data are presented in this section to demonstrate the effectiveness of the PSR method for SAR motion compensation.

4.4.3.1   Simulations

The parameters of the simulated radar and scene are as follows. The center frequency of the radar is 10 GHz, the pulse repetition frequency is 1 kHz, the aperture of the real antenna is 1 m, the bandwidth of the transmitted signal is 150 MHz, and hence the nominal range resolution is 1 m. The simulated scene contains 2,856 scatterers at the slant range of 10 km as shown in Figure 4.16. A total of 5,834 pulses are received during the observation time. The Doppler frequency domain is discretized into $K = 5,834$ grids. Considering the constraint condition expressed in (4.24), the number of pulses in a subaperture is set to $N = 1,175$. Thus, the full-aperture can be decomposed into five subapertures. It is assumed that the true azimuth velocity of the platform is 120 m/s during the entire observation time, the true values of the radial acceleration of the platform in these five subapertures are 0.3 m/s^2, -0.3 m/s^2, -0.2 m/s^2, 0.2 m/s^2, and 0.1 m/s^2, respectively. Meanwhile, we assume that the inaccurate values of the azimuth velocity and the radial acceleration measured by the inertial navigation system are 118 m/s and 0 m/s^2, respectively, which are used as the initialized values for different algorithms. The thresholds in Figure 4.15 are set as $\eta_x = 0.01$ and $\eta_y = 0.0001$. The convergence processes of the azimuth velocity estimation and the radial acceleration estimation in different subapertures are shown in Figure 4.17. As shown in Figure 4.17, the estimates of the azimuth velocity and the radial acceleration provided by the PSR method in all the subapertures converge to the true value after several iterations. This implies that in the noise free case, the PSR method can exactly retrieve the azimuth velocity and the radial acceleration of the platform. Figure 4.18 provides the trajectory estimated by the PSR method. One can find that there exists an obvious difference between the actual and ideal trajectories, and the trajectory estimated by the PSR method well fits the actual one.

Next, the noise-corrupted data are generated by adding a white Gaussian noise to the radar echo of the synthetic scene and used to analyze the performance of the PSR method in the noisy environment. Define the relative errors of the azimuth velocity and the radial acceleration as

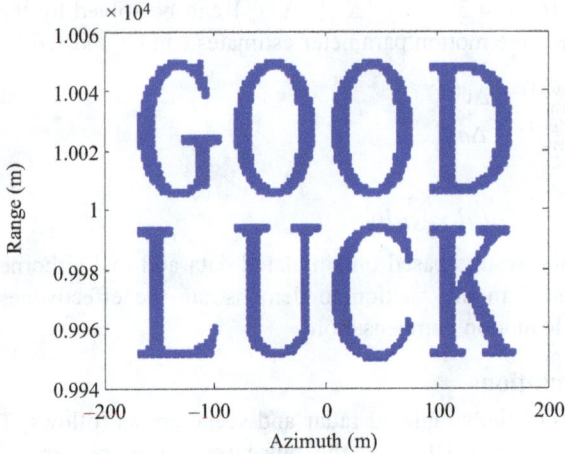

Figure 4.16   *The simulated scene*

$$E_v = \frac{1}{W} \sum_{w=1}^{W} \left| \frac{v_{x;w} - v_{xT;w}}{v_{xT;w}} \right| \times 100\% \tag{4.38}$$

and

$$E_a = \frac{1}{W} \sum_{w=1}^{W} \left| \frac{a_{y;w} - a_{yT;w}}{a_{yT;w}} \right| \times 100\% \tag{4.39}$$

where $v_{xT;w}$ and $a_{yT;w}$ denote the true values of the azimuth velocity and the radial acceleration in the wth subaperture, respectively. Figure 4.19 shows the relative errors of the estimated results obtained by the PSR method with different signal-noise-ratio (SNR) values. One can see that the relative errors are smaller than 5% when SNR ≥ -14 dB, which indicates the robustness of the PSR method against the noise. The imaging results of four motion compensation algorithms are shown in Figure 4.20. The images in the first row of Figure 4.20 are the results of the MD algorithm. The degradation in the reconstructed images shows that MD cannot accurately retrieve the platform motion parameters in the case of low SNR. From the images in the second row of Figure 4.20, one can see that the focusing performance of PGA is seriously deteriorated. The reason is that the simulated scene consisting of multiple scatterers with uniform scattering behavior is far from the condition required by PGA, in which some strong scatterers are assumed to be isolated from the others. The images in the third row of Figure 4.20 are obtained by MEA under the minimum-entropy criterion. The images in the fourth row of Figure 4.20 are formed by the PSR method, which are competitive with the imaging results of MEA in terms of the image quality.

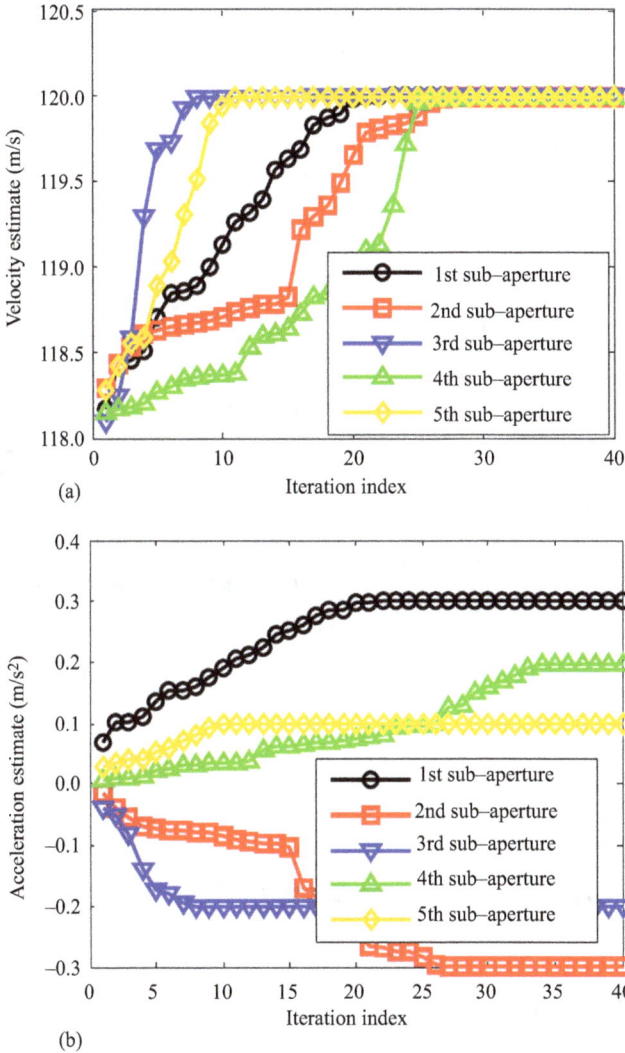

Figure 4.17   The convergence processes of the PSR method in different
subapertures: (a) azimuth velocity estimate versus iteration index
and (b) radial acceleration estimate versus iteration index

Two different metrics, that is, the target-to-background ratio (TBR) and the
image contrast (IC), are taken into account for quantitative evaluation of the above
four algorithms. TBR is defined by [20]

$$\text{TBR} = 20\log_{10}\left(\frac{\max_{a,b\in T_{\text{arget}}}\left|\widehat{\Theta}(a,b)\right|}{\frac{1}{I_B}\sum_{a,b\in B_{\text{ack}}}\left|\widehat{\Theta}(a,b)\right|}\right) \qquad (4.40)$$

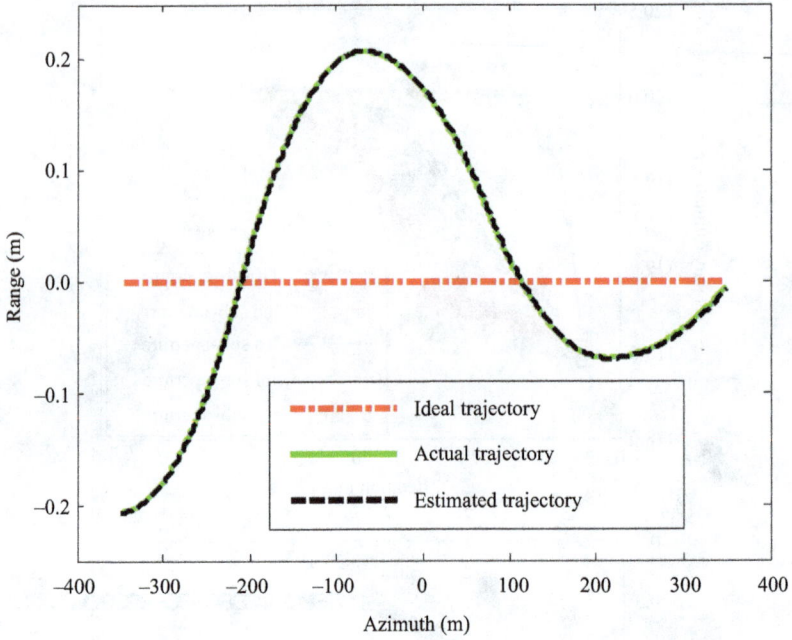

Figure 4.18   *The actual and estimated platform trajectories*

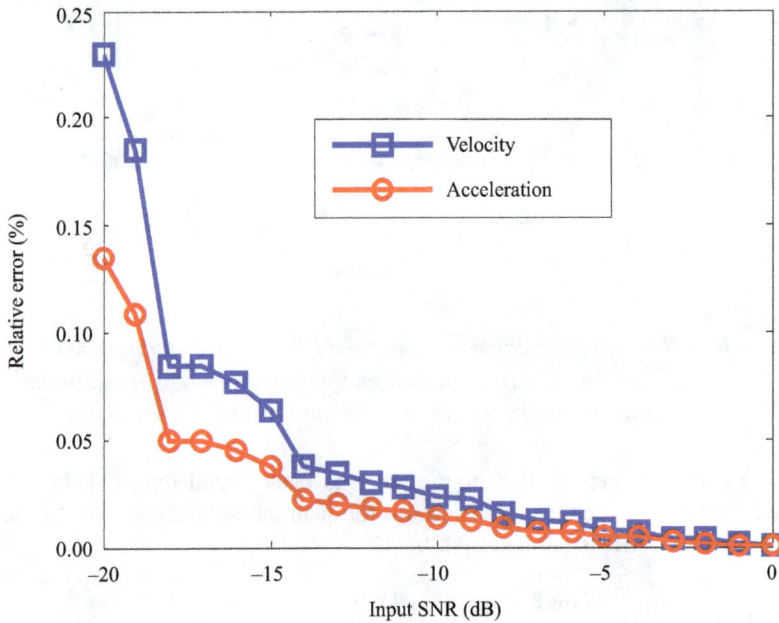

Figure 4.19   *Relative error of motion parameters versus SNR*

Figure 4.20 The images in the first row are formed by MD. The images in the second row are formed by PGA. The images in the third row are formed by MEA. The images in the fourth row are formed by the PSR method. (a) SNR = −20 dB for the first column, (b) SNR = −10 dB for the second column, and (c) SNR = 0 dB for the third column

where T_{arget} and B_{ack} denote the pixel indices in the target region and the background region, respectively, I_B is the number of background pixels, $\widehat{\Theta}(a, b)$ denotes the corresponding pixel in the imaging result. TBR is usually used to determine the energy concentration of the target region in comparison to the background. In this experiment, the resolution cells that contain the scatterers are viewed as the target region. A large value of TBR is obviously desired. As a contrast, a small value of TBR indicates that the target region and the background region have similar amplitude levels, which results in difficult extraction of targets from the

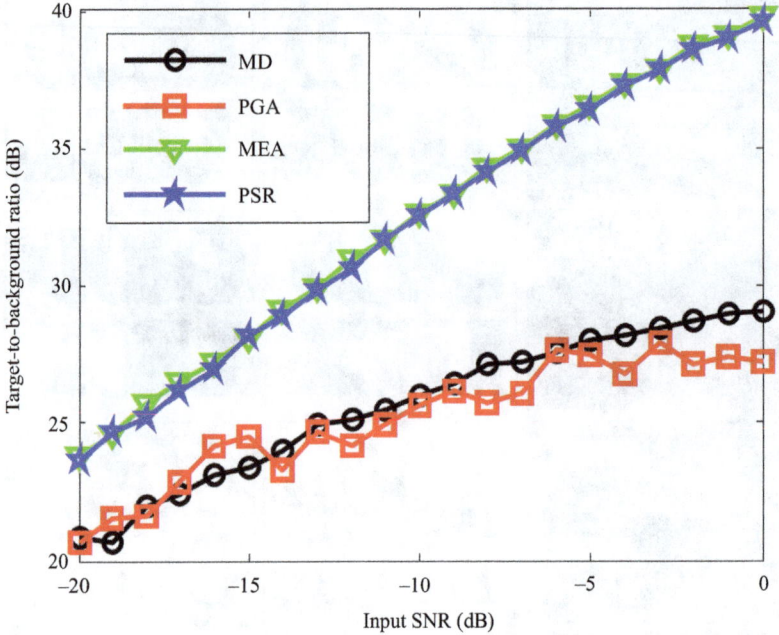

Figure 4.21   TBR versus SNR

background. Another indicator of image quality is the image contrast (IC) [26,27], which is defined by

$$IC = \frac{\text{mean}(|\Theta|)^2 - [\text{mean}(|\Theta|)]^2}{[\text{mean}(|\Theta|)]^2} \tag{4.41}$$

In Figure 4.21 and Figure 4.22, TBR and IC versus varying value of SNR are depicted, respectively, which clearly shows that the MEA and the PSR method yield better image quality.

The running time of these four algorithms are compared in Table 4.2. All the experiments are implemented by using the nonoptimized MATLAB® code in a general PC (Intel Core i7 Duo CPU 5500U). One can find from Table 4.2 that MD takes the least time because it only contains the simple correlation operations, and the PSR method is more computationally efficient than PGA and MEA. Figure 4.20, Figure 4.21, Figure 4.22, and Table 4.2 demonstrate that the PSR method can achieve satisfactory performance of SAR motion compensation with an affordable computational complexity.

4.4.3.2   Experiments on real radar data

In this section, real airborne SAR data are utilized to validate the PSR method. The main radar parameters are as follows. The radar wavelength is 0.031 m, the

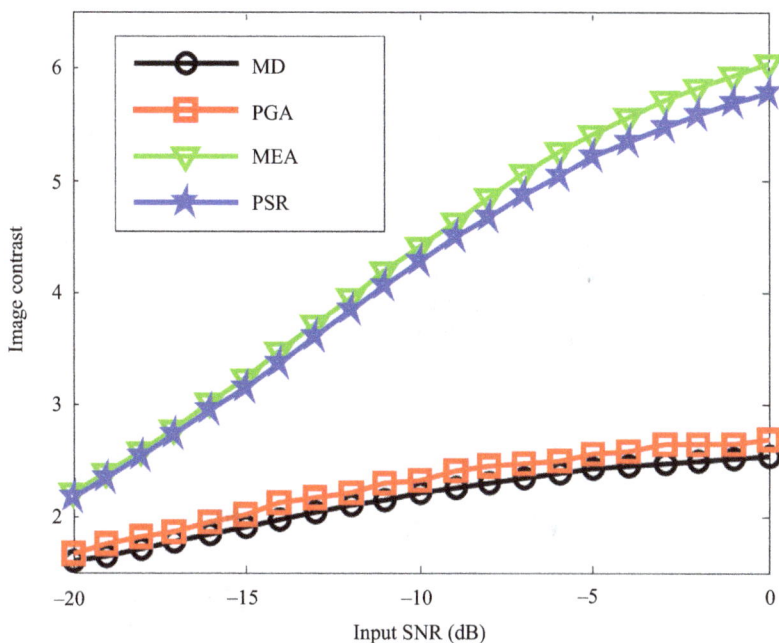

Figure 4.22   IC versus SNR

nearest distance between the radar and the scene is 37.2 km, the pulse repetition frequency is 1 kHz, the real antenna aperture is 0.44 m, the pulse width is 15 μs, the bandwidth of the transmitted signal is 200 MHz, and hence, the nominal range resolution is 0.75 m. The raw data are consisted of 8,192 pulses, in each of which 4,096 range-dimensional measurements are sampled. According to (4.24), the range migration can be ignored when $N < 1,707$. Therefore, the full-aperture is decomposed into five subapertures, each of which contains 1,600 sequential pulses. The Doppler frequency domain is discretized into $K = 8,192$ grids. In the PSR method, the velocity of the radar platform provided by the inertial navigation system (138 m/s) is used as the initial value of the azimuth velocity, and the initial radial acceleration is set as 0 m/s^2. The convergence processes of the azimuth velocity estimation and the radial acceleration estimation in different subapertures are plotted in Figure 4.23, which shows the stable convergence of the PSR method. The results of the motion parameter estimation in every subaperture obtained by the PSR method are shown in Figure 4.24, where one can find that the motion status of the platform obviously changes during the full synthetic aperture. MD, strip-map PGA, and iterative MEA are also carried out for comparison. The full-aperture imaging results obtained by these four algorithms are shown in Figure 4.25, and the zoom-in images containing an urban region are highlighted in Figure 4.26. It can be found from Figure 4.25 and Figure 4.26 that

Table 4.2   Running time of different motion compensation algorithms

Algorithm	MD	PGA	MEA	The PSR method
Time (s)	11.86	394.27	425.15	376.82

the PSR method is superior over MD and PGA in terms of imaging quality. Moreover, the IC values of the images in Figure 4.25, and the running time of these algorithms are listed in Table 4.3. Compared to MEA, the PSR method consumes less computational time and provides competitive image quality. From all the above experiments, we can conclude that the PSR method is reliable to produce satisfactory accuracy of SAR motion compensation at an affordable computational cost.

4.5   Application to ISAR imaging of aircrafts

In this section, the PSR method is applied to ISAR imaging of uniformly rotating aircrafts.

4.5.1   Problem formulation

In the scenario of ISAR imaging of aircrafts, the complicated motion of the maneuvering target consists of translational and rotational parts. Assume that the translational motion compensation has been already achieved, then the target can be considered as uniformly rotating as shown in Figure 4.27. Such a model is generally reasonable when the radar observation duration is not too long. The instantaneous range migration of the scatterer P located at (x_p, y_p) can be written as [28,29]

$$
\begin{aligned}
r(t; \omega_0) &= d \sin(\theta_p + \omega_0 t) \\
&= x_p \sin(\omega_0 t) + y_p \cos(\omega_0 t)
\end{aligned}
\tag{4.42}
$$

where t is the slow time, ω_0 is the rotation rate of the target, $d = \sqrt{x_p^2 + y_p^2}$ and $\theta_p = \tan^{-1}(y_p/x_p)$. Here ω_0 is considered constant during the short observation duration. Thus the baseband ISAR echo reflected from the scatterer P can be expressed as [30,31]

$$
\begin{aligned}
s(\tau, t; \omega_0) &= \sigma \cdot a\left(\tau - \frac{2 \cdot r(t; \omega_0)}{c}\right) \cdot \exp\left[-j\frac{4\pi}{\lambda} r(t; \omega_0)\right] \\
&= \sigma \cdot a\left(\tau - 2 \cdot \frac{x_p \sin(\omega_0 t) + y_p \cos(\omega_0 t)}{c}\right) \\
&\quad \cdot \exp\left[-j\frac{4\pi}{\lambda}(x_p \sin(\omega_0 t) + y_p \cos(\omega_0 t))\right]
\end{aligned}
\tag{4.43}
$$

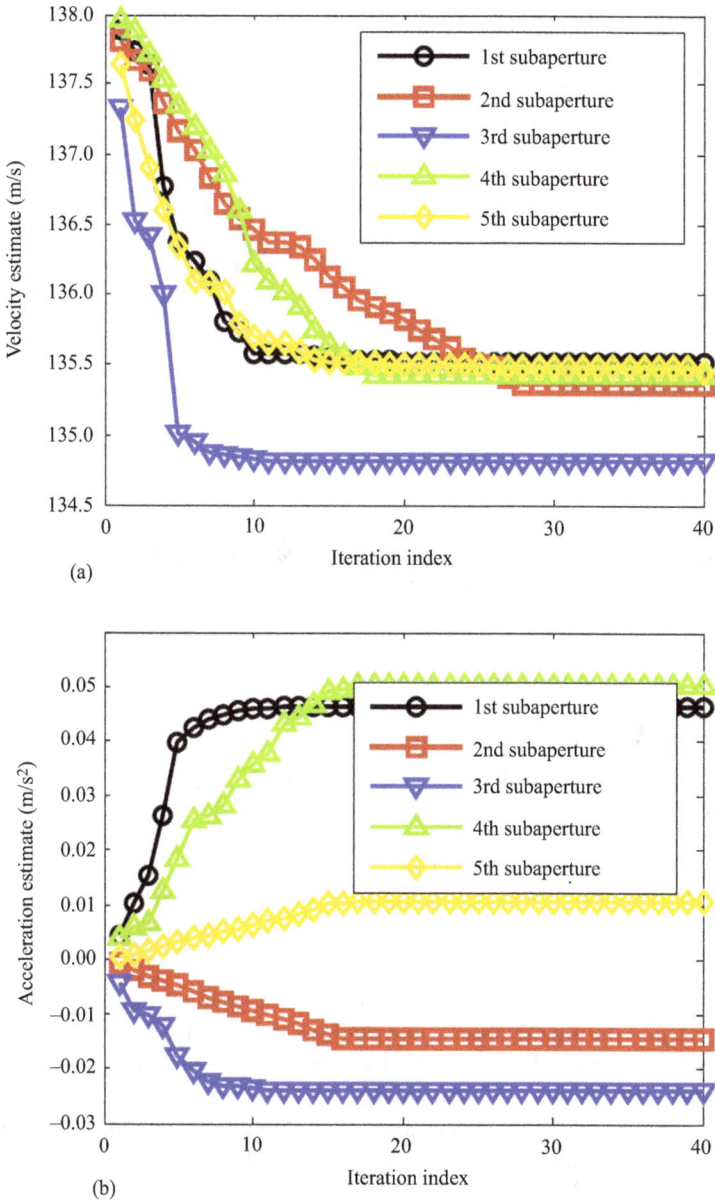

Figure 4.23   *The convergence process of the PSR method in every subaperture: (a) velocity estimate versus iteration index and (b) acceleration estimate versus iteration index*

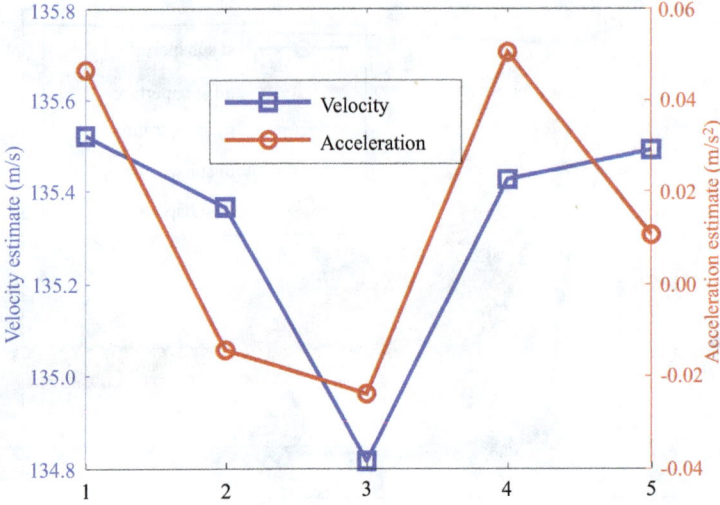

Figure 4.24   The motion parameter estimates obtained by the PSR method in every subaperture

where τ is the fast time, c is the speed of light, $a(\cdot)$ denotes the complex profile of the range-compressed signal, λ is the radar wavelength, and σ is proportional to the scatterer reflectivity. Since the observation duration is short, we have $\sin(\omega_0 t) \approx \omega_0 t$ and $\cos(\omega_0 t) \approx 1 - 0.5(\omega_0 t)^2$. Thus (4.43) can be approximated as

$$s(\tau, t; \omega_0) \approx \sigma \cdot a\left(\tau - \frac{2 \cdot r(t; \omega_0)}{c}\right) \cdot \exp\left[-j\frac{4\pi}{\lambda}\left(y_p + x_p \omega_0 t - \frac{1}{2} y_p \omega_0^2 t^2\right)\right]$$

(4.44)

It is clear from (4.44) that (1) the range cell migration may occur if the variance of $r(t; \omega_0)$ during the observation duration is larger than the range resolution; and (2) the last term of the phase in (4.44) induces the time-varying Doppler frequency and causes a blurred ISAR image.

In general, a real target such as aircraft or ship contains a number of dominant scatterers. Thus, from (4.43), the practical ISAR echo can be expressed as

$$s(\tau, t; \omega_0) = \sum_i \sigma_i \cdot a\left(\tau - 2 \cdot \frac{x_i \sin(\omega_0 t) + y_i \cos(\omega_0 t)}{c}\right)$$
$$\cdot \exp\left[-j\frac{4\pi}{\lambda}(x_i \sin(\omega_0 t) + y_i \cos(\omega_0 t))\right]$$

(4.45)

where i denotes the scatterer index.

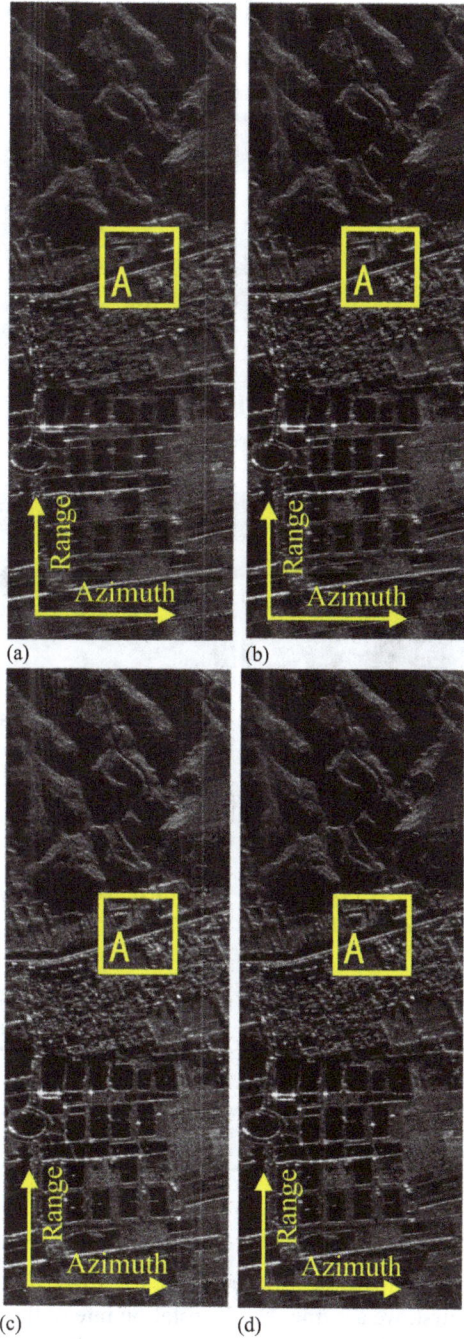

Figure 4.25   Imaging results of the entire scene obtained by (a) MD, (b) PGA, (c) MEA, and (d) the PSR method

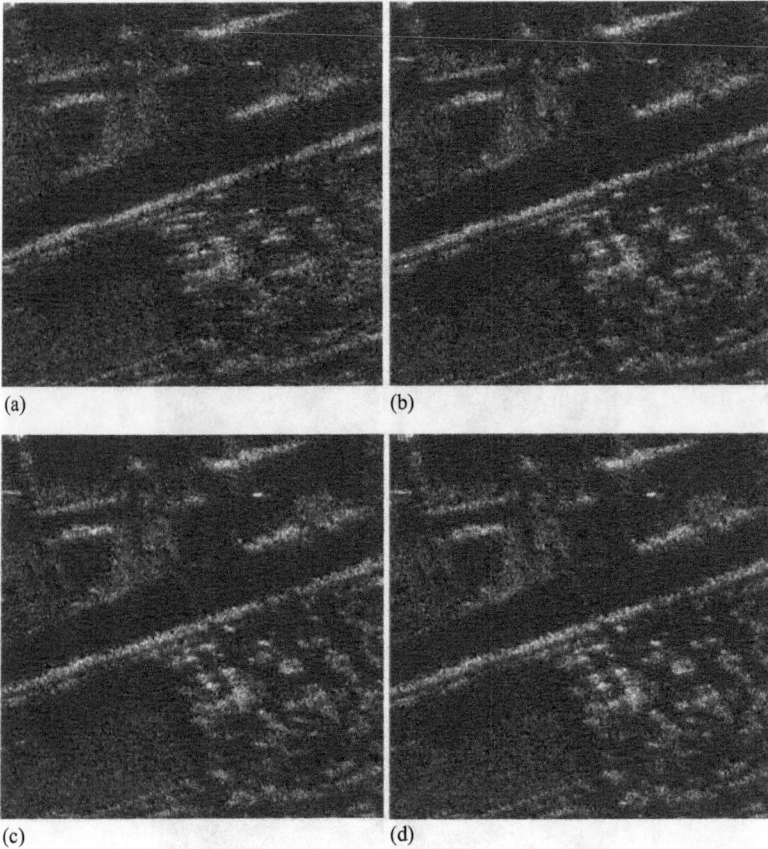

Figure 4.26   Zoom-in images of region "A" in Figure 4.25 obtained by (a) MD, (b) PGA, (c) MEA, and (d) the PSR method

Table 4.3   Image contrasts of the image results shown in Figure 4.25

Algorithm	MD	PGA	MEA	The PSR method
IC	1.1552	1.2518	1.2536	1.2528
Time (s)	14.73	3,105.83	3,290.63	1,634.11

4.5.2   Algorithm description

In this section, we formulate the PSR method for ISAR imaging of uniformly rotating targets [6]. First, we assume that the rotation rate of the target is known and formulate the algorithm. Then we deal with the case when the rotation rate of the target is unknown.

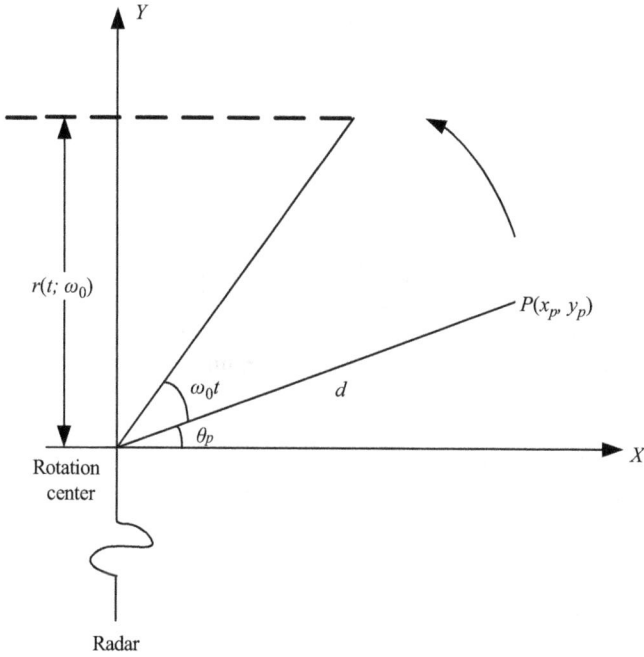

Figure 4.27   The geometry of ISAR imaging

4.5.2.1   With known rotation rate of the target

The target spatial domain is discretized as $K \times L$ positions, where K is the number of discretized range bins and L is the number of discretized cross-range bins. If a scatterer with the reflectivity $\sigma = 1$ is located at spatial coordinates (k, l), the ISAR data reflected from it can be directly derived from (4.43) as

$$s_{k,l}(\tau, t; \omega_0) = a\left(\tau - \frac{2 \cdot r_{k,l}(t; \omega_0)}{c}\right) \cdot \exp\left[-j\frac{4\pi}{\lambda}r_{k,l}(t; \omega_0)\right] \tag{4.46}$$

where

$$\begin{aligned} r_{k,l}(t; \omega_0) &= l\rho_x \sin(\omega_0 t) + k\rho_y \cos(\omega_0 t) \\ &\approx k\rho_y + l\rho_x\omega_0 t - \frac{1}{2}k\rho_y\omega_0^2 t^2 \end{aligned} \tag{4.47}$$

where the target rotation rate ω_0 is assumed known, and ρ_x and ρ_y are the cross-range and range resolutions, respectively. By discretizing τ and t as $\tau = \frac{1}{F_s}\left[-\frac{M}{2}, -\frac{M}{2} + 1, \ldots, \frac{M}{2} - 1\right]$ and $t = \frac{1}{F_p}\left[-\frac{N}{2}, -\frac{N}{2} + 1, \ldots, \frac{N}{2} - 1\right]$, respectively, where F_s is the radar sampling rate and F_p is the pulse repetition frequency, the discrete form of (4.46) can be described as an $M \times N$ matrix $\mathbf{S}_{k,l}$, whose (m, n)th element is given by

$$\mathbf{S}_{k,l}(m, n; \omega_0) = a\left(\frac{m}{F_s} - \frac{2}{c} \cdot r_{k,l}\left(\frac{n}{F_p}; \omega_0\right)\right) \cdot \exp\left[-j\frac{4\pi}{\lambda} r_{k,l}\left(\frac{n}{F_p}; \omega_0\right)\right]$$

(4.48)

where $r_{k,l}(\cdot)$ follows (4.47), $m = 1, 2, \ldots, M$, $n = 1, 2, \ldots, N$.

From (4.48), one can see that given the radar system parameters and the known ω_0, $\mathbf{S}_{k,l}$ is only dependent on (k, l) that indicates the spatial position where the scatterers are located. Thus finding which $\mathbf{S}_{k,l}$ contributes to the received ISAR echo can reflect the scatterer positions, and the projection coefficients of the ISAR echo on these $\mathbf{S}_{k,l}$ indicate the scatterer reflectivities. Moreover, in most ISAR applications, the target spatial domain is sparse, that is, the number of dominant scatterers is much smaller than the number of the discretized spatial positions.

Define the vectorization of $\mathbf{S}_{k,l}$ (the result is denoted by $\mathbf{S}_{k,l}^{vec}$) as a basis-signal corresponding to the spatial position (k, l). Here the vectorization of $\mathbf{S}_{k,l}$ is carried out by stacking the columns of $\mathbf{S}_{k,l}$ one underneath the other in sequence, that is, the $MN \times 1$ basis-signal $\mathbf{S}_{k,l}^{vec}$ is generated by

$$\mathbf{S}_{k,l}^{vec} \triangleq \underbrace{\left[\mathbf{S}_{k,l}(1,1), \mathbf{S}_{k,l}(2,1), \ldots, \mathbf{S}_{k,l}(M,1)\right.}_{\text{1st column of } \mathbf{S}_{k,l}}, \underbrace{\mathbf{S}_{k,l}(1,2), \mathbf{S}_{k,l}(2,2), \ldots, \mathbf{S}_{k,l}(M,2)}_{\text{2nd column of } \mathbf{S}_{k,l}}, \ldots,$$

$$\underbrace{\mathbf{S}_{k,l}(1,N), \mathbf{S}_{k,l}(2,N), \ldots, \mathbf{S}_{k,l}(M,N)}_{N\text{th column of } \mathbf{S}_{k,l}}\Big]^{T}$$

(4.49)

For $k = 1, 2, \ldots, K$ and $l = 1, 2, \ldots, L$, we generate KL basis-signals that respectively correspond to all discretized spatial positions, and we stack them as a $MN \times KL$ dictionary matrix $\mathbf{\Phi}(\omega_0)$:

$$\mathbf{\Phi}(\omega_0) \triangleq \left[\mathbf{S}_{1,1}^{vec}, \mathbf{S}_{2,1}^{vec}, \ldots, \mathbf{S}_{K,1}^{vec}, \mathbf{S}_{1,2}^{vec}, \mathbf{S}_{2,2}^{vec}, \ldots, \mathbf{S}_{K,2}^{vec}, \ldots, \mathbf{S}_{1,L}^{vec}, \mathbf{S}_{2,L}^{vec}, \ldots, \mathbf{S}_{K,L}^{vec}\right]$$

(4.50)

where the $(K(l-1)+k)$ th column of $\mathbf{\Phi}(\omega_0)$ corresponds to the ISAR data reflected from the possible scatterer located at (k, l)th discretized position in the target spatial domain. Thus the received ISAR echo $\mathbf{x}$ can be expressed as

$$\mathbf{x} = \sum_{k=1}^{K}\sum_{l=1}^{L} \mathbf{S}_{k,l}^{vec} \cdot \theta_{k,l} + \mathbf{e} = \mathbf{\Phi}(\omega_0) \cdot \mathbf{\theta} + \mathbf{e}$$

(4.51)

where $\mathbf{x}$ is a $MN \times 1$ vector, $\mathbf{\theta} = \left[\theta_{1,1}, \theta_{2,1}, \ldots, \theta_{K,1}, \theta_{1,2}, \theta_{2,2}, \ldots, \theta_{K,2}, \ldots, \theta_{1,L}, \theta_{2,L}, \ldots, \theta_{K,L}\right]^{T}$ is a $KL \times 1$ vector indicating the complex reflectivities at all discretized spatial positions, and $\mathbf{e}$ denotes the additive noise. Moreover, $\mathbf{\theta}$ is a sparse vector, that is, it has a few large coefficients and many small coefficients because the number of dominant scatterers is much smaller than the number of discretized spatial positions. From this observation, the ISAR imaging problem is equivalent to solving θ from the measurement vector $\mathbf{x}$ with a predesigned dictionary $\Phi(\omega_0)$. This is the typical problem of sparse signal recovery:

$$\widehat{\theta} = \arg\min_{\theta} \|\theta\|_0 \quad \text{s.t.} \quad \|\mathbf{x} - \Phi(\omega_0) \cdot \theta\|_2^2 \leq \varepsilon \tag{4.52}$$

where ε is the error threshold. The problem in (4.52) can be solved by using the OMP algorithm [25], which is also provided in Chapter 1. Once θ is solved, an ISAR image of size $K \times L$ is obtained.

4.5.2.2 With unknown rotation rate of the target

In Section 4.5.2.1, the target rotation rate ω_0 is assumed known, but this is not the case for noncooperative targets. In what follows, we deal with the case of unknown rotation rate of the target. Given a candidate rotation rate ω, one can create the dictionary $\Phi(\omega)$ according to (4.48)–(4.50) and solve (4.52) by replacing $\Phi(\omega_0)$ with $\Phi(\omega)$, and the solution of (4.52) is denoted by $\theta(\omega)$. Now the problem of interest is to search for a parameter/quantity to measure whether $\theta(\omega)$ correctly represents the true target spatial domain. Similar to Section 4.4.3, image contrast is also used here to evaluate the quality of ISAR images. Different from (4.41), here the image contrast of $\theta(\omega)$ is viewed as the function of ω and expressed as

$$C_\theta(\omega) = \frac{\text{mean}\left(|\theta(\omega)|^2\right) - [\text{mean}(|\theta(\omega)|)]^2}{[\text{mean}(|\theta(\omega)|)]^2} \tag{4.53}$$

If the candidate ω is equal to the true ω_0, the solution of $\theta(\omega)$ presents a clear ISAR image that correctly represents the true target spatial domain, so a larger C_θ is obtained. Contrarily, if the candidate ω is apart from the true ω_0, the solution of $\theta(\omega)$ presents a blurred ISAR image, so C_θ becomes smaller. This implies that the C_θ produced by the true ω_0 is larger than the that produced by a wrong candidate ω. Therefore, the correct rotation rate can be determined by a search on ω such that $C_\theta(\omega)$ reaches the maximum.

In summary, the PSR method for ISAR imaging can be carried out by following steps.

Step 1 Given a candidate rotation rate ω, create the dictionary $\Phi(\omega)$ according to (4.48)–(4.50).

Step 2 Solve $\theta(\omega)$ in (4.52) by inputting $\mathbf{x}$ and $\Phi(\omega)$ into the OMP algorithm, and compute $C_\theta(\omega)$ according to (4.53).

Step 3 Solve for the true ω_0 by searching all possible value of ω such that the maximum of $C_\theta(\omega)$ is reached, that is, $\widetilde{\omega}_0 = \arg\max_{\omega} \{C_\theta(\omega)\}$, and the correct ISAR image is obtained as $\theta(\widetilde{\omega}_0)$.

Note that, in this section, the PSR method for ISAR imaging is implemented by parameter searching, which is different from the alternating iterations used in Section 4.3 and Section 4.4, that is, iteratively updating the dictionary and the sparse solution. The reason is that the parametric dictionary for ISAR imaging in (4.50) is difficult to be represented in analytical form. This section provides an example about how to apply the PSR method to the case where the analytical expression of the dictionary is unavailable.

4.5.3   Experimental results

Some experiments on the simulated MIG-25 data [32], which have been widely used to evaluate the performances of ISAR imaging algorithms, are provided to verify the effectiveness of the PSR method. The simulated radar is stepped frequency type, the radar bandwidth is 512 MHz that is synthesized by a pulse burst consisting of 64 pulses, the pulse repetition frequency is 15 kHz, and the radar carrier frequency is 9 GHz. The result of the inverse Fourier transform on 64 pulses in a burst is taken as the range signal profile $a(\cdot)$. For computational convenience, the original pulse burst repetition frequency in [32] is halved, that is, only the pulse bursts with odd indexes are picked from the original data. The solution of (4.52) is carried out by the OMP algorithm, in which the iterations are terminated if the relative recovery error is smaller than $0.05\|\mathbf{x}\|_2^2$. Let $M = 40$, $K = 80$, and $L = 100$, that is, 40 range samples are taken, and the target spatial domain is discretized as 80×100 positions.

An additive noise is added to the original MIG-25 data, and the signal-noise-ratio (SNR) is set to 5 dB. When N, the number of the used pulse bursts, varies from 50 to 70, $C_\theta(\omega)$ in (4.53) versus various rotation rate candidate ω is plotted in Figure 4.28, where each point represents the average of 100 trials. It can be seen that the true rotation rate of the aircraft can be obtained (about 8.5°/s) by finding the maximum of $C_\theta(\omega)$. For a fixed rotation rate candidate, the image contrast value goes down with increased number of used pulse bursts. The reason is that the more measurements are used, the more detailed characteristics are recovered by the OMP algorithm. Moreover, the more pulse bursts are used, the larger gradient of the contrast curve is. This means that the rotation rate estimation accuracy can be improved by increasing the number of measurements. The reason is that the rotational motion effect becomes serious in a long observation duration, and therefore the wrong rotation rate candidate induces more serious blurring in the ISAR image.

When $N \geq 50$, substituting $\widetilde{\omega}_0 = 8.5°/s$ into (4.47), one can calculate that the range cell migration occurs and the high order phase in terms of slow time cannot be ignored. This implies that the ISAR signal during N pulse bursts cannot be seen as a sum of some sinusoids. The imaging result of the algorithm of time-frequency analysis [33] by using 50 pulse bursts is shown in Figure 4.29(a), where the image focusing is accomplished by first compensating the high order phase error and then performing Fourier transform in terms of slow time. It can be seen that the image is blurred due to the poor resolution and the high sidelobe. The imaging result of the algorithm of time-frequency analysis [33] by using 100 pulse bursts is shown in Figure 4.29(b), where the resolution is improved but the high-sidelobe still exists. By substituting $\widetilde{\omega}_0 = 8.5°/s$ into (4.48)–(4.50) and solving the ISAR image $\theta(\omega_0)$ by (4.52), a

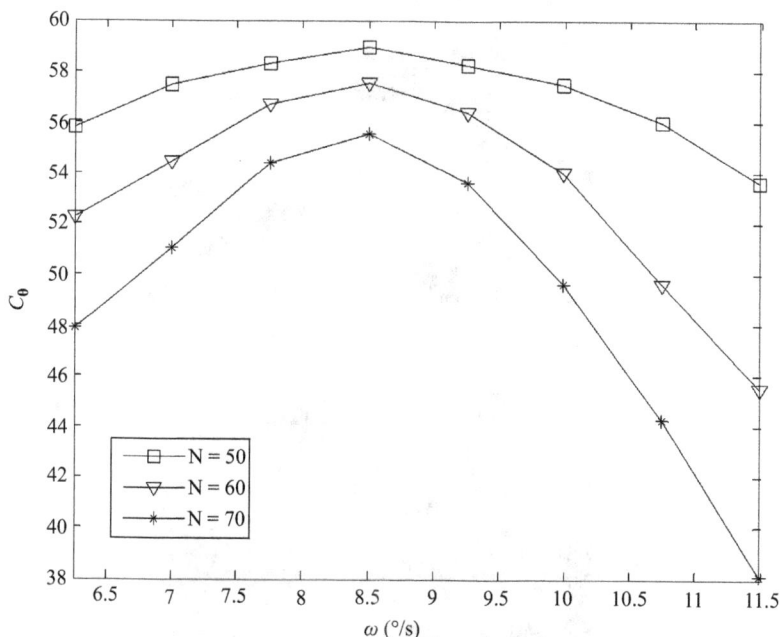

Figure 4.28 $C_\theta(\omega)$ versus various rotation rate candidate ω for different number of used pulse bursts

clear ISAR image is obtained by the PSR method as shown in Figure 4.29(c), which improves both range and cross-range resolutions and removes sidelobe artifacts.

Assume that the number of the used pulse bursts N is set to be equal to 60 and the SNR varies from −5 dB to infinitude. Figure 4.30 shows the C_θ versus various rotation rate candidate ω, where each point represents the average of 100 trials. It is clear that the true rotation rate of the aircraft can be obtained (about 8.5°/s) by searching the maximum of $C_\theta(\omega)$ for moderate or high SNR. When SNR = −5 dB, by using the rotation rate estimate 8.5°/s, the imaging result of the PSR algorithm is given in Figure 4.31. Compared with Figure 4.29(c), although some unwanted artificial points occur around dominant scatterers due to the increased noise energy level, the resulting ISAR image is still acceptable for target classification and recognition.

4.6   Conclusion

In this chapter, a PSR method was presented to eliminate the negative effect of model uncertainty on radar imaging quality. A parameter vector formulating the model uncertainty is embedded into the basis-signal dictionary so that the

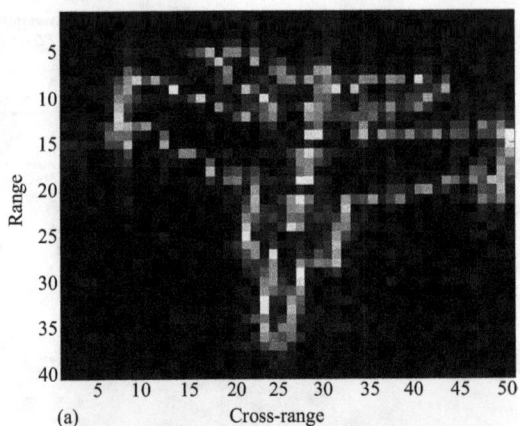

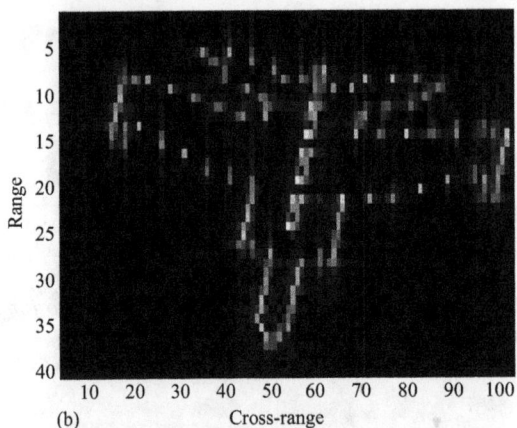

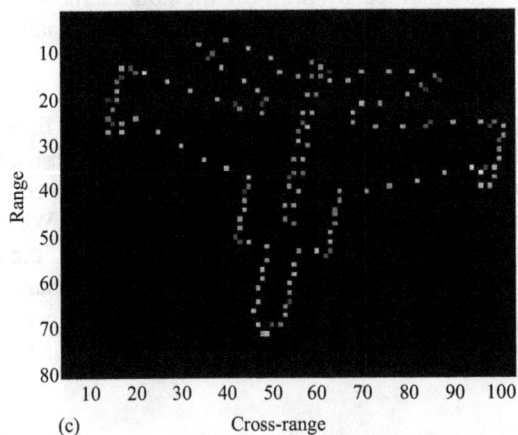

Figure 4.29   Imaging results when SNR = 5 dB: (a) by using the chirplet
decomposition algorithm on 50 pulse bursts, (b) by using the chirplet
decomposition algorithm on 100 pulse bursts, and (c) by using the
PSR method on 50 pulse bursts

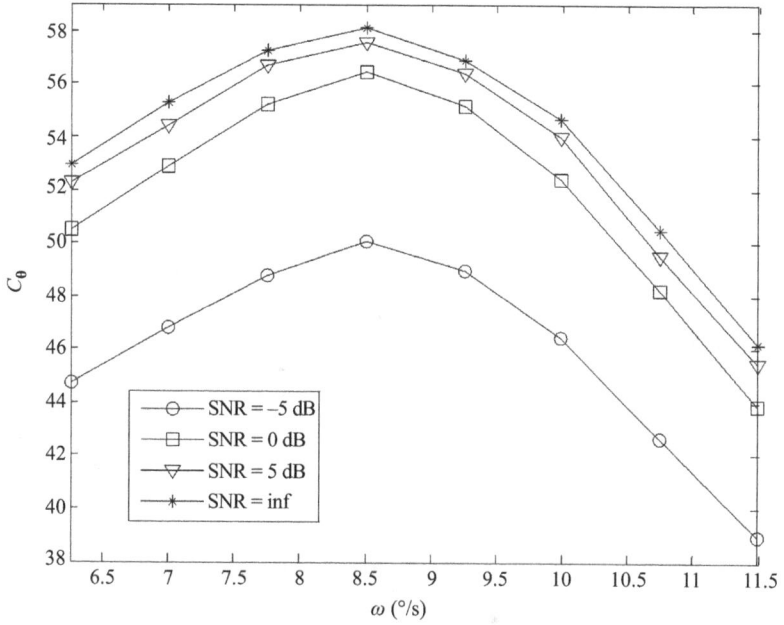

Figure 4.30 C_θ versus various rotation rate candidate ω for different SNR

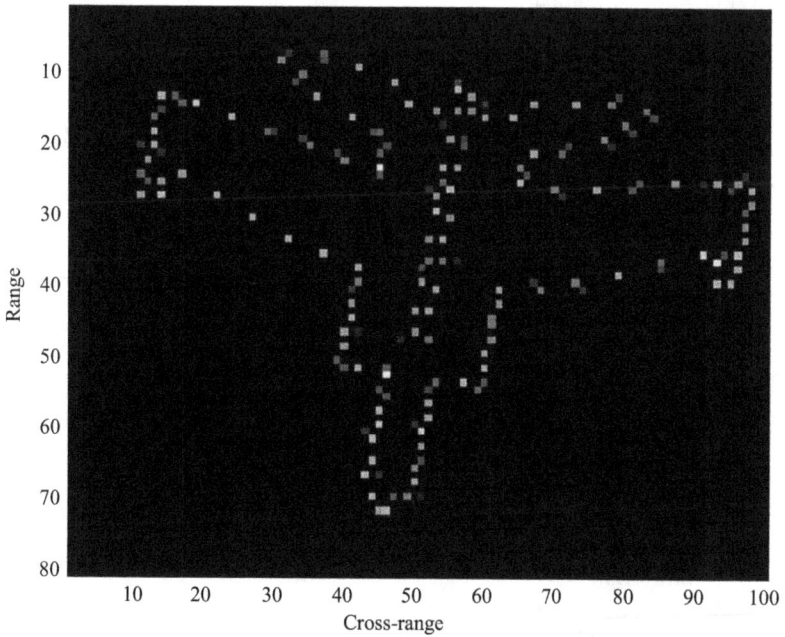

Figure 4.31   Imaging result when SNR $= -5$ dB and 60 pulse bursts are used

dictionary is adjustable with the varying value of the parameter vector. Both the spare radar image and the parametric dictionary can be solved by the alternating iterations or the parameter searching scheme. The PSR method has been applied to SAR refocusing of moving targets, SAR motion compensation, and ISAR imaging of maneuvering aircrafts. Experimental results demonstrate that the PSR method is superior over some widely used algorithms in terms of the radar imaging quality at an affordable computational complexity. The model uncertainty considered in this chapter is mainly caused by the unknown relative motion between the radar and the target. It is worth investigating the applications of the PSR method in other radar tasks with model uncertainty. For example, in the case of through-wall radar imaging or ground penetrating radar imaging, the model uncertainty lies in the unknown characteristics of the propagation channel. By designing a proper parametric dictionary that is closely related to the propagation characteristics, the PSR is expected to be able to improve the quality of through-wall radar imaging and ground penetrating radar imaging. One may also extend the PSR method to the problems of radar detection and classification with model uncertainty, in which the parametric dictionary is expected to make the features corresponding to different hypotheses more separable.

References

[1] Bamler R. "A comparison of range-Doppler and wavenumber domain SAR focusing algorithms." *IEEE Transactions on Geoscience Remote Sensing*. 1992; 30(4): 706–713.

[2] Raney R. K., Runge H., Bamler R., Cumming I. G., and Wong F. H. "Precision SAR processing using chirp scaling." *IEEE Transactions on Geoscience Remote Sensing*. 1994; 32(4): 786–799.

[3] Cafforio C., Prati C., and Rocca F. "SAR data focusing using seismic migration and techniques." *IEEE Transactions on Aerospace Electronic Systems*. 1991; 27(2): 194–207.

[4] Chen Y., Li G., Zhang Q., and Sun J. "Refocusing of moving targets in SAR images via parametric sparse representation." *Remote Sensing*. 2017; 9(8): 795; doi:10.3390/rs9080795.

[5] Chen Y., Li G., Zhang Q., Zhang Q., and Xia X.-G. "Motion compensation for airborne SAR via parametric sparse representation." *IEEE Transactions on Geoscience and Remote Sensing*. 2017; 55(1): 551–562.

[6] Li G., Zhang H., Wang X., and Xia X.-G. "ISAR 2-D imaging of uniformly rotating targets via matching pursuit." *IEEE Transactions on Aerospace and Electronic Systems*. 2012; 48(2): 1838–1846.

[7] Ender J. H. G. "On compressive sensing applied to radar." *Signal Processing*. 2010; 90(5): 1402–1414.

[8]   Rao W., Li G., Wang X., and Xia X.-G. "Adaptive sparse recovery by parametric weighted L1 minimization for ISAR imaging of uniformly rotating targets." *IEEE Journal of Selected Topics in Applied Earth Observations and Remote Sensing.* 2013; 6(2): 942–952.

[9]   Li G. and Varshney P. K. "Micro-Doppler parameter estimation via parametric sparse representation and pruned orthogonal matching pursuit." *IEEE Journal of Selected Topics in Applied Earth Observations and Remote Sensing.* 2014; 7(12): 4937–4948.

[10] Tošić I. and Frossard P. "Dictionary learning." *IEEE Signal Processing Magazine.* 2011; 28(2): 27–38.

[11] Li G., Xia X.-G., Xu J., and Peng Y.-N. "A velocity estimation algorithm of moving targets using single antenna SAR." *IEEE Transactions on Aerospace and Electronic Systems.* 2009; 45(3): 1052–1062.

[12] Zhang Y., Sun J., Lei P., Li G., and Hong W. "High-resolution SAR based ground moving target imaging with defocused ROI data." *IEEE Transactions on Geoscience and Remote Sensing.* 2016; 54(2): 1062–1073.

[13] Blumensath T. and Davies M. "Iterative hard thresholding for compressed sensing." *Applied and Computational Harmonic Analysis.* 2009; 27(3): 265–274.

[14] Cumming L. G. and Wong F. H. *Digital processing of synthetic aperture radar data: Algorithms and implementation.* Norwood, MA: Artech House, 2005.

[15] Wyholt A. "SAR image focus errors due to incorrect geometrical positioning in fast factorized back-projection." Licentiate thesis. Chalmers University of Technology, Gothenburg, Sweden, 2008.

[16] Carrara W. G., Goodman R. S., and Majewski R. M. *Spotlight synthetic aperture radar signal processing algorithms.* Norwood, MA: Artech House, 1995.

[17] Samczynski P. and Kulpa K. "Concept of the coherent autofocus map-drift technique." *Proceedings of International Radar Symposium (IRS).* Krakow, Poland: IEEE; 2006, pp. 1–4.

[18] Van Rossum W. L., Otten M. P. G., and Van Bree R. J. P. "Extended PGA for range migration algorithms." *IEEE Transactions on Aerospace and Electronic Systems.* 2006; 42(2): 478–488.

[19] Kragh T. J. "Monotonic iterative algorithm for minimum-entropy auto-focus." *Proceedings of Adaptive Sensor Array Processing (ASAP) Workshop.* Lexington, MA, USA, pp. 1–6, 2006.

[20] Onhon N. O. and Cetin M. "A sparsity-driven approach for joint SAR imaging and phase error correction." *IEEE Transactions on Image Processing.* 2012; 21(4): 2075–2088.

[21] Xing M., Jiang X., Wu R., Zhou F., and Bao Z. "Motion compensation for UAV SAR based on raw radar data." *IEEE Transactions on Geoscience and Remote Sensing.* 2009; 47(8): 2870–2883.

[22] Kir J. C. "Motion compensation for synthetic aperture radar." *IEEE Transactions on Aerospace and Electronic Systems.* 1975; 11(3): 338–348.

[23] Moreira J. R. "A new method of aircraft motion error extraction from radar raw data for real time motion compensation." *IEEE Transactions on Geoscience and Remote Sensing.* 1990; 28(4): 620–626.

[24] Tello Alonso M., Lopez-Dekker P., and Mallorqui J. J. "A novel strategy for radar imaging based on compressive sensing." *IEEE Transactions on Geoscience and Remote Sensing.* 2010; 48(12): 4285–4295.

[25] Tropp J. and Gilbert A. C. "Signal recovery from partial information via orthogonal matching pursuit." *IEEE Transactions on Information Theory.* 2007; 53(12): 4655–4666.

[26] Oliver C. and Quegan S. *Understanding synthetic aperture radar images.* Norwood, MA: Artech House, 1998.

[27] Berizzi F. and Corsini G. "Autofocusing of inverse synthetic aperture radar images using contrast optimization." *IEEE Transactions on Aerospace and Electronic Systems.* 1996; 32(3): 1185–1191.

[28] Chen C. C. and Andrews H. C. "Target motion induced radar imaging." *IEEE Transactions on Aerospace and Electronic Systems.* 1980; AES-16(1): 2–14.

[29] Wang J. and Kasilingam D. "Global range alignment for ISAR." *IEEE Transactions on Aerospace and Electronic Systems.* 2003; 39(1): 351–357.

[30] Munoz-Ferreras J. M. and Perez-Martinez F. "Uniform rotational motion compensation for inverse synthetic aperture radar with non-cooperative targets." *IET - Radar, Sonar & Navigation.* 2008; 2(1): 25–34.

[31] Chen V. C. and Qian S. "Joint time-frequency transform for radar range-Doppler imaging." *IEEE Transactions on Aerospace and Electronic Systems.* 1998; 34(2): 486–499.

[32] See http://airborne.nrl.navy.mil/~vchen/tftsa.html

[33] Wang G. and Bao Z. "Inverse synthetic aperture radar imaging of maneuvering targets based on chirplet decomposition." *Optical Engineering.* 1999; 38(9): 1534–1541.

Chapter 5

Poisson disk sampling for high-resolution and wide-swath SAR imaging

5.1 Introduction

Modern applications of synthetic aperture radar (SAR) demand both high spatial resolution and wide imaging swath. Unfortunately, traditional SAR imaging algorithms based on the Nyquist sampling theorem and the matched filtering can hardly achieve high-resolution and wide-swath simultaneously, since there is a trade-off between these two demands [1–4]. Existing wide-swath SAR imaging systems usually adopt the flexible digital beamforming or waveform coding techniques with phased arrays, at the cost of increased system complexity [5,6]. It is worth investigating how to achieve wider imaging swath of single-antenna SAR while keeping high resolution. If this solution is found, the imaging swath of the phased array SAR systems can be further increased.

In a single-antenna SAR, high azimuth resolution is dependent on a large Doppler bandwidth. Thus, high pulse repetition frequency (PRF) is required for achieving high azimuth resolution according to the Nyquist sampling theorem. On the other hand, the swath width in the range direction is limited by PRF, as shown in Figure 5.1. From Figure 5.1(a), it is obvious that the sampling window length is required to be smaller than the pulse repetition interval (PRI) to make sure that the radar echo is collected unambiguously and completely. To have a wider imaging swath in the range direction, a larger PRI is required. A simple scheme to increase PRI is the uniform under-sampling strategy, as shown in Figure 5.1(b). However, uniform under-sampling will cause the well-known aliasing phenomenon in the azimuth direction.

Inspired by the stochastic sampling strategy widely used in computer graphics, in this chapter, we attempt to adopt the Poisson disk sampling pattern instead of the uniform sampling pattern in SAR pulse sampling. Under the framework of the Poisson disk sampling, an iterative shrinkage thresholding like (IST-like) algorithm is presented for high-resolution and wide-swath SAR image formation [7]. Simulations and experimental results on real SAR data demonstrate the feasibility of utilizing the Poisson disk sampling pattern for high-resolution and wide-swath imaging with single-antenna SAR systems.

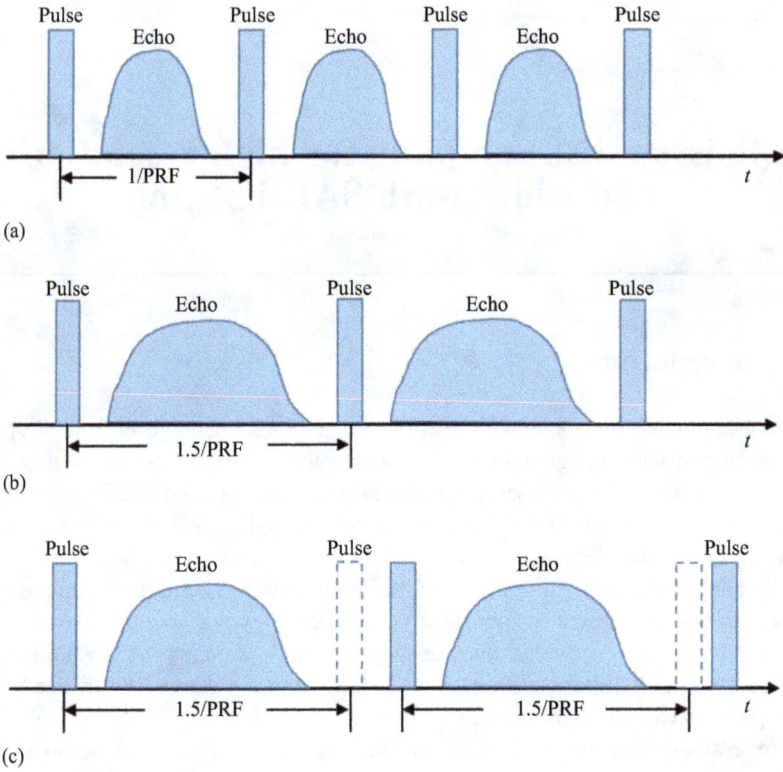

Figure 5.1   Pulse sampling schemes in the azimuth direction: (a) uniform
             sampling pattern; (b) uniform under-sampling pattern;
             and (c) Poisson disk sampling pattern

5.2   Tradeoff between high-resolution and wide-swath in SAR imaging

Consider the scenario of the stripmap SAR, where the direction of the antenna beam is fixed during the platform flight. The SAR echo reflected from a scatterer located at (x, y) can be expressed as [2]

$$
\begin{aligned}
s(\tau, t) \\
= \iint \sigma(x, y) \cdot p\left(\tau - \frac{2R(x, y, t)}{c}\right) \cdot w_a\left(t - \frac{x}{V}\right) \cdot \exp\left\{-j4\pi \frac{R(x, y, t)}{\lambda}\right\} dxdy + N(\tau, t)
\end{aligned}
$$

$$(5.1)$$

where τ is the fast time, t is the slow time, $\sigma(x, y)$ is the backscattering coefficient at the spatial position (x, y), w_a is the azimuth window function, V is the equivalent velocity of the SAR platform, $p(\tau)$ is the transmitted chirp signal, $R(x, y, t)$ is the slant distance between the radar and the scatterer, λ is the radar wavelength, c is the speed of light, and $N(\tau, t)$ is the additive noise at the receiver.

By discretizing the observed scene, the discrete form of the received echo can be expressed as

$$s(\tau_{n_r}, t_{n_a}) = \sum_{m_r=1}^{M_r} \sum_{m_a=1}^{M_a} \sigma(x_{m_a}, y_{m_r}) \cdot a[\tau_{n_r}, t_{n_a}, x_{m_a}, y_{m_r}] + N(\tau_{n_r}, t_{n_a}) \quad (5.2)$$

where $s(\tau_{n_r}, t_{n_a})$ is the sample at the n_rth range sample and the n_ath pulse of the original signal $s(\tau, t)$, M_r and M_a are the numbers of the discrete cells in range and azimuth directions, respectively, $N(\tau_{n_r}, t_{n_a})$ is the discretized additive noise, and $a[\tau_{n_r}, t_{n_a}, x_{m_a}, y_{m_r}]$ is defined by

$$a\left[\tau_{n_r}, t_{n_a}, x_{m_a}, y_{m_r}\right]$$
$$= p\left(\tau_{n_r} - \frac{2R(x_{m_a}, y_{m_r}, t_{n_a})}{c}\right) \cdot w_a\left(t_{n_a} - \frac{x_{m_a}}{V}\right) \cdot \exp\left\{-j4\pi \frac{R(x_{m_a}, y_{m_r}, t_{n_a})}{\lambda}\right\}$$
$$(5.3)$$

Record the discretized data $s(\tau_{n_r}, t_{n_a})$ and the reflectivity map $\sigma(x_{m_a}, y_{m_r})$ in two matrices $\mathbf{Y} \in \mathbb{C}^{M \times N_r}$ and $\mathbf{X} \in \mathbb{C}^{M_a \times M_r}$, respectively, where M is the number of the transmitted pulses in the azimuth direction, N_r is the number of samples in the range direction. Then (5.2) can be rewritten as

$$\mathbf{y} = \mathbf{A} \cdot \mathbf{x} + \mathbf{N} \quad (5.4)$$

where $\mathbf{y} = \text{vec}(\mathbf{Y})$ is the vectorization of $\mathbf{Y}$, $\mathbf{x} = \text{vec}(\mathbf{X})$ is the vectorization of $\mathbf{X}$, $\mathbf{N}$ is the vectorization of additive noise, and $\mathbf{A}$ is defined by

$$\mathbf{A} = \begin{bmatrix} a[\tau_1, t_1, x_1, y_1] & \cdots & a[\tau_1, t_1, x_{M_a}, y_{M_r}] \\ a[\tau_1, t_2, x_1, y_1] & \cdots & a[\tau_1, t_2, x_{M_a}, y_{M_r}] \\ \vdots & \ddots & \vdots \\ a[\tau_1, t_M, x_1, y_1] & \cdots & a[\tau_1, t_M, x_{M_a}, y_{M_r}] \\ \vdots & \ddots & \vdots \\ a[\tau_{N_r}, t_M, x_1, y_1] & \cdots & a[\tau_{N_r}, t_M, x_{M_a}, y_{M_r}] \end{bmatrix} \quad (5.5)$$

For the stripmap mode SAR, the relation between the azimuth resolution and the Doppler bandwidth can be expressed as [2]

$$\rho_a \approx \frac{V_g}{B_a} \quad (5.6)$$

where ρ_a is the azimuth resolution, B_a is the Doppler bandwidth, V_g is the velocity of the beam footprint on the ground, and the relation between V_g and V can be expressed as

$$V \approx \sqrt{V_g \cdot V_s} \quad (5.7)$$

where V_s represents the physical velocity of the SAR platform. It can be seen that high azimuth resolution demands wide Doppler bandwidth. According to the Nyquist sampling theorem, PRF must be larger than the Doppler bandwidth to avoid the azimuth ambiguity, that is,

$$PRF \geq (1 + \alpha_{os})B_a \tag{5.8}$$

where α_{os} is the oversampling ratio in the azimuth direction.

The upper limit of the sampling window time is $1/PRF - T_p$, where T_p is the width of the transmitted pulse. To guarantee that the echo of whole range swath can be collected unambiguously and completely, the relation between PRF and the range swath width W is required to satisfy [4]:

$$
\begin{aligned}
W &\leq \left(T_R - 2T_p\right)\frac{c}{2} \\
&= \left(\frac{1}{PRF} - 2T_p\right)\frac{c}{2}
\end{aligned}
\tag{5.9}
$$

where $T_R = \frac{1}{PRF} = PRI$.

In order to improve the azimuth resolution of the SAR system, a wider Doppler bandwidth is required, which demands higher PRF based on the Nyquist sampling theorem. However, the width of range swath is restricted by PRF according to (5.9). Thus, it is difficult to satisfy high-resolution and wide-swath simultaneously. To achieve a wider imaging swath in the range direction, a larger PRI is required. A simple scheme to increase PRI is the uniform under-sampling strategy. Nevertheless, the aliasing phenomenon will occur in the azimuth direction when the uniform sampling frequency is below the Doppler bandwidth according to the Nyquist sampling theorem. As a result, the aliasing errors caused by uniform under-sampling will create false patterns that seriously deteriorate the quality of SAR images.

5.3   Poisson disk sampling scheme

5.3.1   Sampling formulation

Different from uniform sampling, stochastic sampling is another class of sampling strategies, in which the samples are no longer taken with a uniform interval. The stochastic sampling pattern can be used for antialiasing since it is capable of converting the structured aliasing errors to the nonstructured noise [8]. The antialiasing capability of stochastic sampling was investigated since the 1960s [9–11]. In the 1980s, several stochastic sampling patterns, for example, jittered sampling and Poisson disk sampling, were adopted in computer graphics for antialiasing [12,13]. As reported in [14], the frequency response function of Poisson disk sampling is superior to that of jittered sampling in terms of antialiasing performance. This inspires us to adopt the Poisson disk sampling pattern instead of the uniform under-sampling pattern in the SAR pulse sampling scheme along the azimuth direction, as shown in Figure 5.1(c).

The samples of the Poisson disk sampling pattern can be expressed as [15,16]

$$p_{l+1} = p_l + \underbrace{\Delta}_{\text{deterministic}} + \underbrace{\delta_l}_{\text{random}} \tag{5.10}$$

where p_l and p_{l+1} are the lth and the $(l+1)$th samples, respectively, Δ is the desired minimum interval of all the Poisson disk sampling pulses, and δ_l is a random variable.

From (5.10), it is clear that Poisson disk sampling is capable of ensuring that the interval between any two adjacent samples is larger than the desired value. Therefore, Poisson disk sampling has the potential to achieve wider-swath imaging by setting the desired interval to be longer than the Nyquist sampling interval. By performing Poisson disk sampling in the azimuth direction, the nonuniformly under-sampled pulses are acquired.

Quantitatively speaking, to make sure that the signals reflected from all the targets in the range swath are collected completely, the interval between any two adjacent pulses must follow $\Delta t \geq \frac{2W}{c} + 2T_p$ according to (5.9). By setting the minimum pulse interval as $\frac{2W}{c} + 2T_p$, the Poisson disk samples in the azimuth direction can be designed as

$$t_{m+1} = t_m + \Delta t_{\min} + \delta_t \tag{5.11}$$

where $\Delta t_{\min} = \frac{2W}{c} + 2T_p$, t_{m+1} and t_m are the $(m+1)$th and the mth instants of pulse transmission, respectively, W is the desired width of the range swath, and δ_t is an exponentially distributed random variable defined by

$$\delta_t \sim exp(\eta_t) = \begin{cases} \eta_t \exp(-\eta_t \delta_t), & \delta_t \geq 0 \\ 0, & \delta_t < 0 \end{cases} \tag{5.12}$$

where $exp(\eta_t)$ denotes the exponential distribution with the parameter η_t. The existence of the random variable δ_t in (5.11) provides the potential of converting the aliasing of high-frequency components to noise-like components, which can avoid the structured aliasing caused by under-sampling in the azimuth direction. The average frequency of the pulse transmission is given by

$$\eta = \frac{1}{E(t_{m+1} - t_m)} = \frac{1}{\Delta t_{\min} + E(\delta_t)} = \frac{1}{\Delta t_{\min} + 1/\eta_t} \tag{5.13}$$

From (5.13), it can be seen that the average frequency of the pulse transmission η is dependent on the values of η_t and $\Delta t_{\min}$. As $\eta_t \to \infty$, the Poisson sampling pattern becomes a uniform sampling pattern with the sampling frequency $1/\Delta t_{\min}$. To achieve a wider imaging swath in the range direction, it is necessary to ensure that $\Delta t_{\min}$ is larger than the Nyquist sampling interval T_R. According to the SAR geometry, the vertical beamwidth of SAR determines the size of footprint in the range direction, that is, the theoretical upper bound of the range swath width. Therefore, a too-large value of $\Delta t_{\min}$ is unnecessary and impractical since the swath width cannot exceed the footprint size in the range direction. Determining the value of Δt_{min} is application-dependent. A method of selecting the parameter η_t is suggested in [17] in order to ensure that all the features are not missed during the sampling process with a high probability. According to the strategy suggested in [17], here η_t is set to $2 / T_R$.

With Poisson disk sampling, the width of SAR imaging swath can be expressed as follows:

$$W \leq \left(\Delta t_{min} - 2T_p\right) \frac{c}{2} \tag{5.14}$$

Since Δt_{min} is larger than T_R in (5.9), the Poisson disk sampling strategy can achieve a wider imaging swath than the traditional Nyquist sampling strategy.

5.3.2 Comparison between Poisson disk sampling and jittered sampling

Another widely used stochastic sampling pattern is jittered sampling. The samples of the jittered sampling pattern can be expressed as [18]

$$p_l = \underbrace{l \cdot T_d}_{\text{deterministic}} + \underbrace{\varsigma_l}_{\text{random}} \tag{5.15}$$

where l is an integer, T_d denotes the uniform sampling interval, ς_l is a random variable following uniform distribution on the interval $[-\beta T_d/2, \beta T_d/2]$, and the parameter $0 \leq \beta \leq 1$ determines the size of the perturbation around the regular sampling grids $\{0, T_d, 2T_d, 3T_d, \ldots\}$.

In the jittered sampling pattern, the sampling instants are determined by randomly perturbing the regular samples, while in the Poisson disk sampling pattern, the distance between any two adjacent samples is larger than a certain minimum interval. In what follows, the antialiasing effects of these two sampling patterns are analyzed.

Figure 5.2 shows the magnitude spectrums of several sampling patterns with the same average sampling density. The spectrum of the uniform sampling pattern in Figure 5.2(a) is composed of a series of Dirac delta functions. This frequency response causes spectral replications in the frequency domain, which will result in structured aliasing at sub-Nyquist sampling rate. Figure 5.2(b) shows the magnitude spectrum of the Poisson disk sampling, where Δ equals to $T_d/2$ and the mathematical expectation of the random variable δ_l is $T_d/2$. It can be seen that the magnitude spectrum of the Poisson disk sampling has only a single Dirac delta function at the origin, surrounded by the noise-like sidelobes. Such a frequency response function ensures that the Poisson disk sampling pattern can avoid the spectral replications that occur in the case of a uniform sampling pattern. That is to say, aliasing phenomenon can be significantly reduced by adopting the Poisson disk sampling strategy. As for more theoretical analysis about the antialiasing capability of the Poisson disk sampling, we refer the readers to [17]. Figure 5.2(c) and (d) demonstrates the magnitude spectrums of the jittered sampling pattern with $\beta = 1$ and $\beta = 0.5$, respectively. One can find that the antialiasing capability of the

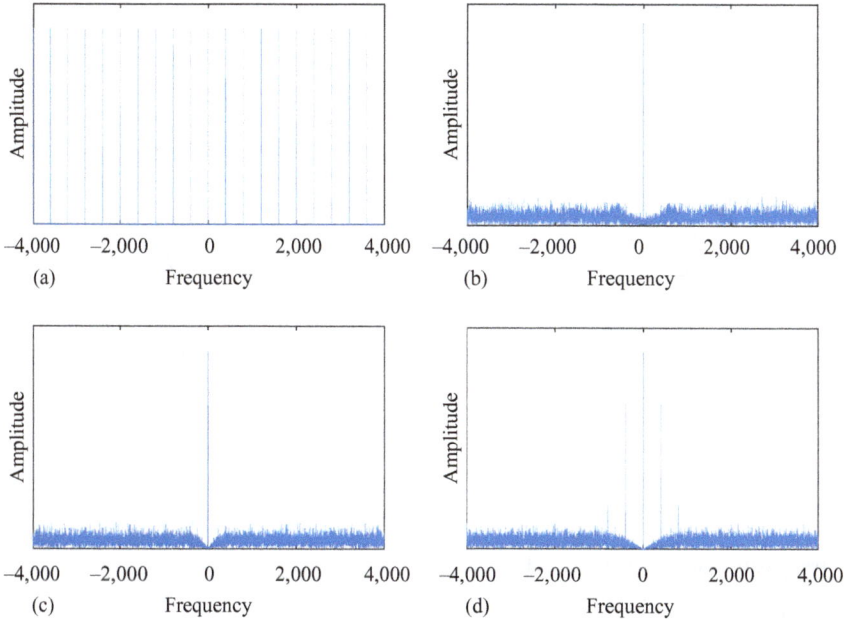

Figure 5.2   Magnitude spectrums of different sampling patterns: (a) uniform sampling pattern; (b) Poisson disk sampling; (c) jittered sampling pattern with $\beta = 1$; and (d) jittered sampling pattern with $\beta = 0.5$

jittered sampling pattern is sensitive to the value of the parameter β. When β equals to 1, the spectral of jittered sampling is similar to that in Figure 5.2(b). Nevertheless, in this case, the theoretical minimum sampling interval is close to zero, which cannot be applied to wide-swath SAR imaging according to (5.9). In Figure 5.2(d), there exist some lower Dirac delta functions except for the one at the origin, which will also cause the aliasing phenomenon. From the above observation, it can be concluded that the Poisson disk sampling is more suitable for high-resolution and wide-swath SAR imaging than the jittered sampling.

5.3.3   Comparison between Poisson disk sampling and random selection from uniform samples

Random selection from uniform samples is a common strategy of data collection adopted in the field of compressed sensing (CS) [28]. This sampling pattern randomly selects a part of the uniform sampling grids which satisfy the Nyquist sampling theorem. Though this sampling pattern shows remarkable performances in sparse signal recovery, it contributes nothing to wide-swath SAR imaging since its minimum sampling interval is required to be the same as that of the uniform sampling pattern for avoiding the aliasing.

The comparison of several sampling patterns is shown in Figure 5.3. Figure 5.3(a) illustrates the sampling result of the uniform sampling pattern with PRF = 600 Hz and

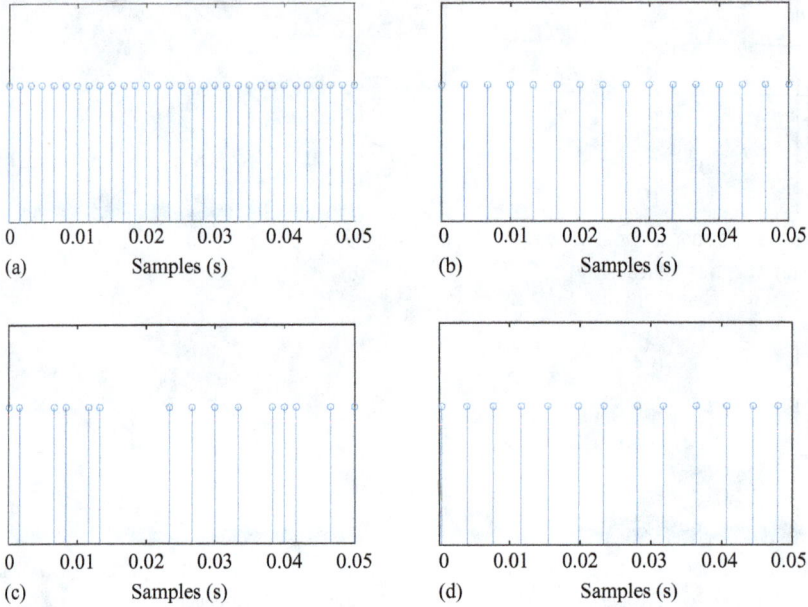

Figure 5.3   *Comparison of different sampling patterns: (a) uniform sampling with sampling frequency of 600 Hz; (b) uniform under-sampling with sampling frequency of 300 Hz; (c) random selection from uniform samples; and (d) Poisson disk sampling pattern*

Figure 5.3(b) shows the result of uniform under-sampling with the sampling rate PRF/2 = 300 Hz, that is to say, the sampling interval length between any two adjacent samples in Figure 5.3(b) is twice that in Figure 5.3(a). Figure 5.3(c) illustrates the random selection from the uniform samples in Figure 5.3(a). Although this sampling pattern can reduce the number of samples, it cannot ensure that the distance between any two adjacent samples is larger than that in Figure 5.3(a). For instance, the distance between the first two samples in Figure 5.3(c) is the same as the sampling interval length in Figure 5.3(a). The sampling pattern in Figure 5.3(d) is the Poisson disk sampling, which ensures that the interval length between any two adjacent samples is at least two times larger than that in Figure 5.3(a), which achieves almost twice the swath width in the range direction. Different from the result of Figure 5.3(b), the distribution of samples in Figure 5.3(d) is nonuniform.

5.4   SAR imaging algorithm with Poisson disk sampled data

The CS algorithms can be used for signal recovery from the Poisson disk sampled data. In [16], Poisson disk sampling was applied to compressed magnetic resonance imaging as the optimal sparse sampling strategy. In [15], the Poisson disk sampling was adopted in the azimuth direction to obtain a wider imaging swath with under-sampled SAR data. However, during the process of range migration correction (RMC) of this method in [15], only the range walk is corrected but the range curvature is ignored. Therefore, the method in [15] is only applicable to the scenarios of middle-level resolution SAR imaging.

In this section, by combining the Poisson disk sampling and the iterative shrinkage thresholding (IST) algorithm [7], we present an IST-like algorithm for high-resolution and wide-swath SAR imaging. By inducing the inverse operator of the traditional Chirp Scaling algorithm (CSA), the SAR imaging problem is converted to a sparse reconstruction problem. IST, which is an efficient algorithm for sparse reconstruction [19,20], is adopted to form the final SAR image from the Poisson disk sampled data. Different from the method in [15], the presented IST-like algorithm corrects both the range walk and the range curvature, providing the high-resolution SAR imaging capability. Moreover, the computational complexity of the IST-like algorithm is affordable.

5.4.1 Inverse CSA operator

The CSA is widely used for SAR image formation thanks to its good balance between efficiency and accuracy [2,21]. The procedure of CSA consists of three main steps: (1) the chirp scaling operation, (2) the range compression, and (3) the azimuth compression and the residual phase correction. The formulation of CSA can be expressed as follows:

$$\widetilde{\mathbf{X}} = H(\mathbf{Y}) = \text{IFFT}_t(\text{IFFT}_\tau(\text{FFT}_\tau((\mathbf{F}_t \cdot \mathbf{Y}) \odot \mathbf{S}_c) \odot \mathbf{P}_\tau) \odot \mathbf{P}_t) \tag{5.16}$$

where $\widetilde{\mathbf{X}}$ is the reconstructed SAR image matrix, $H(\cdot)$ is the equivalent CSA operator, $\mathbf{Y}$ is the matrix of the raw SAR data defined in (5.4), FFT_τ is the fast Fourier transform (FFT) along the range direction, IFFT_τ and IFFT_t are the inverse fast Fourier transforms (IFFT) along the range direction and the azimuth direction, respectively, $\mathbf{S}_c$ is the matrix of the chirp scaling operation, $\mathbf{P}_\tau$ and $\mathbf{P}_t$ are the matrices of matched filtering along range and azimuth directions, respectively, and $\mathbf{F}_t$ is the nonuniform Fourier transform matrix, whose definition is given later. Specifically, $\mathbf{S}_c$ is expressed as

$$\mathbf{S}_c(\tau, f_t; R_s) = \exp\left[-j\pi\gamma_e(f_t; R_s)a(f_t)\left(\tau - \frac{2R_s(1 + a(f_t))}{c}\right)^2\right] \tag{5.17}$$

where f_t is the Doppler frequency and R_s is the distance between the center of the observed scene and the radar flight path. The chirp scaling factor $a(f_t)$ is expressed as

$$a(f_t) = \frac{1}{\sqrt{1 - (f_t/f_{tM})^2}} - 1 \tag{5.18}$$

where $f_{tM} = 2V/\lambda$. $\gamma_e(f_t; R_s)$ is the equivalent frequency modulation rate and it is expressed

$$\frac{1}{\gamma_e(f_t; R_s)} = \frac{1}{\gamma} + R_s \cdot \frac{2\lambda}{c^2} \cdot \frac{(f_t/f_{tM})^2}{\left[1 - (f_t/f_{tM})^2\right]^{3/2}} \tag{5.19}$$

where γ is the chirp rate of the transmitted signal. $\mathbf{P}_\tau$ and $\mathbf{P}_t$ are expressed as

$$\mathbf{P}_\tau(f_\tau, f_t; R_s) = \exp\left[-j\pi \frac{1}{\gamma_e(f_t; R_s)[1 + a(f_t)]}f_\tau^2\right] \cdot \exp\left[j\frac{4\pi R_s a(f_t)}{c}f_\tau\right] \tag{5.20}$$

and

$$\mathbf{P}_t(\tau, f_t; R_B) = \exp\left[-j\frac{2\pi}{\lambda}c\tau\left(1 - \sqrt{1 - (f_t/f_{tM})^2}\right) + j\Theta_\Delta(f_t; R_B)\right] \tag{5.21}$$

where f_τ is the range frequency, R_B is the distance between the range cell under consideration and the radar flight path, $\Theta_\Delta(f_t; R_B)$ is the residual phase induced by chirp scaling operation, which is expressed as

$$\Theta_\Delta(f_t; R_B) = \frac{4\pi}{c^2}\gamma_e(f_t; R_s)a(f_t)[1 + a(f_t)](R_B - R_s)^2 \tag{5.22}$$

The Poisson disk sampling pattern produces the nonuniform pulse samples in the azimuth direction. As a result, the regular FFT cannot be adopted in the azimuth direction. Instead, the nonuniform digital Fourier transform $\mathbf{F}_t$ is used in (5.16) to map the raw data into the range-Doppler domain. Specifically speaking, $\mathbf{F}_t$ is consisting of M column vectors:

$$\mathbf{F}_t = [\boldsymbol{\alpha}_1, \boldsymbol{\alpha}_2, \dots, \boldsymbol{\alpha}_M] \tag{5.23}$$

where the mth column is defined by

$$\boldsymbol{\alpha}_m = \left[\exp\left(-j2\pi t_m f_t^{(1)}\right), \dots, \exp\left(-j2\pi t_m f_t^{(N_a)}\right)\right]^T \tag{5.24}$$

In (5.24), t_m is the mth Poisson disk sample in the azimuth direction given by (5.11), $m = 1, 2, \dots, M$, and M is the total number of the Poisson disk samples, $f_t^{(n)} = (n-1)/(N_a T_R)$ means the nth Doppler frequency bin, $n = 1, 2, \dots, N_a$, and N_a is the number of the uniform frequency grids in the Doppler domain.

By performing the CSA operator on the raw data $\mathbf{Y}$, we can obtain the reconstructed scattering behavior of the observed scene, which are approximate to $\mathbf{X}$, the true scattering coefficients [22,23], that is,

$$H(\mathbf{Y}) = \tilde{\mathbf{X}} \approx \mathbf{X} \tag{5.25}$$

From (5.16), it is clear that the CSA operator is invertible [22]. Performing the inverse of CSA operator on the true scattering coefficients of the scene, we can obtain the approximation of the raw data $\mathbf{Y}$. The inverse CSA operator G can be explicitly expressed by

$$\widetilde{\mathbf{Y}} = G(\mathbf{X}) = \mathbf{I}_t \cdot \left(\text{IFFT}_\tau \left(\text{FFT}_\tau \left(\text{FFT}_t(\mathbf{X}) \odot \mathbf{P}_t^* \right) \odot \mathbf{P}_\tau^* \right) \odot \mathbf{S}_c^* \right) \tag{5.26}$$

where $\widetilde{\mathbf{Y}}$ is the approximated echo data matrix generated by G and FFT_t is the fast Fourier transform along the azimuth direction. $\mathbf{I}_t$ is the nonuniform inverse Fourier transform matrix:

$$\mathbf{I}_t = [\boldsymbol{\beta}_1, \boldsymbol{\beta}_2, \dots, \boldsymbol{\beta}_{N_a}] \tag{5.27}$$

where the nth column is given by

$$\boldsymbol{\beta}_n = \frac{1}{N_a} \left[\exp\left(j2\pi t_1 f_t^{(n)} \right), \dots, \exp\left(j2\pi t_M f_t^{(n)} \right) \right]^T \tag{5.28}$$

where $t_1, \dots, t_M$ are the Poisson disk samples as expressed in (5.11).

5.4.2   The IST-like algorithm for SAR imaging

As the Poisson disk sampling pattern can avoid the spectral replications, only noise-like sidelobes rather than the intolerable aliasing phenomenon will appear in the formed image. The amplitude of noise-like caused by the Poisson disk sampling is usually much lower than the amplitudes of dominant targets, especially in the sparse scenes. For example, in applications of sea surface surveillance with SAR, the scattering coefficients of ships are dominant, which are much greater than both the noise-like sidelobes caused by the Poisson disk sampling and the reflectivity of the water surface. Therefore, the sparse reconstruction algorithms can be applied to suppress the noise-like sidelobes and reconstruct the image of the dominant targets. As for more complicated nonsparse scenes, the sparsity-driven methods can also be applied to retrieve the main components since the Poisson disk sampling produces weak noise-like sidelobes rather than aliasing.

Based on the inverse CSA operator in (5.26), SAR imaging can be accomplished by solving the following problem:

$$\widehat{\mathbf{X}} = \arg\min_{\mathbf{X}} \left\{ \|\mathbf{Y} - G(\mathbf{X})\|_F^2 + \rho \|\mathbf{X}\|_1 \right\} \tag{5.29}$$

where $\widehat{\mathbf{X}}$ is the recovered image, and ρ is the regularization parameter. Here the problem in (5.29) is solved by the IST-like algorithm. The detailed procedure of the IST-like algorithm for SAR imaging is described in Table 5.1.

In Table 5.1, the input variables include the raw data $\mathbf{Y}$, the regularization parameter ρ, the threshold parameter μ, and the maximum number of iterations $I_{\max}$. First, the scene image $\mathbf{X}^{(0)}$, the residual matrix $\mathbf{R}^{(0)}$, and the recovery error $\varepsilon^{(0)}$ are initialized. At every iteration, the scene image $\mathbf{X}^{(i)}$, the residual matrix $\mathbf{R}^{(i)}$, and the recovery error $\varepsilon^{(i)}$ are updated sequentially. The shrinkage thresholding operator $\text{Th}_\rho(\cdot)$ is adopted for each entry of a matrix to update $\mathbf{X}^{(i)}$. $\text{Th}_\rho(\cdot)$ is expressed as

Table 5.1   *The IST-like algorithm*

Input:
    $\mathbf{Y}, \rho, \mu, I_{\max}$.
Initialization:
    $\mathbf{X}^{(0)} = \mathbf{0}, \mathbf{R}^{(0)} = \mathbf{Y}, \varepsilon^{(0)} = \left\|\mathbf{R}^{(0)}\right\|_F^2 + \rho\left\|\mathbf{X}^{(0)}\right\|_1$.
Iteration:
    for $i = 1$ to $I_{\max}$

1. Update $\mathbf{X}^{(i)} = \mathrm{Th}_\rho(\mathbf{X}^{(i-1)} + H(\mathbf{R}^{(i-1)}))$.
2. Update $\mathbf{R}^{(i)} = \mathbf{Y} - G(\mathbf{X}^{(i)})$.
3. Update $\varepsilon^{(i)} = \left\|\mathbf{R}^{(i)}\right\|_F^2 + \rho\left\|\mathbf{X}^{(i)}\right\|_1$.
4. Compute $\xi^{(i)} = \left|\varepsilon^{(i)} - \varepsilon^{(i-1)}\right|/\varepsilon^{(i-1)}$.
5. If $\xi^{(i)} \geq \mu$, return to step (1); else, terminate the iterations.

    end for
Output:
    $\mathbf{X}^{(i)}$

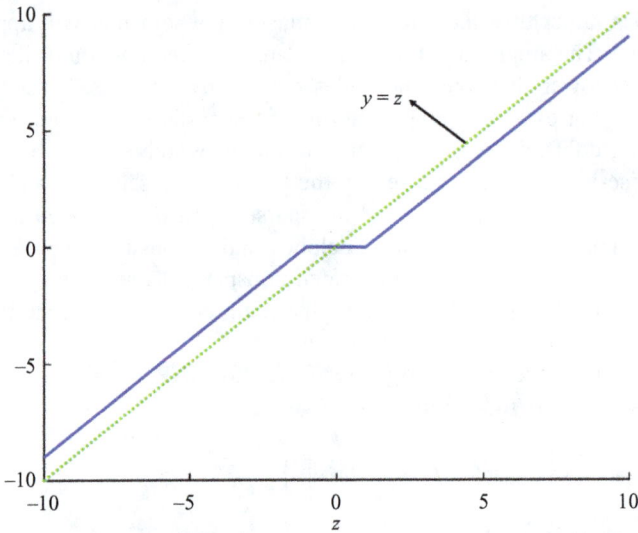

Figure 5.4   *The shrinkage thresholding function $Th_\rho(z)$*

$$Th_\rho(z) = \begin{cases} \mathrm{sgn}(z)(|z| - \rho), & |z| \geq \rho \\ 0, & |z| < \rho \end{cases} \qquad (5.30)$$

where sgn($\cdot$) is the sign function. The shrinkage thresholding function $Th_\rho(z)$ is plotted and compared with the function of $y = z$ in Figure 5.4. It is obvious than the thresholding operator sets the small coefficients to zero for enforcing the sparsity of the signal and slightly decreases the magnitudes of the large coefficients. The iterative process is terminated when the recovery error is smaller than the prede-signed threshold μ or the maximum number of iterations $I_{\max}$ is reached. The values of ρ, μ, and $I_{\max}$ are application-dependent and can be selected empirically.

The above algorithm is called IST-like because the thresholding operator $Th_{\rho}(\cdot)$ is similar to that in the IST algorithm [19,20]. The thresholding strategy in the original IST algorithm is deduced by the gradient descent method, as described in (1.16) of Chapter 1. Another viewpoint on (1.16) is that the current solution is updated as the weighted sum of the solution estimated at the previous iteration and the correlation between the residual and the dictionary. In the area of radar signal processing, the correlation between the residual and the dictionary can be regarded as the matched filtering. Step 1 in Table 5.1 follows this viewpoint, in which $\mathbf{X}^{(i-1)}$ is the image solution estimated at the previous iteration and $H(\mathbf{R}^{(i-1)})$ is essentially matched filtering according to (5.16). This supports the reasonability of the IST-like algorithm presented in Table 5.1 to some extent, although without the strict derivation. In Step 1 in Table 5.1, $\mathbf{X}^{(i-1)}$ and $H(\mathbf{R}^{(i-1)})$ are equally weighted. One may further consider the weighted sum of them as done in the original IST algorithm.

5.4.3 Analysis of computational complexity

Since the Poisson disk samples in azimuth direction are nonuniform, $\mathbf{F}_t$ and $\mathbf{I}_t$ are used in (5.16) and (5.26) instead of the regular FFT and IFFT. Accordingly, the efficient algorithms of the nonuniform FFT (NUFFT) and the nonuniform IFFT (NUIFFT) are needed. The core idea of NUFFT/NUIFFT is to compute an over-sampled FFT/IFFT of the original signal samples and then perform an accurate interpolation for desired sample positions [24,25]. The performance of NUFFT relies heavily on the accuracy of interpolation. Here we adopt the min–max inter-polation method to realize NUFFT and NUIFFT in azimuth direction. More details about the interpolation strategy can be found in [24].

As mentioned above, M is the number of Poisson disk samples in the azimuth direction, N_a is the number of the uniform frequency grids in the Doppler domain, and N_r is the number of the uniform samples in the range direction. The operator $H(\cdot)$ in (5.16) contains N_r NUFFTs and N_r IFFTs along the azimuth direction, while the operator $G(\cdot)$ in (5.26) contains N_r NUIFFTs and N_r FFTs along the azimuth direction. In addition, both $H(\cdot)$ and $G(\cdot)$ contain N_a FFTs and N_a IFFTs along the range direction. The computational complexities of NUFFT and NUIFFT along the azimuth direction are $O(N_a \log N_a + JN_a)$ and $O(N_a \log N_a + JM)$, respectively, where J is the length of the interpolator. In SAR imaging scenario, N_a, N_r, and M are usually on the order of thousands, while J is at most tens. Thus, the total computational cost of the IST-like algorithm is of the order $O(I \cdot N_T \log N_T)$, where $N_T = N_a N_r$, and I is the actual number of iterations to recover the SAR image. The exact value of I is difficult to be confirmed through theoretical analysis. As demonstrated by the experiments on real SAR data, the value of I is also about tens.

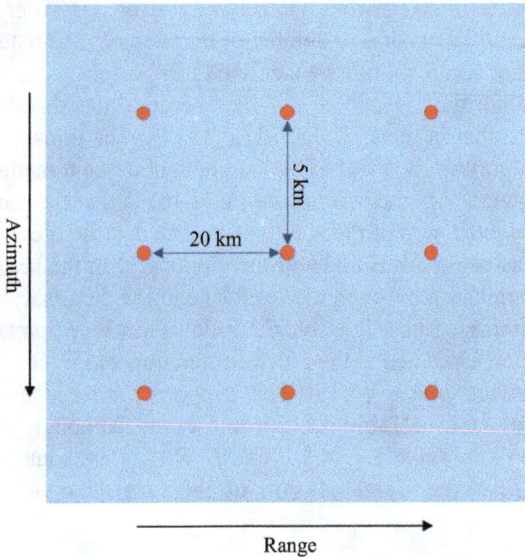

Figure 5.5   The simulated scene

As for the memory cost of the IST-like algorithm, we need to storage the input, the output, the parameter matrices including $\mathbf{S}_c$, $\mathbf{P}_\tau$, and $\mathbf{P}_t$, and the interpolation coefficients. In summary, the memory cost of the IST-like algorithm is of the order $O(N_T)$.

5.5   Experimental results

In this section, simulations and experimental results on real SAR data collected by RADARSAT-1 and GAOFEN-3 satellites are provided to demonstrate the effectiveness of the combination of the Poisson disk sampling strategy and the IST-like algorithm.

5.5.1   Simulations

The simulated scene contains nine scatterers with unit amplitude and random phases. The spatial distribution of these scatterers is shown in Figure 5.5. The range-direction distance between the nearest and the farthest scatterers is 40 km. The main parameters of the simulated SAR system are listed in Table 5.2, where the parameter PRF means the pulse repetition frequency required by the Nyquist sampling theorem. The upper limit of the range swath width in this SAR system is 26.7 km according to (5.9). Therefore, if the traditional uniform sampling pattern is adopted in azimuth direction, the radar echo reflected from part of the range swath cannot be completely collected, that is, the recovered SAR image cannot cover the whole range swath. The image generated by the traditional CSA algorithm with the

Table 5.2   Main parameters of SAR systems

Parameters	Simulations	RADARSAT-1	GAOFEN-3
Slant range of scene center: R_s (km)	850	1,016.7	954.7
Equivalent radar velocity: V (m/s)	7,100	7,062	7,136.4
Bandwidth: B_r (MHz)	20	30.111	240
Sampling frequency: F_s (MHz)	24	32.317	266.67
Pulse duration: T_p (µs)	50	41.74	45
Radar center wavelength: λ (m)	0.057	0.057	0.056
Pulse repetition frequency: PRF (Hz)	3,600	1,733	3,533
Beam squint angle: θ (°)	5	3.44	0

uniform pulse samples are shown in Figure 5.6(a). It is clear from Figure 5.6(a) that the nearest and the farthest scatterers are missing, and only the three targets in the middle of the range swath are well reconstructed, which is consistent with the above analysis.

To achieve a wider imaging swath, the Poisson disk sampling pattern is adopted. The desired range swath width W is set to 53.4 km, which doubles the swath of the original SAR system. The Poisson disk sampling pattern is designed according to (5.14) and the number of Poisson disk samples is 3,307. For comparison, the imaging results produced by the method proposed in [15] and the IST-like algorithm presented in this chapter with the Poisson disk sampled data are shown in Figure 5.6(b) and (c), respectively. We can see that the IST-like algorithm can provide slightly better imaging quality than the traditional CSA with a doubled range swath width and fewer measurements. In addition, there exists a geometric distortion in Figure 5.6(b). The reason is that the method in [15] performs RMC in the time domain, which is not accurate enough. Specifically, in Figure 5.6(b), wrong target positioning is obvious for some scatterers that are far from the scene center, that is, the scatterers actually located in the same range cell are reconstructed to different range cells. In contrast, the IST-like algorithm correctly locates all the scatterers distributed in a wide range swath as shown in Figure 5.6(c). The range profiles of the scatterer located at the scene center generated by different algorithms are compared in Figure 5.7. It is apparent that the imaging result of the IST-like algorithm has fewer sidelobes than the method in [15] without broadening the main lobe.

5.5.2   Experiment on RADARSAT-1 data

The real SAR data collected by RADARSAT-1 and GAOFEN-3 satellites are used to evaluate different algorithms. RADARSAT-1 and GAOFEN-3 were launched in 1995 and 2016, respectively, and both of them work at C-band.

The RADARSAT-1 data were collected over the English Bay, Canada. The main parameters of this space-borne SAR are listed in Table 5.2. In the scene to be imaged, several vessels are sparsely distributed on the sea surface. The number of uniform pulses of the raw data is 1,536. Since the raw data have been already

(a)   Range

(b)   Range

(c)   Range

Figure 5.6   *Imaging results of different imaging methods for the scene containing*
nine scatterers: (a) the traditional CSA method with uniform pulse
samples; (b) the method in [15] with the Poisson disk sampled data;
and (c) the IST-like algorithm with the Poisson disk sampled data

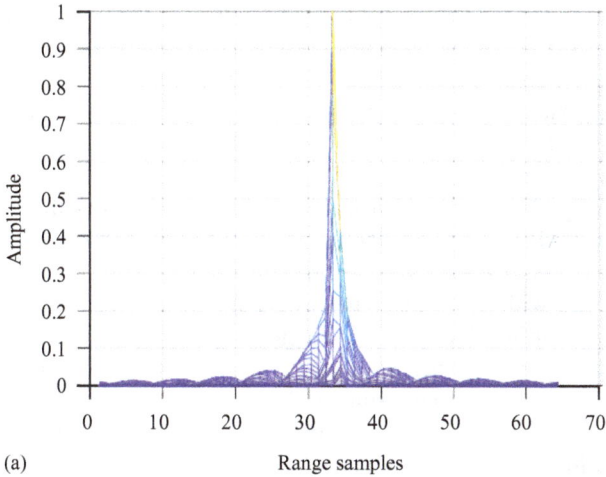

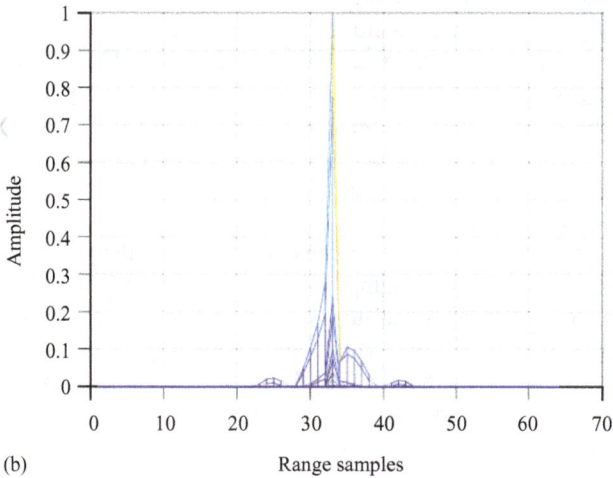

Figure 5.7   The range profiles of the scatterer located at the scene center obtained by (a) the method in [15] and (b) the IST-like algorithm

sampled at the uniform PRF, we need to generate the Poisson disk sampled data from the raw data. First, the raw data in the azimuth direction is interpolated to increase the sampling rate with 30 times, and then the upsampled data are resampled according to the Poisson disk sampling strategy given in (5.11). Specially, we set the minimum distance between any two adjacent Poisson disk sampled pulses in azimuth direction as 1.5/PRF so that the width of the range swath is increased by 1.5 times in comparison to the case of uniformly sampled pulses.

Figure 5.8 shows the imaging results of different methods. The imaging result of the traditional CSA with all the uniformly sampled 1,536 pulses is shown in Figure 5.8(a). Figure 5.8(b) provides the imaging result of the sparsity-driven method in [26] with 1,024 pulses randomly selected from the uniformly sampled pulses. Figure 5.8(c) and (d) illustrates the imaging results obtained by using the method in [15] and the IST-like algorithm with the generated Poisson disk sampled data, respectively. The comparison in Figure 5.8 demonstrates the advantages of sparsity-driven SAR imaging algorithms in comparison with the traditional CSA. The dominant targets are accurately recovered by the sparsity-driven algorithms in Figure 5.8(b)–(d), with less measurement data and lower sidelobes.

In order to further compare the imaging performances of the above three sparsity-driven algorithms, that is, the methods in [26] and [15] and the IST-like algorithm, the detailed imaging results of the third vessel (from top to bottom) are shown in Figure 5.9. Since the traditional range compression is retained in the process of the method in [15], the sidelobes in the range direction are obvious in Figure 5.9(b). On the other hand, the method in [26] and the IST-like algorithm perform the sparse reconstruction in the 2D range-azimuth domain, which provides the SAR images with fewer sidelobes than the method in [15], as shown in Figure 5.9(a) and (c). In addition, though the method in [26] can provide high-resolution images, it cannot achieve wide-swath SAR imaging since the random selection sampling is unable to guarantee a larger distance between two adjacent pulses than that required by the Nyquist sampling theorem. In contrast, the IST-like algorithm can increase the range swath width by 1.5 times thanks to the designed Poisson disk sampling pattern.

Table 5.3 gives the quantitative comparison of the method in [26] and the IST-like algorithm in terms of the imaging quality. Here three measures are considered: (1) image entropy; (2) peak signal-to-noise ratio (PSNR) relative to the imaging result of the method in [26]; and (3) structural similarity index (SSIM) relative to the imaging result of the method in [26]. Entropy is a statistical measure that can characterize the texture information of an image. PSNR is widely used to measure the absolute errors between two images, while SSIM considers image degradation as the changes in structural information [27]. The results in Figure 5.9 and Table 5.3 demonstrate that the IST-like algorithm can provide comparable imaging performance to the method in [26]. Besides, the IST-like algorithm can widen the range swath width while the method in [26] cannot. That is to say, the most attractive advantage of the IST-like algorithm is that it is able to guarantee both high-resolution and wide-swath simultaneously.

The running time of the traditional CSA and the IST-like algorithm are compared in Table 5.4. All the experiments were conducted on the same computer with a 4-core 3.5-GHz CPU and 16 GB memory. Since the size of the scene to be imaged is relatively small, the IST-like algorithm only takes several seconds to form the final image. The computational complexity of the IST-like algorithm is about 10 times larger than that of the traditional CSA.

Figure 5.10(a) plots the quantitative imaging performance versus the number of iterations. The imaging result of the method in [26] with a full set of the pulses is

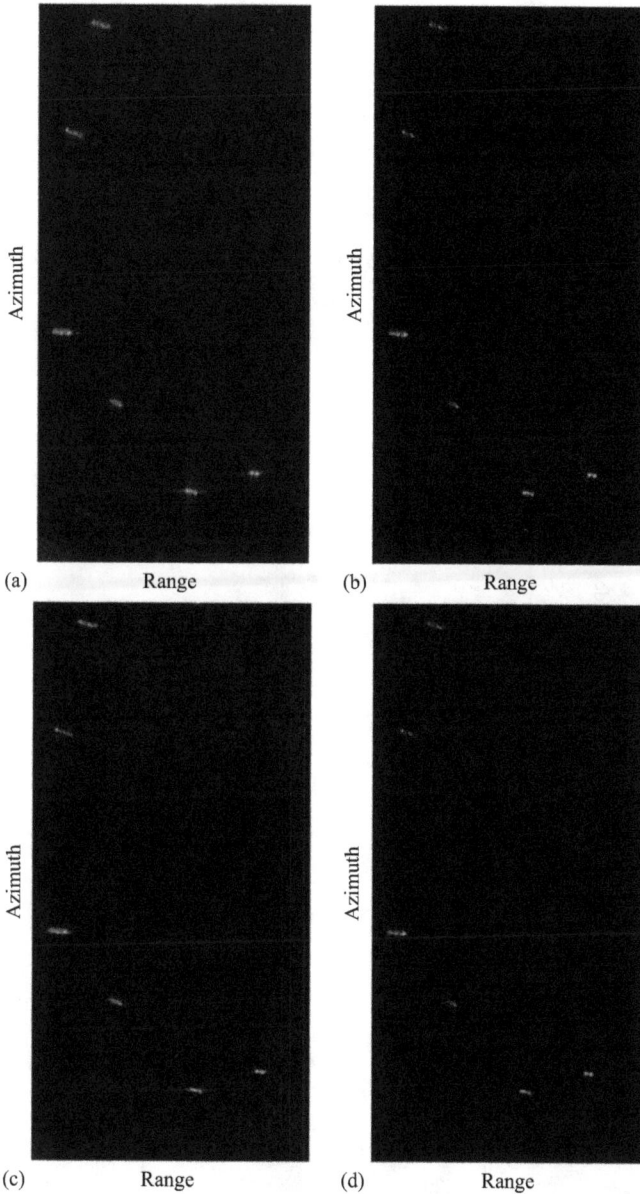

Figure 5.8 Imaging results on the RADARSAT-1 data: (a) traditional CSA method with uniformly sampled pulses; (b) the method in [26] with randomly selected pulses; (c) the method in [15] with Poisson disk sampled data; and (d) the IST-like algorithm with Poisson disk sampled data

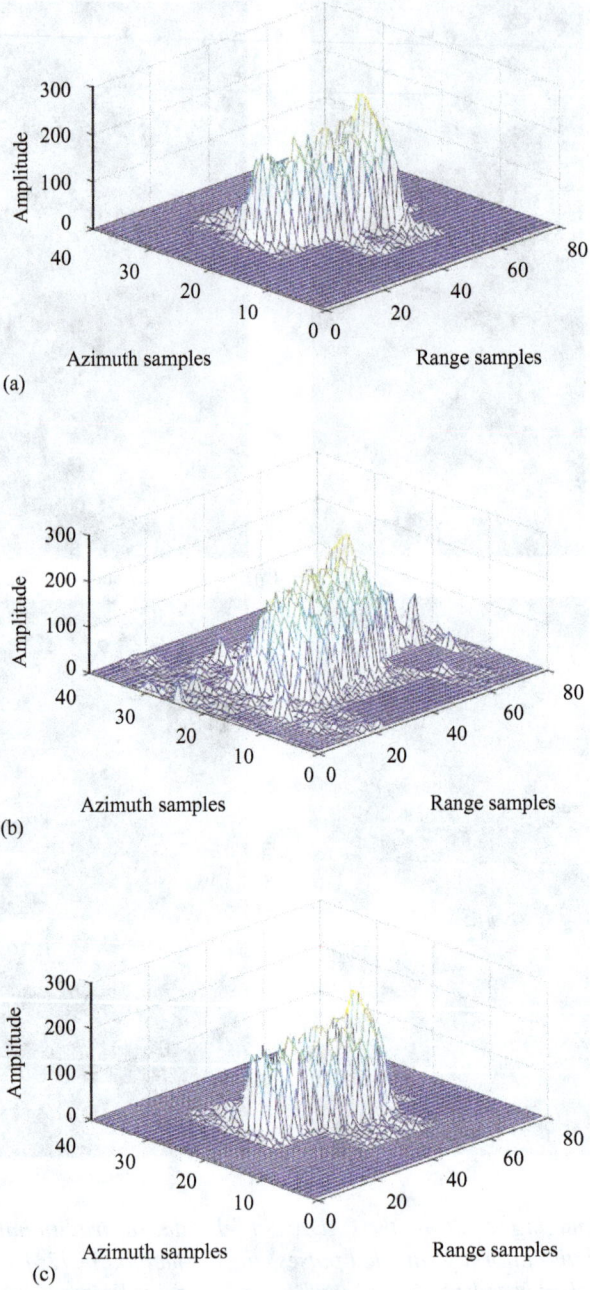

Figure 5.9   The imaging results of a vessel: (a) the method in [26]; (b) the method in [15]; and (c) the IST-like algorithm

Table 5.3 Comparison of two algorithms in terms of imaging quality

Data	Method	Entropy	PSNR	SSIM
RADARSAT-1	The method in [26]	0.086	/	/
	The IST-like algorithm	0.081	43.291	0.997
GAOFEN-3	The method in [26]	6.109	/	/
	The IST-like algorithm	5.974	20.915	0.412

Table 5.4 Comparison of running times of two algorithms

Data	Method	Running time (s)
RADARSAT-1	CSA	0.53
	The IST-like algorithm	5.02
GAOFEN-3	CSA	11.22
	The IST-like algorithm	278.01

adopted as the reference image to compute the values of PSNR and SSIM. It can be seen that the IST-like algorithm converges quickly after several iterations as both PSNR and SSIM of the formed SAR image become stable. This also gives a suggestion on setting the value of $I_{\max}$. The quantitative imaging performance versus the minimum sampling interval of the Poisson disk samples is plotted in Figure 5.10(b). Given the observation duration or the length of the synthetic aperture, as the minimum sampling interval length increases, both PSNR and SSIM gradually deteriorate. Thus, there exists a trade-off between the SAR image quality and the range swath width.

5.5.3 Experiment on GAOFEN-3 data

The GAOFEN-3 data were acquired over the Xiamen city, China. The main parameters of this space-borne SAR are listed in Table 5.1. The size of the raw data is 4,184 × 6,656. It can be calculated that the range resolution is 0.625 m and the range curvature is about 9.352 m. Therefore, the range curvature must be considered during the range migration correction, which makes the method in [15] unusable since it ignores the range curvature. In contrast, the IST-like algorithm ensures the accurate range migration correction. Here the Poisson disk data are sampled according to (5.14) so that the range swath width is 1.5 times that of the original uniform sampling pattern. The number of the Poisson disk sampled pulses is 2,133. The comparison among the imaging results of the three algorithms is given in Figure 5.11. Figure 5.11(a) and (b) demonstrates the imaging results of the traditional CSA with the full data and the method in [26] with part of pulses randomly selected from the uniform pulses. Though the method in [26] can obtain high-resolution images with less pulses, it cannot realize wide-swath imaging due to the fact that the minimum interval between any two adjacent

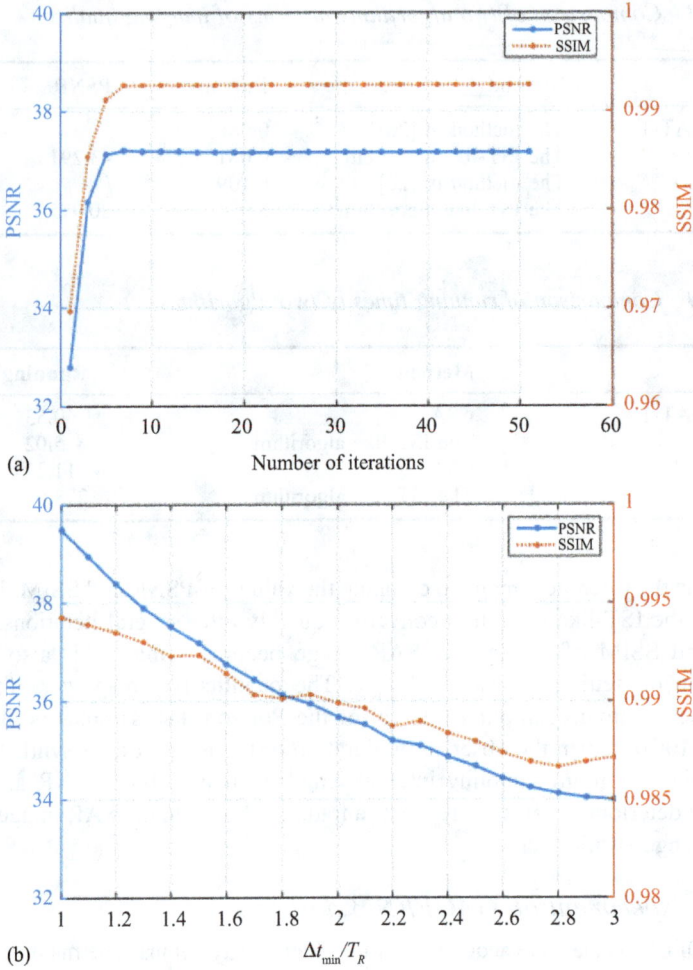

Figure 5.10   Quantitative imaging quality of the IST-like algorithm on the RADARSAT-1 data: (a) PSNR and SSIM versus the number of iterations; and (b) PSNR and SSIM versus the minimum sampling interval

pulses is the same as the original PRI required by the Nyquist sampling theorem. The imaging result of the IST-like algorithm is shown in Figure 5.11(c) and its quantitative measure is given in Table 5.3. It is clear from Figure 5.11 and Table 5.3 that the IST-like algorithm can achieve the comparable imaging quality with the traditional CSA and the method in [26], at the same time, has the potential to enlarge the width of the range swath.

The quantitative imaging performance versus the number of iterations is shown in Figure 5.12(a). Similar to the experimental result on RADARSAT-1 data, the IST-like algorithm also converges well on GAOFEN-3 data. In addition, the scene

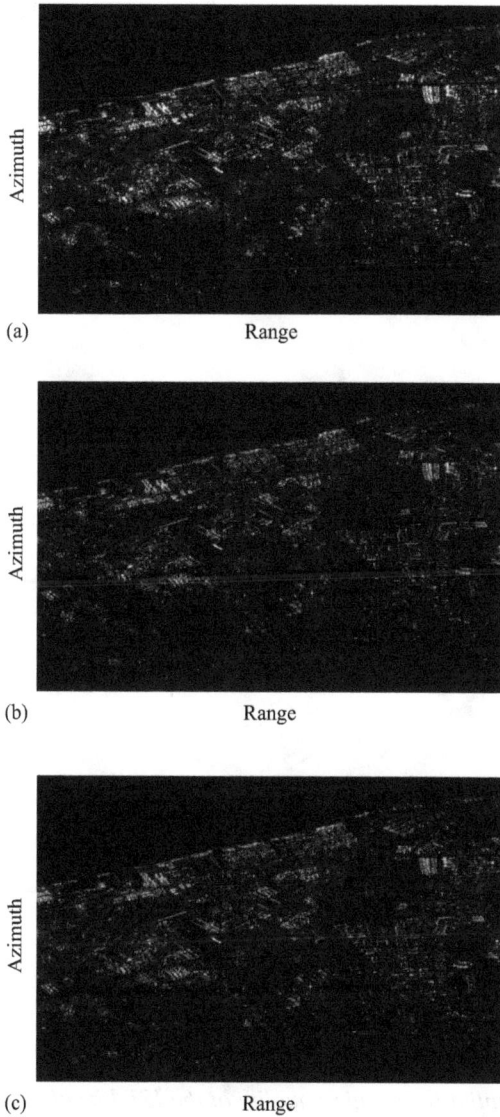

*Figure 5.11   Imaging results on GAOFEN-3 data: (a) traditional CSA method
with full 4,184 pulses; (b) the method in [26] with randomly selected
pulses; and (c) the IST-like algorithm with Poisson disk sampled data*

observed by GAOFEN-3 data is more complicated than that observed by
RADARSAT-1, so here more iterations are required to recover the final image. In
Figure 5.12(b), the relation between the imaging performance and the minimum
sampling interval of the Poisson disk samples is provided, which shows a similar

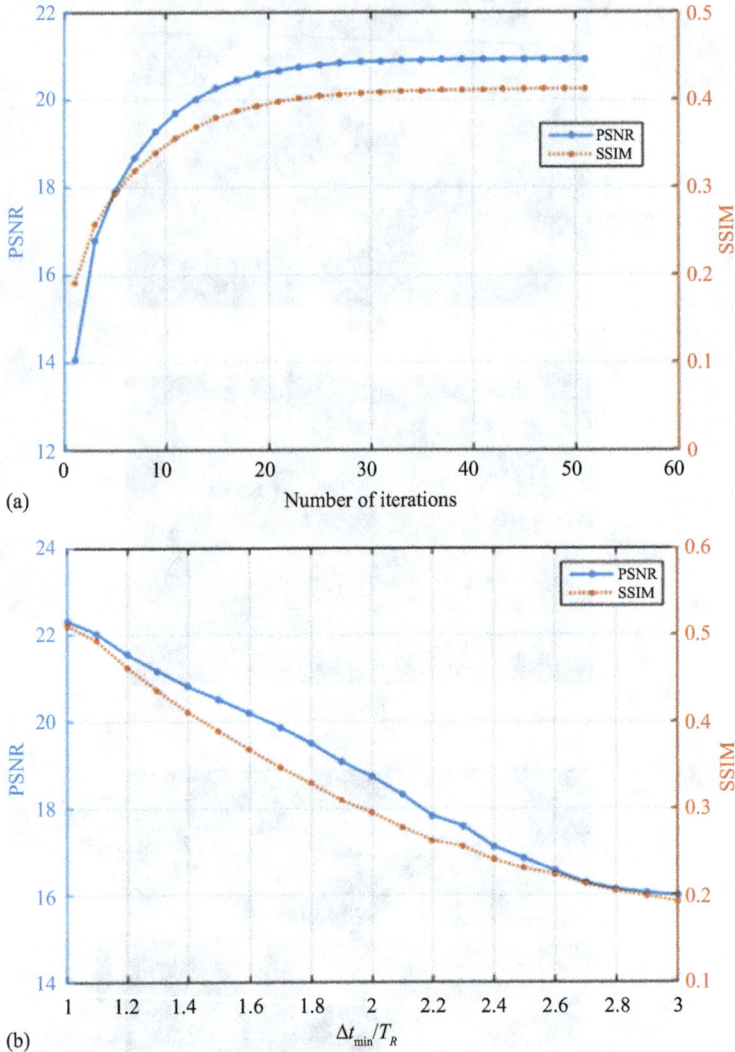

Figure 5.12   *Quantitative imaging quality of the IST-like algorithm on the GAOFEN-3 data: (a) PSNR and SSIM versus the number of iterations; and (b) PSNR and SSIM versus the minimum sampling interval*

trend to Figure 5.10(b). The trade-off between the imaging performance and the range swath width must be taken into account for the design of the Poisson disk sampling pattern in practical SAR systems.

The running time of two imaging algorithms on GAOFEN-3 data is compared in Table 5.4. The IST-like algorithm is much more time consuming than the traditional CSA. The computational burden of the IST-like algorithm comes from two aspects: (1) at each iteration of the IST-like algorithm, each of the operations $H(\cdot)$ and $G(\cdot)$ has the comparable complexity with the traditional CSA, and (2) the number of iterations required by the IST-like algorithm may greatly increase for complicated scenes.

5.6   Conclusion

In this chapter, we investigated how to achieve high-resolution and wide-swath of SAR imaging simultaneously. The Poisson disk sampling pattern was adopted in the azimuth direction to broaden the range swath of SAR since it can ensure that the interval between any two adjacent pulses is larger than that required by the Nyquist sampling theorem. Then the IST-like algorithm was presented to reconstruct the high-resolution SAR image from the Poisson disk sampled pulses. The experiments on simulations and real SAR data collected by RADARSAT-1 and GAOFEN-3 have demonstrated that compared to the traditional CSA based on the uniformly sampled pulses and the sparsity-aware algorithms based on the randomly selected pulses, the combination of the Poisson disk sampling and the IST-like algorithm can increase the range swath width by at least 1.5 times, without the obvious sacrifice of the SAR imaging quality. Besides the increased complexity of the imaging algorithm, another cost of applying the Poisson disk sampling pattern to SAR imaging is the loss of the signal-to-noise ratio, since less pulses are acquired given a synthetic aperture time. It is worth theoretically analyzing the trade-off between the SAR imaging performance and the minimum interval of the Poisson disk samples. Another issue worth studying is the effect of the Poisson disk sampling pattern on multi-antenna SAR systems for high-resolution and wide-swath imaging.

References

[1]   Curlander J. C. and Mcdonough R. N. *Synthetic aperture radar: Systems and signal processing.* New York, NY: Wiley, 1991.

[2]   Cumming I. G. and Wong F. H. *Digital signal processing of synthetic aperture radar data: Algorithms and implementation.* Norwood, MA: Artech House, 2005.

[3]   Wiley C. A. "Synthetic aperture radars." *IEEE Transactions on Aerospace and Electronic Systems.* 1985; 21(3): 440–443.

[4]   Currie A. and Brown M. A. "Wide-swath SAR." *IEE Proceedings F (Radar and Signal Processing).* 1992; 139(2): 122–135.

[5]   Gebert N., Krieger G., and Moreira A. "Digital beamforming on receive: Techniques and optimization strategies for high-resolution wide-swath SAR imaging." *IEEE Transactions on Aerospace and Electronic Systems.* 2009; 45(2): 564–592.

[6]   Gebert N. and Krieger G. "Azimuth phase center adaptation on transmit for high-resolution wide-swath SAR imaging." *IEEE Geoscience and Remote Sensing Letters.* 2009; 6(4): 782–786.

[7]   Yang X., Li G., Sun J., Liu Y., and Xia X.-G. "High-resolution and wide-swath SAR imaging via Poisson disk sampling and iterative shrinkage thresholding." *IEEE Transactions on Geoscience and Remote Sensing.* 2019; 57(7): 4692–4704.

[8]   Zhang X. H. *Stochastic sampling for antialiasing in computer graphics.* Ph.D. dissertation. Stanford University, Stanford, CA, USA, 2005.

[9]   Shapiro H. S. and Silverman R. A. "Alias-free sampling of random noise." *SIAM Journal on Applied Mathematics.* 1960; 8(2): 225–236.

[10]   Beutler F. J. "Alias-free randomly timed sampling of stochastic processes." *IEEE Transactions on Information Theory.* 1970; 16(2): 147–152.

[11]   Masry E. "Alias free sampling: An alternative conceptualization and its applications." *IEEE Transactions on Information Theory.* 1978; 24(3): 317–324.

[12]   Cook R. L. "Stochastic sampling in computer graphics." *ACM Transactions on Graphics.* 1986; 5(1): 51–72.

[13]   Dippé M. A. Z. and Wold E. H. "Antialiasing through stochastic sampling." *ACM SIGGRAPH Computer Graphics.* 1985; 19(3): 69–78.

[14]   Mitchell D. P. "Generating antialiased images at low sampling densities." *ACM SIGGRAPH Computer Graphics.* 1987; 21(4): 65–72.

[15]   Sun J. P., Zhang Y. X., Tian J. H., and Wang J. "A novel spaceborne SAR wide-swath imaging approach based on Poisson disk-like nonuniform sampling and compressive sensing." *Science China Information Sciences.* 2012; 55(8): 1876–1887.

[16]   Vasanawala S. S., Alley M. T., Hargreaves B. A., Barth R. A., Pauly J. M., and Lustig M. "Improved pediatric MR imaging with compressed sensing." *Radiology.* 2010; 256(2): 607–616.

[17]   Dippé M. A. Z. and Wold E. H. "Antialiasing through stochastic sampling." *ACM SIGGRAPH Computer Graphics.* 1985; 19(3): 69–78.

[18]   Hennenfent G. and Herrmann F. J. "Simply denoise: Wavefield reconstruction via jittered undersampling." *Geophysics.* 2008; 73(3): v19–v28.

[19]   Daubechies I., Defrise M., and De Mol C. "An iterative thresholding algorithm for linear inverse problems with a sparsity constraint." *Communications on Pure and Applied Mathematics.* 2004; 57(11): 1413–1457.

[20]   Beck A. and Teboulle M. "A fast iterative shrinkage-thresholding algorithm for linear inverse problems." *SIAM Journal on Imaging Sciences.* 2009; 2(1): 183–202.

[21]   Raney R. K., Runge H., Bamler R., Cumming I. G., and Wong F. H. "Precision SAR processing using chirp scaling." *IEEE Transactions on Geoscience and Remote Sensing.* 1994; 32(4): 786–799.

[22]   Bhattacharya S., Blumensath T., Mulgrew B., and Davies M. "Fast encoding of synthetic aperture radar raw data using compressed sensing." *Proceedings*

of IEEE/SP Workshop on Statistical Signal Processing. Madison, WI, USA: IEEE; 2007, pp. 448–452.

[23] Khwaja A. S., Ferro-Famil L., and Pottier E. "SAR raw data simulation using high precision focusing methods." *Proceedings of European Radar Conference (EuRAD)*. Paris, France: IEEE; 2005, pp. 33–36.

[24] Fessler J. A. and Sutton B. P. "Nonuniform fast Fourier transforms using min-max interpolation." *IEEE Transactions on Signal Processing.* 2003; 51(2): 560–574.

[25] Greengard L. and Lee J. Y. "Accelerating the nonuniform fast Fourier transform." *SIAM Review.* 2004; 46(3): 443–454.

[26] Fang J., Xu Z., Zhang B., Hong W., and Wu Y. "Fast compressed sensing SAR imaging based on approximated observation." *IEEE Journal of Selected Topics in Applied Earth Observations and Remote Sensing.* 2014; 7(1): 352–363.

[27] Wang Z., Bovik A. C., Sheikh H. R., and Simoncelli E. P. "Image quality assessment: From error measurement to structural similarity." *IEEE Transactions on Image Processing.* 2004; 13(4): 600–612.

[28] Ender J. H. G. "On compressive sensing applied to radar." *Signal Processing.* 2010; 90(5): 1402–1414.

Chapter 6

When advanced sparse signal models meet coarsely quantized radar data

6.1 Introduction

The original compressed sensing (CS) algorithms were designed to recover sparse signals from continuous-valued measurements. In practice, however, the measurements must be quantized, that is, each measurement is mapped from a finite set to a discrete value via an analog-to-digital converter (ADC). To deal with the quantized measurements, several methods of quantized compressive sensing (QCS) have been proposed in [1,2]. An extreme case of QCS is so-called 1-bit CS [3], where only the signs of the real-valued measurements are maintained and used for follow-on processing. The coarse quantization is appealing due to its simplified ADC hardware implementation. In addition, the coarse quantization significantly alleviates the burden of data transmission and storage. In the case of 1-bit quantization, ADC only needs to take the form of a comparator to zero with very low cost [4]. Moreover, it is shown in [5] that the 1-bit measurement is optimal in presence of heavy noise, that is, the noise can be neglected as long as it does not alter the signs of the measurements.

Existing literatures have shown the feasibility of sparsity-driven signal and image reconstructions from coarsely quantized data. In [6], it is demonstrated that the convex optimization is able to accurately recover sparse vector from 1-bit measurements. In [7,8], the Bayesian methods are proposed for sparse signal recovery from coarsely quantized data. Only the additive Gaussian noise is considered in [7], while the combined effect of the quantization error and the additive Gaussian noise is imposed in [8]. Another alternate scheme of QCS is to enforce the consistent reconstruction. This condition implies that, if the reconstructed sparse signal is remeasured by the same measurement system and the regenerated measurements are quantized by the same ADC, the quantized results should be the same as the original quantized measurements. The algorithms based on the consistent reconstruction include the matching sign pursuit (MSP) [9], the restricted-step shrinkage (RSS) [10], the quantized iterative hard thresholding (QIHT) algorithm [11], and binary iterative hard thresholding (BIHT) [12].

It is attractive for radar imaging applications to quantize the raw radar measurements with few-bit ADC, even 1-bit ADC. The resulting benefits include inexpensive hardware and less cost of data transmission and storage. Conventional radar imaging methods based on matched filtering have been used to process the coarsely quantized radar data [13–15]. However, these methods suffer from the resolution loss and ghost target images due to the imbalances of amplitude and phase between the coarsely quantized I/Q data. There have been some attempts to suppress the ghost target images by developing the sparsity-driven algorithms based on the consideration of the basic sparse signal models and the quantized radar data [7,16,17].

In this chapter, we investigate how to adapt the advanced sparse signal models to the coarsely quantized data and present two algorithms for enhancing the radar imaging quality with coarsely quantized data.

1. The first algorithm is called parametric quantized iterative hard thresholding (PQIHT) [18]. The PQIHT algorithm aims to refocus moving targets from the coarsely quantized synthetic aperture radar (SAR) echo, by combing the original quantized iterative hard thresholding (QIHT) algorithm [11] and the parametric sparse representation (PSR) framework [19]. In comparison with the original QIHT algorithm, the PQIHT algorithm makes significant extension to the case of parametric procedure of solution updating. Estimating the motion-related parameter and refocusing the moving target are simultaneously accomplished by a pruned searching procedure. Simulations and experiments on real SAR data demonstrate that the images of moving targets produced by the PQIHT algorithm based on coarsely quantized data and even 1-bit data are comparable with that reconstructed from high-precision data.

2. The second algorithm is called enhanced-binary iterative hard thresholding (E-BIHT) [20]. The E-BIHT algorithm aims to enhance the quality of 1-bit radar imaging of stationary targets by combining the BIHT algorithm [12] with the two-level block sparsity model [21]. Compared to the original BIHT algorithm, the key modification is that the nonlinear sparse approximation in BIHT is replaced with a two-level block sparse approximation. Simulations and experiments based on measured radar data demonstrate that, in comparison with existing 1-bit CS algorithms, the E-BIHT algorithm offers better imaging results with higher target-to-clutter-ratio (TCR).

6.2   Parametric quantized iterative hard thresholding for SAR refocusing of moving targets with coarsely quantized data

In this section, the PQIHT algorithm [18] is presented for SAR refocusing of moving targets with coarsely quantized data, by combing the QIHT algorithm [11] and the PSR framework [19]. The detailed formulation of the PSR method and some of its applications are also introduced in Chapter 4.

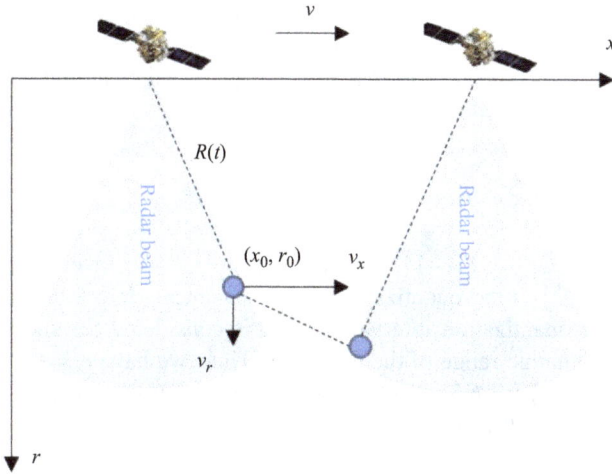

Figure 6.1   The geometry of side-looking SAR

6.2.1   Signal model

Consider the geometry of side-looking SAR, as shown in Figure 6.1. The x and r axes denote the cross-range and down-range directions, respectively. A moving target is denoted by the blue point in Figure 6.1, and its initial location is (x_0, r_0). Assume that the SAR platform moves along the cross-range direction at a constant velocity v. The velocities of the moving target in down-range and cross-range directions are denoted by v_r and v_x, respectively. The transmitted signal of the SAR system is a linear frequency modulated (LFM) signal with a modulation rate K_r, expressed as

$$s_t(t, t_s) = \text{rect}\left(\frac{t}{T_p}\right) \cdot \exp(j\pi K_r t^2 + j2\pi f_0 t) \tag{6.1}$$

where $\text{rect}(\cdot)$ represents the rectangular function, T_p denotes the pulse width, f_0 is the carrier frequency, t and t_s are the fast time and the slow time, respectively. Then, the baseband echo reflected from this moving scatterer with the scattering coefficient σ can be written as

$$
s_r(t, t_s) = \sigma \cdot \text{rect}\left(\frac{t - 2R(t_s)/c}{T_p}\right) \cdot \text{rect}\left(\frac{t}{T_a}\right)
$$
$$
\cdot \exp\left\{ j\pi K_r \left(t - \frac{2R(t_s)}{c}\right)^2 - j4\pi f_0 \frac{R(t_s)}{c} \right\} \tag{6.2}
$$

where T_a is the synthetic aperture time, c is the speed of light, and $R(t_s) = \sqrt{(r_0 + v_r t_s)^2 + (v t_s - x_0 - v_x t_s)^2}$ is the slant range from the radar platform to the target at slow time t_s. The scattering coefficient σ is assumed to be constant during the observation time duration.

It is assumed that the analog SAR signal is quantized by a b-bit quantizer $Q_b(\cdot)$ at the platform, expressed as

$$Q_b(x) = l_i \quad \text{if } x \in [r_i, r_{i+1}) \text{ for } i = 1, 2, \ldots, 2^b \tag{6.3}$$

where $\{r_i | i = 1, 2, \ldots, 2^b + 1\}$ is the quantization level sequence and $\{l_i | i = 1, 2, \ldots, 2^b\}$ is the quantization value sequence. Here a uniform quantizer with a constant quantization interval $\Delta_q = \frac{2A}{2^b}$ is considered for simplicity, where $[-A, A]$ is the dynamic range of the quantizer. Thus, we have $r_i = -A + (i - 1)\Delta_q$ and $l_i = \frac{1}{2}(r_{i+1} + r_i)$ in (6.3). With the quantizer $Q_b(\cdot)$, the quantized data of the received signal in (6.2) can be expressed as

$$s_r^{(q)}(t, t_s) = Q_b(s_r(t, t_s)) = s_r(t, t_s) + n(t, t_s) \tag{6.4}$$

where $n(t, t_s)$ denotes the quantization error.

Before extracting the region of interest (ROI) containing the moving target, some conventional SAR imaging algorithms, such as the back projection algorithm [22], the range-Doppler algorithm [23], the chirp scaling algorithm [24], and the ω-K algorithm [25], can be carried out on the quantized data to obtain the focused image of the static scene. In what follows, the effect of performing the ω-K algorithm [25] on the quantized signal reflected from a moving target (i.e., $s_r^{(q)}(t, t_s)$ in (6.4)) is analyzed.

After performing matched filtering and the modified Stolt interpolation in the two-dimensional (2D) frequency domain, as the ω-K algorithm operates [25], the 2D frequency signal of the quantized data in (6.4) becomes:

$$S_1^{(q)}(f_r, f_a) = \sigma \cdot W_r(f_r) \cdot W_a(f_a) \cdot \exp\left\{-j\frac{2\pi f_a \theta}{v_e^2}\right\}$$
$$\cdot \exp\left\{-j\frac{4\pi(x_0 v_r + r_0(v - v_x))}{c v_e}\sqrt{(f_0 + f_r)^2 + \left(\frac{cf_a}{2}\right)^2 \cdot \left(\frac{1}{v^2} - \frac{1}{v_e^2}\right)}\right\}$$
$$+ \mathrm{FT}_{2D}(f(n(t, t_s))) \tag{6.5}$$

where $v_e^2 = (v - v_x)^2 + v_r^2$, $\theta = x_0(v - v_x) - r_0 v_r$, f_r and f_a denote the down-range and cross-range frequencies, respectively, W_r and W_a are the down-range and cross-range envelope functions, respectively, $\mathrm{FT}_{2D}(\cdot)$ denotes the operation of taking 2D Fourier transform, and the function $f(\cdot)$ contains the operations of the ω-K algorithm on the quantization error $n(t, t_s)$.

The first item in (6.5) is similar to the 2D frequency signal obtained by the ω-K algorithm base on the precise data, as detailed in [19]. The second item in (6.5) is the result produced by performing the corresponding operations on the quantization error $n(t, t_s)$, which leads to higher side-lobes in the refocused SAR image. Taking the 2D inverse Fourier transform of (6.5), the regular SAR image $s_1^{(q)}(t, t_s)$ is obtained:

$$s_1^{(q)}(t, t_s) = \sigma \cdot W_r\left(t - \frac{2R_1(t_s)}{c}\right) \cdot W_a\left(t_s - \frac{\theta}{v_e^2}\right) \cdot \exp\left\{-\frac{j4\pi R_1(t_s)}{\lambda}\right\}$$

$$+ f(n(t, t_s)) \tag{6.6}$$

where

$$R_1(t_s) = \sqrt{\left(\frac{x_0 v_r + r_0(v - v_x)}{v_e}\right)^2 - \frac{(t_s - \theta/v_e^2)^2}{1/v^2 - 1/v_e^2}} \tag{6.7}$$

We can see from (6.5) and (6.6) that, when the target is stationary, the phase of the signal in 2D frequency domain only contains the first-order terms of f_r and f_a, and the coupling term of them is eliminated. Thus, the image based on the quantized data can be focused well except for higher side-lobes induced by the quantization error. As for the moving target, because of the motion-induced error, high-order phase residual still exists after performing the conventional ω-K algorithm, which can lead to blurring and wrong position of the moving target in the reconstructed image, as shown in (6.6). Hence, extra phase compensation is required for moving target refocusing.

According to (6.5) and the analysis in [19], when the measurement data have high-precision, the filter used to compensate motion-induced error of the moving target should be designed as

$$H_a(f_r, f_a) = \exp\left\{j\left(\frac{4\pi R_{\text{ref}}}{c}\sqrt{(f_0 + f_r)^2 + \left(\frac{cf_a}{2}\right)^2 \cdot \left(\frac{1}{v^2} - \alpha\right)} - \frac{4\pi R_{\text{ref}}(f_0 + f_r)}{c}\right)\right\} \tag{6.8}$$

where $\alpha = 1/(v_r^2 + (v - v_x)^2)$ and R_{ref} is the reference slant range. This filter is dependent on the motion-related parameter α. By multiplying the first term of (6.5) by (6.8) and taking 2D inverse Fourier transform, the focused image of a moving target can be achieved if the filter used to compensate the motion-induced phase error in (6.8) is well designed with an accurate estimate of α, as discussed in [19] and [26]. However, in most cases of SAR refocusing of moving targets, the true value of the parameter α is unknown in advance. Thus, the parameter α should be estimated during the refocusing procedure.

With the coarsely quantized echo data, even if the motion-related parameter α is estimated correctly, the ROI sub-image containing the moving target cannot be well focused using the above process because of quantization-induced error $n(t, t_s)$. Here the operations in the BIHT algorithm are used to compensate the quantization-induced error.

The ROI subimage containing the moving target is cropped from the image of the entire scene and referred to as $s_{1\_\text{ROI}}^{(q)}(\bar{t}, \bar{t}_s)$, where $\bar{t}_s$ and $\bar{t}$ denote the equivalent slow time and fast time of the ROI data, respectively. By taking 2D Fourier transform of $s_{1\_\text{ROI}}^{(q)}(\bar{t}, \bar{t}_s)$ with size of $n_r \times n_a$, the 2D frequency signal of the quantized ROI data can be expressed as

$$S_{1\_\text{ROI}}(\bar{f}_r, \bar{f}_a) = \sigma \cdot W_r(\bar{f}_r) \cdot W_a(\bar{f}_a)$$

$$\cdot \exp\left\{ -j\frac{2\pi \bar{f}_a \theta}{v_e^2} - j\frac{4\pi(x_0 v_r + r_0(v - v_x))}{c v_e} \sqrt{(f_0 + \bar{f}_r)^2 + \left(\frac{c\bar{f}_a}{2}\right)^2 \cdot \left(\frac{1}{v^2} - \frac{1}{v_e^2}\right)} \right\}$$

$$+ \text{FT}_{2D}\left(n_{\text{ROI}}^{(q)}(\bar{t}, \bar{t}_s)\right)$$

(6.9)

where $\bar{f}_r \in [-0.5 f_s/n_r, 0.5 f_s/n_r]$ denotes the down-range frequency and $\bar{f}_a \in [-0.5 \text{PRF}/n_a, 0.5 \text{PRF}/n_a]$ denotes the cross-range frequency of the ROI data, f_s and PRF represent the sample frequency and the pulse repetition rate, respectively, $n_{\text{ROI}}^{(q)}(\bar{t}, \bar{t}_s)$ is the noise of the ROI data extracted from $f(n(t, t_s))$ in (6.6). Thus, the filter used to compensate the motion-induced error of the ROI data should be designed as

$$H_a^{(q)}(\bar{f}_r, \bar{f}_a) = \exp\left\{ j\left(\frac{4\pi R_{\text{ref}}}{c} \sqrt{(f_0 + \bar{f}_r)^2 + \left(\frac{c\bar{f}_a}{2}\right)^2 \cdot \left(\frac{1}{v^2} - a\right)} - \frac{4\pi R_{\text{ref}}(f_0 + \bar{f}_r)}{c} \right) \right\}$$

(6.10)

By multiplying (6.9) by (6.10) with an accurate estimate of α and performing 2D inverse Fourier transform subsequently, the refocused ROI image of the moving target can be obtained as

$$\Theta = \sigma \cdot \text{sinc}\left\{ B_a\left(\bar{t}_s - \frac{x_0(v - v_x) - r_0 v_r}{v_e^2} \right) \right\} \cdot \text{sinc}\left\{ B_r\left(\bar{t} - \frac{2}{c} \cdot \frac{x_0 v_r + r_0(v - v_x)}{v_e} \right) \right\}$$

$$\cdot \exp\left\{ -j4\pi \frac{x_0 v_r + r_0(v - v_x)}{v_e \lambda} \right\} + \text{IFT}_{2D}\left[\text{FT}_{2D}\left(n_{\text{ROI}}^{(q)}(\bar{t}, \bar{t}_s) \right) \odot H_a^{(q)}(\bar{f}_r, \bar{f}_a) \right]$$

(6.11)

where B_a and B_r represent the bandwidths in cross-range and down-range directions, respectively, $\text{IFT}_{2D}(\cdot)$ denotes the operation of taking 2D inverse Fourier transform.

It can be seen from (6.11) that the quality of SAR refocusing of moving targets is influenced by both motion-induced error and quantization-induced error. Therefore, in order to obtain the focused image of a moving target, two operations are required: (1) suppression of the quantization-induced error and (2) accurate estimation of the motion-related parameter α. This is the key idea of the PQIHT algorithm and it is presented in Section 6.2.2.

6.2.2   Description of the PQIHT algorithm

In this section, we formulate the PQIHT algorithm for moving target refocusing based on coarsely quantized SAR data.

From (6.9)–(6.11) we can deduce that the image formation of the moving target from the ROI data $s_{1\_ROI}^{(q)}(\bar{t},\bar{t}_s)$ with a certain value of α can be expressed as the following formulation:

$$\Theta = G_\alpha(s_{1\_ROI}^{(q)}) = \Psi_r^{-1} \cdot \left(\left(\Psi_r \cdot s_{1\_ROI}^{(q)} \cdot \Psi_a \right) \odot H_\alpha^{(q)} \right) \cdot \Psi_a^{-1} \tag{6.12}$$

where Θ denotes the reconstructed image of the moving target, $\Psi_r \in \mathbb{C}^{n_r \times n_r}$ and $\Psi_a \in \mathbb{C}^{n_a \times n_a}$ are the Fourier transform matrices in down-range and cross-range directions, respectively, Ψ_r^{-1} and Ψ_a^{-1} are the inverse matrices of Ψ_r and Ψ_a, respectively, and $G_\alpha(\cdot)$ is called a sparse transform. From (6.12), it is clear that the transform $G_\alpha(\cdot)$ is invertible. Thus, the quantized data $s_{1\_ROI}^{(q)}$ can be expressed by the inverse transform of $G_\alpha(\cdot)$:

$$s_{1\_ROI}^{(q)} = G_\alpha^{-1}(\Theta) = \Psi_r^{-1} \cdot \left(\left(\Psi_r \cdot \Theta \cdot \Psi_a \right) \odot \left(H_\alpha^{(q)} \right)^* \right) \cdot \Psi_a^{-1} \tag{6.13}$$

In order to ensure the quality of the moving target refocusing, we need to deal with both motion-induced error and quantization-induced error. The compensation of the motion-induced error lies in the estimation of the parameter α, as discussed in [19]. According to analysis in [11], the quantization error can be suppressed by forcing the consistent reconstruction, which requires that the quantization of the reconstructed data well approximate $s_{1\_ROI}^{(q)}$. Thus, the desired solution of (6.13) is given by

$$(\widehat{\Theta}, \widehat{\alpha}) = \arg \min_{\Theta, \alpha} \left\{ \mathcal{E} + \lambda_1 \|\Theta\|_1 \right\} \tag{6.14}$$

where $\widehat{\alpha}$ is the estimation of the motion-related parameter, $\mathcal{E}$ corresponds to the quantization consistency between $G_\alpha^{-1}(\Theta)$ and $s_{1\_ROI}^{(q)}$, described as [11]:

$$\mathcal{E} = \sum_{i=1}^{2^b} \sum_{m=1}^{n_r \times n_a} \Delta_q \cdot \left| \left\{ \text{sign}\left(\left[G_\alpha^{-1}(\Theta) \right]_m - r_i \right) \cdot \text{sign}\left(\left[s_{1\_ROI}^{(q)} \right]_m - r_i \right) \right\}_- \right| \tag{6.15}$$

where the representation $[\cdot]_m$ means selecting the mth element of the vector, and the operator $(\cdot)_-$ is defined as element-wise negative operator, which projects any positive component to zero while maintaining the negative ones, that is,

$$(x)_- = \frac{x - |x|}{2} \tag{6.16}$$

The quantization consistency $\mathcal{E}$ in the objective function (6.15) is similar to that in the original QIHT algorithm. However, different from QIHT that can only handle stationary targets, a parametric operation $G_\alpha^{-1}(\cdot)$ with a motion-related parameter α is utilized here to represent echoes from moving targets. In this way, the PQIHT algorithm in (6.14) is able to adaptively achieve moving target refocusing, with a motion-related parameter optimization procedure involved. The original QIHT algorithm can be viewed as a special case of the PQIHT algorithm for stationary targets.

Define the reconstructed scene at the kth iteration of the PQIHT algorithm as $\widehat{\Theta}^k$. Given a candidate value of α, calculating the gradient of the objective function in (6.15) and thresholding the obtained refocused ROI image can give the update of the solution:

$$\widehat{\Theta}^{k+1} = H_K\left[\widehat{\Theta}^k + \mu G_\alpha\left(s^{(q)}_{1\_ROI} - Q_b\left(G_\alpha^{-1}(\widehat{\Theta}^k)\right)\right)\right] \tag{6.17}$$

where $\mu > 0$ controls the gradient step size, and the hard thresholding operator $H_K(\cdot)$ is defined to maintain the K biggest coefficients while setting others to zero. It can be seen from (6.17) that the solution updating is a function of α.

According to (6.14) and (6.17), we can conclude that different candidate values of α will generate different results of the refocused image of the moving target. The estimation of α can be achieved by searching for the maximum image contrast, which can be expressed as

$$\widehat{\alpha} = \arg\max_\alpha C\left(\widehat{\Theta}(\alpha)\right)$$

$$C\left(\widehat{\Theta}(\alpha)\right) \triangleq \left\{\frac{\text{mean}\left[\left|\widehat{\Theta}(\alpha)\right|^2\right] - \left[\text{mean}\left(\left|\widehat{\Theta}(\alpha)\right|\right)\right]^2}{\left[\text{mean}\left(\left|\widehat{\Theta}(\alpha)\right|\right)\right]^2}\right\} \tag{6.18}$$

where $\widehat{\Theta}(\alpha)$ is the image result reconstructed by the QIHT algorithm with the certain candidate value of α.

The full search over a large range of the candidate value of α is time-consuming. Here a pruned searching method is embedded into the iterative process of QIHT for efficiently estimating the motion-related parameter α. When the solution is updated as shown in (6.17), the image contrast calculated with a candidate value of α can to some extent reflect the correctness of the candidate. That is, the image contrast will be larger when the candidate value of α is closer to the true value. From this consideration, at each iteration of the solution updating, some candidate values of α that lead to small image contrasts are removed from the candidate set since they are less likely to be around the true value. Assume that there are n_k candidates at the kth iteration of QIHT, and only $n_k/2$ candidate values of α are maintained and plugged into the next iteration after the pruning process based on image contrast comparison. As the iterative process proceeds, the size of the candidate set of α is continuously reduced. Thus, the whole iterative process of the original QIHT algorithm is only performed for the candidate of α that is very close to the true value. For those wrong candidates of α, only part of iterations of QIHT are carried out. In this way, the computational efficiency is remarkably improved in comparison to full search on the imaging results generated by performing the whole iterative process of QIHT for a large candidate range of α. When the pruning process is terminated, the accurate estimate of α is obtained, and accordingly, the corresponding refocused image of the moving target $\widehat{\Theta}(\alpha)$ is also obtained. The steps of the PQIHT algorithm are summarized in Table 6.1.

6.2.3 Experimental results

In this section, experiments based on simulated and real SAR data are carried out to demonstrate the effectiveness of the PQIHT algorithm. In all of these experiments, the maximal number of iterations $I_{\max}$ is set to 100, and the error tolerance ε is set to 0.001.

6.2.3.1 Simulations

In our simulations, the point-like scattering model is adopted and an air-borne radar with the bandwidth of 300 MHz and the carrier frequency of 10 GHz is used. The

Table 6.1   The PQIHT algorithm

Input:

The ROI image containing the moving target $s^{(q)}_{1\_\text{ROI}}(\bar{t}, \bar{t}_s)$, the gradient step size μ, the error tolerance ε, the maximal number of iterations $I_{\max}$, and the sparsity K.

Initialization:

$k = 0$, the vector $\mathbf{A}^0$ including all the possible candidate values of α, and the initial scene $\widehat{\Theta}^0 = \mathbf{0}$.

Iteration:

(1) For each candidate value $\tilde{\alpha}$ in vector $\mathbf{A}^k$, calculate:

$H^{(q)}_{\tilde{\alpha}}$, according to (6.10);

$G^{-1}_{\tilde{\alpha}}(\widehat{\Theta}^k(\tilde{\alpha})) = \text{IFT}_{2D}(\text{FT}_{2D}(\widehat{\Theta}^k(\tilde{\alpha})) \odot (H^{(q)}_{\tilde{\alpha}})^*)$;

$\widehat{\Theta}^{k+1}(\tilde{\alpha}) = H_K[\widehat{\Theta}^k(\tilde{\alpha}) + \mu G_{\alpha}(s^{(q)}_{1\_\text{ROI}} - Q_b(G^{-1}_{\alpha}(\widehat{\Theta}^k(\tilde{\alpha}))))]$;

$C(\widehat{\Theta}^{k+1}(\tilde{\alpha}))$ according to (6.18).

End For.

(2) Pruning operation:

Let $n_k = \text{card}(\mathbf{A}^k)$.

If $n_k = 1$

Let $\widehat{\alpha}$ denote the only entry in $\mathbf{A}^k$.

If $\|\widehat{\Theta}^{k+1}(\widehat{\alpha}) - \widehat{\Theta}^k(\widehat{\alpha})\|_2 / \|\widehat{\Theta}^k(\widehat{\alpha})\|_2 \leq \varepsilon$ or $k > I_{\max}$

Stop the iterative process;

Else

Let $\mathbf{A}^{k+1} = \mathbf{A}^k$ and $k = k + 1$, and return to Step (1)

End If

Else

Remove $n_k/2$ candidates corresponding to relatively small image contract values in $\{C(\widehat{\Theta}^{k+1}(\tilde{\alpha})), \tilde{\alpha} \in \mathbf{A}^k\}$ from $\mathbf{A}^k$, and the remaining candidates of α are composed as $\mathbf{A}^{k+1}$, let $k = k + 1$, and return to Step (1).

End If

Output:

$\widehat{\alpha}$ and $\widehat{\Theta}^k(\widehat{\alpha})$

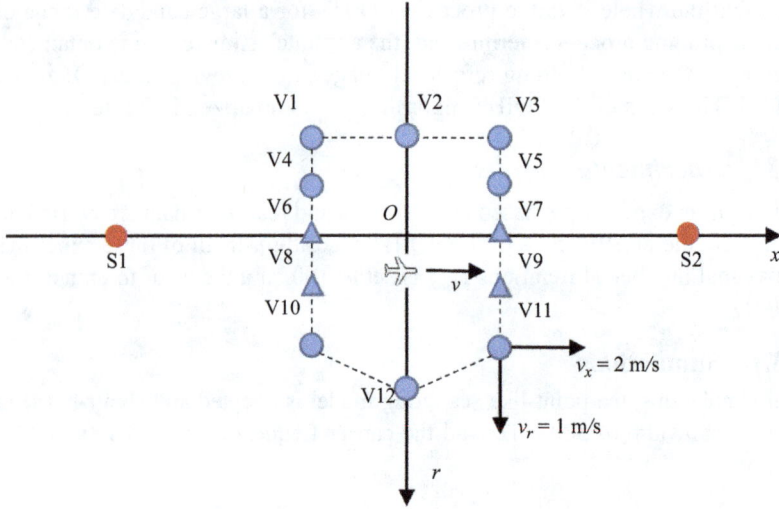

*Figure 6.2   The simulated imaging scene, with red and blue points denoting static
target and moving scattering point, respectively*

pulse width of the transmitted LMF signal is 2.2 µs and the pulse repetition frequency is set to 3 KHz. The radar platform is assumed to move along the cross-range direction with a constant velocity of 150 m/s. The simulated scene is shown in Figure 6.2, where the scattering coefficients of the point-like targets denoted by circles and triangles are supposed to be 1 and 0.5, respectively. The scene to be imaged consists of two stationary point-like targets denoted by S1 and S2 and a moving target comprised of 12 scattering points denoted by V1–V12 with constant velocities of 1 m/s and 2 m/s along down-range and cross-range directions, respectively, that is, $v_r = 1$ m/s, $v_x = 2$ m/s. The reference slant range is 10 km. The two stationary targets are at the same range cell with 40 m separated in the cross-range direction. The center of the moving target lies between the two stationary targets. The additive Gaussian white noise is added to the measurements, wherein the signal-to-noise ratio (SNR) is 15 dB. Since the velocities of the moving target are relatively small compared to the velocity of the radar platform in airborne and spaceborne SAR applications, the true value of the motion-related parameter α should be close to $1/v^2$. Therefore, the initial candidate value set of α can be chosen around this value.

The regular image of the entire scene obtained by the ω-K algorithm based on 2-bit quantized data is shown in Figure 6.3, where the white dashed box indicates ROI containing the moving target. The image of the two stationary targets is well focused, but high side-lobes exist due to the presence of the quantization error. The ROI image containing the moving target is blurred duo to both the motion-induced error and the quantization error.

We first compare the PQIHT method with the original PSR method based on high-precision data, which is presented in [19] and in Chapter 4, in terms of the imaging quality. The image generated by the PSR method with 64-bit quantized

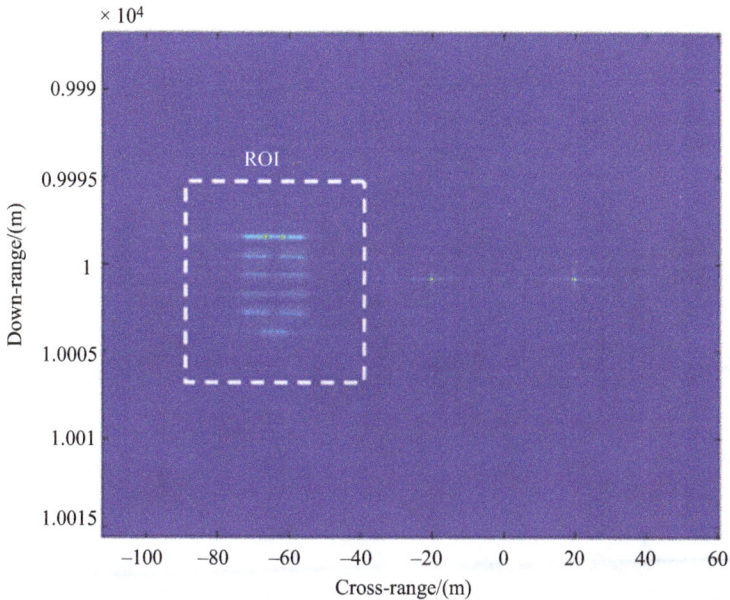

Figure 6.3   SAR image of the entire scene obtained by the ω-K algorithm based on 2-bit quantized data, with a white dashed box denoting the ROI image containing the moving target

data is given in Figure 6.4(a), which serves as the benchmark. The refocusing results of the ROI data obtained by PSR method and the PQIHT method with 2-bit quantized data are shown in Figure 6.4(b) and (c), respectively. The image produced by the PSR method is unsatisfactory because it cannot compensate the quantization-induced error. It is evident from Figure 6.4(d) that the PQIHT method can provide better quality of moving target refocusing with coarsely quantized data, thanks to its capability of compensating the quantization-induced error for each candidate of the motion-related parameter α and accurately estimating the value of α by pruned searching over all of its candidates. Especially, the range and cross-range profiles of the image of the scatterer V12 are plotted in Figure 6.5 and Figure 6.6, respectively, to compare PQIHT and PSR in the coarse quantization case, both of which are able to deal with the motion-induced error. The refocused image of the moving scatterer produced by PQIHT has asymmetric side-lobes lower than -20 dB, which illustrates the positive effect of suppressing the quantization-induced error.

Based on the simulation setting, it can be calculated that the true value of α is 4.5652×10^{-5}. The estimation result of α obtained by PQIHT is 4.5690×10^{-5}, which is very close to the true value and corresponds to the image in Figure 6.4(d). This indicates that the PQIHT algorithm can yield an accurate estimation of α from 2-bit quantized data.

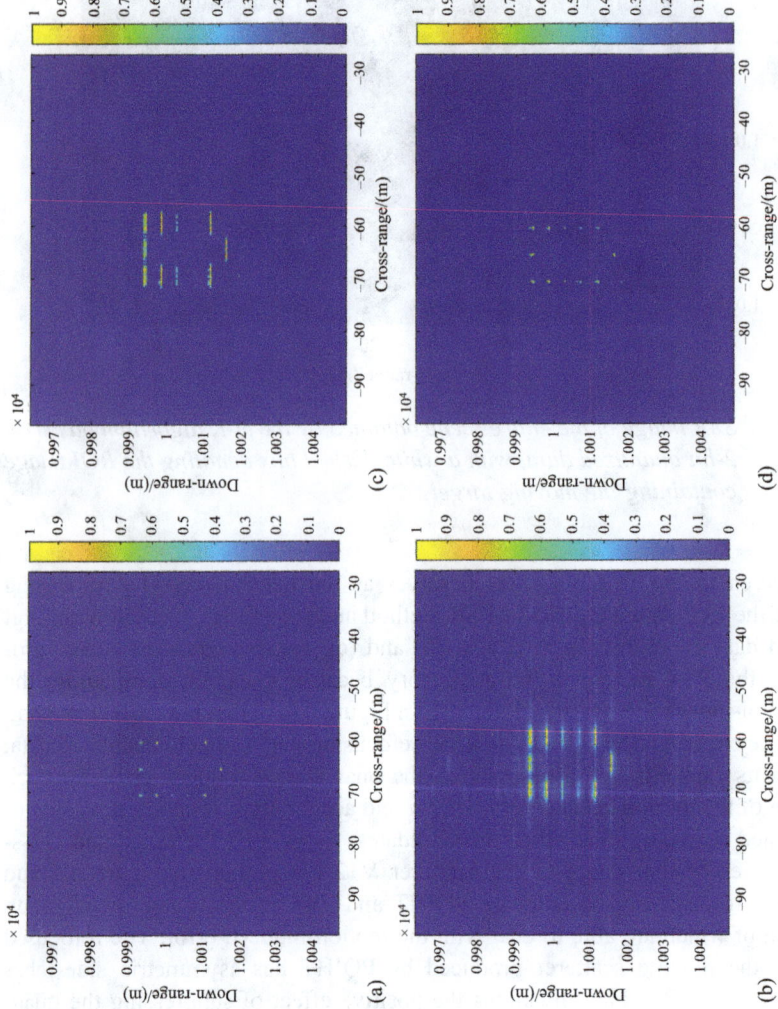

Figure 6.4 Comparison in terms of the quality of moving target refocusing: (a) benchmark: PSR, 64-bit data; (b) PSR, 2-bit data; (c) QIHT method with 2-bit quantized data; and (d) PQIHT, 2-bit data

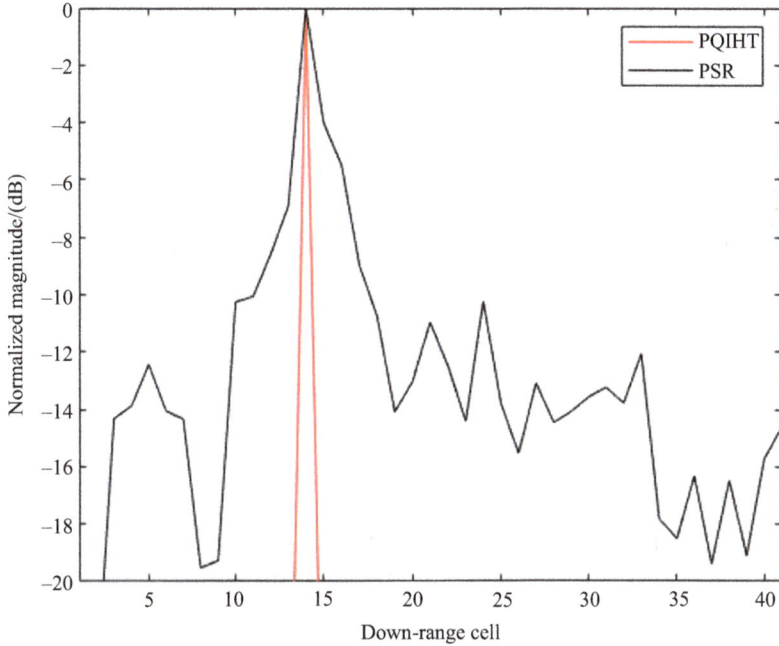

Figure 6.5   Comparison of the range profiles

The image contrasts obtained from the quantized data with different quantization levels are investigated in Table 6.2, where SNR is set to 15 dB. It is clear that the image contrast produced by the PQIHT algorithm is higher than 37 based on coarsely quantized data even 1-bit data, which is close to the benchmark image contrast value 40.94 produced by PSR based on 64-bit data. Thus we can conclude that compared to the PSR method for moving target refocusing with precise data, the PQIHT algorithm can achieve comparable image quality with much less data volume. In contrast, the PSR method fails to generate satisfactory imaging quality by using 8-bit data and more coarsely quantized data.

The image contrasts produced by performing PQIHT and PSR on 2-bit quantized data with different values of SNR are plotted in Figure 6.7, where each point is obtained by averaging over 50 Monte Carlo trials. One can see that PQIHT consistently outperforms PSR in terms of the refocusing quality.

Note that PQIHT is more computationally expensive than PSR, due to the additional operations for suppressing the quantization-induced error. The computational complexity of PQIHT is related to the number of iterations and the initial range of the candidate set of α. In our simulations, the running time of the PQIHT algorithm and the PSR method are 1,728.408 s and 114.573 s, respectively. All the experiments are implemented using the nonoptimized MATLAB® code on a laptop (Intel Core i7-8550U CPU at 1.80 GHz and 16 GB of RAM).

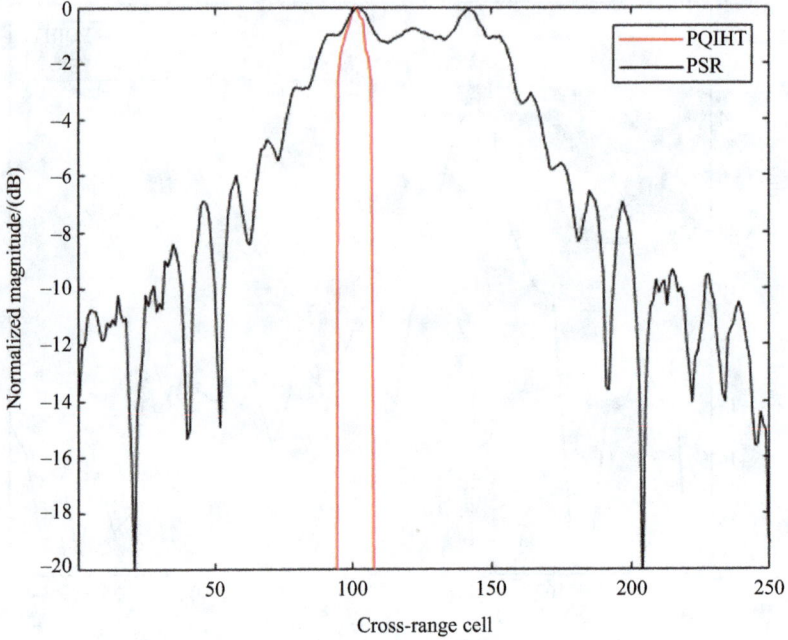

Figure 6.6   Comparison of the cross-range profiles

Table 6.2   The image contrasts of the simulated ROI data

Method	Data	Image contrast	Data volume
PSR	Precise (64-bit)	40.94	$64 \times n_r \times n_a$
PQIHT	1-bit	37.19	$1 \times n_r \times n_a$
	2-bit	37.76	$2 \times n_r \times n_a$
	4-bit	37.86	$4 \times n_r \times n_a$
	8-bit	37.90	$8 \times n_r \times n_a$
PSR	1-bit	0.94	$1 \times n_r \times n_a$
	2-bit	1.93	$2 \times n_r \times n_a$
	4-bit	3.61	$4 \times n_r \times n_a$
	8-bit	4.23	$8 \times n_r \times n_a$
QIHT	1-bit	11.25	$1 \times n_r \times n_a$
	2-bit	11.25	$2 \times n_r \times n_a$
	4-bit	11.37	$4 \times n_r \times n_a$
	8-bit	11.80	$8 \times n_r \times n_a$

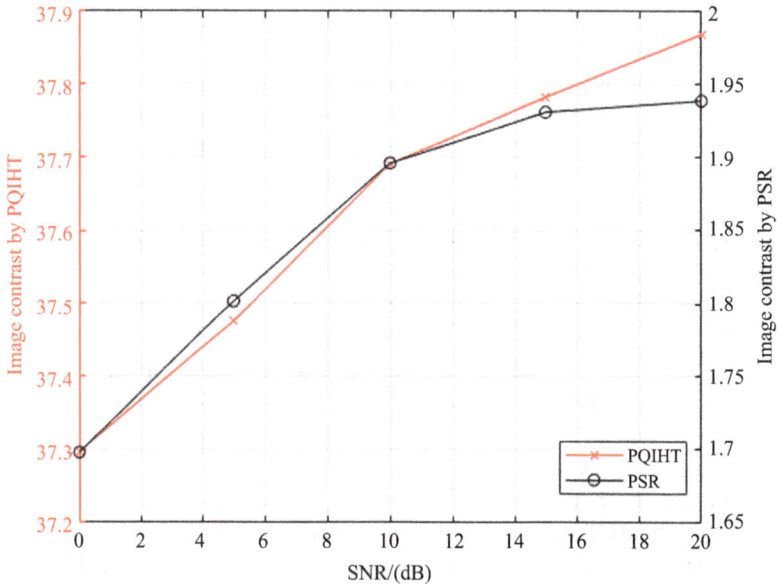

Figure 6.7 Image contrast of the refocused image of the moving target versus SNR

The PQIHT algorithm extends the QIHT algorithm to fit the case of SAR refocusing of moving targets. Both of them are good at suppressing the quantization-induced error, and PQIHT also has the capability of compensating the motion-induced error. Based on the 2-bit quantized echo data of the simulated scene, the image of the moving target produced by QIHT is given in Figure 6.4(c). It can be seen that the QIHT algorithm fails to refocus the moving target image since the motion-induced error is not considered in its signal model. The comparison between Figure 6.4(c) and (d) demonstrates the positive effect of refining the estimation of the motion-related parameter α in the iterative process of PQIHT. The quantitative comparison between QIHT and PQIHT in terms of the image contrast is provided in Table 6.2, which also demonstrates the superiority of PQIHT. The cost PQIHT has to pay is the increased computational complexity, in comparison to the QIHT algorithm. The running time of the QIHT algorithm for generating Figure 6.4(c) is 47.319 s, which is much less than that of the PQIHT algorithm for obtaining Figure 6.4(d) (1728.408 s).

6.2.3.2 Experiments on real SAR data

The real SAR data were collected by the GF-3 satellite. The bandwidth and the pulse width of the transmitted signal are 60 MHz and 35.01 μs, respectively, and the PRF is 2.362 KHz. The down-range and cross-range resolutions are both 5 m. The equivalent velocity of the SAR platform is 7.1463×10^3 m/s. The coarsely quantized data are regenerated from the raw data before the imaging process. The regular SAR image of the sea surface obtained by the conventional ω-K algorithm is shown in Figure 6.8, where three moving ships are indicated by white boxes P1–P3. These ships are defocused, while the sea background is focused well except higher side-lobes induced by the quantization error. We assume that the maximum velocities of the moving ships along down-range and cross-range directions are both 30 kn, that is, 15.42 m/s, which is utilized to determine the range of the candidate values of α. In the following experiments, the range of the candidate value of α is set to $\left[1.9 \times 10^{-8}, 2 \times 10^{-8}\right]$.

The refocused images of these moving ships yielded by three algorithms are shown in Figure 6.9, Figure 6.10, and Figure 6.11, respectively. For comparison, the images produced by the PSR algorithm based on the precise data (64-bit quantized data) serve as the image quality benchmark.

Take Figure 6.9 as an example. The refocused images of the moving ship P1 generated by performing PSR and QIHT on 2-bit quantized data are given in the sub-figures (c) and (d) in Figure 6.9, respectively. These two images are still defocused because PSR and QIHT ignore the quantization-induced error and the motion-induced error, respectively. The error that these two algorithms do not consider accumulate during their iterative processes, resulting in their failure of moving target refocusing. The refocused images of the moving ship P1 yielded by performing PQIHT on 2-bit data and 1-bit data are shown in the sub-figures (e) and (f) in Figure 6.9, respectively. One can easily determine the size and shape of the ship P1 from the well-focused image in Figure 6.9(e). The image reconstructed by PQIHT from 1-bit quantized data, shown in Figure 6.9(f), slightly degenerates in comparison with Figure 6.9(e), but it is still acceptable for target classification

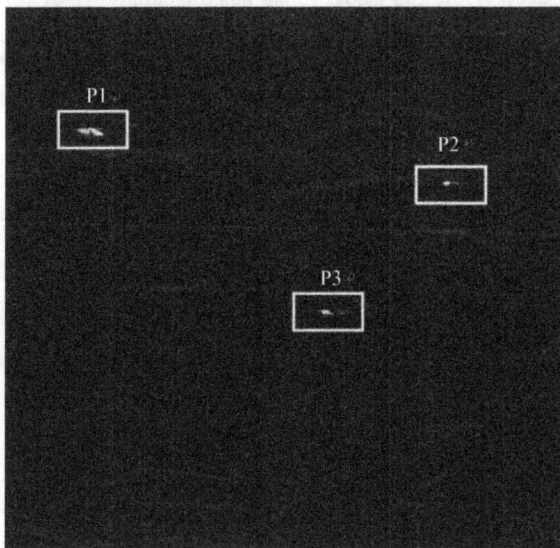

Figure 6.8   Image of the sea surface generated by the ω-K algorithm from 2-bit quantized data

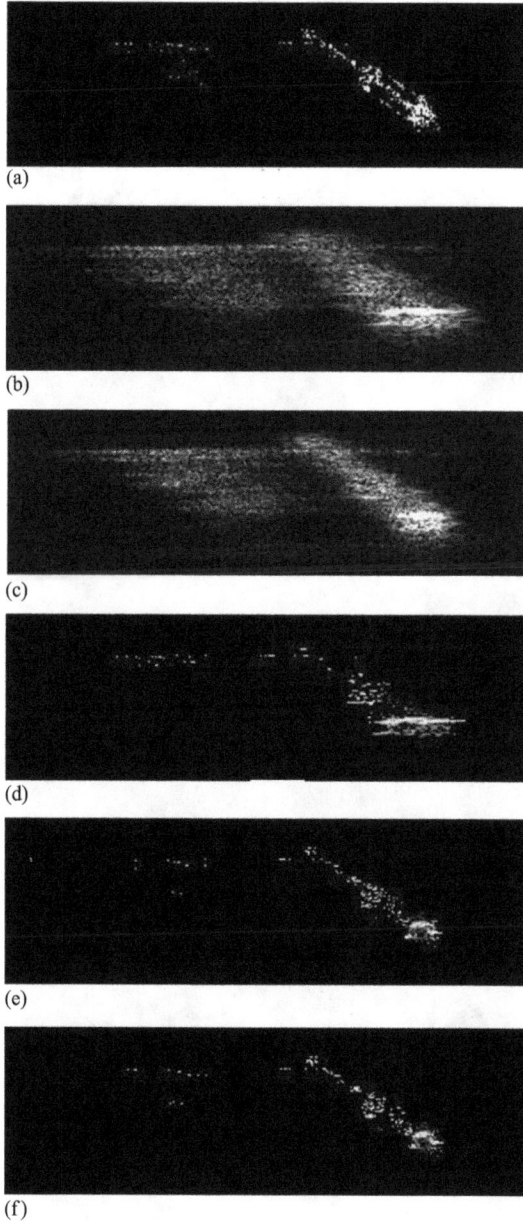

Figure 6.9 Comparison in terms of the imaging quality for moving ship P1: (a) benchmark: PSR, 64-bit data; (b) defocused image directly cropped from Figure 6.8; (c) PSR, 2-bit data; (d) QIHT, 2-bit data; (e) PQIHT, 2-bit data; and (f) PQIHT, 1-bit data

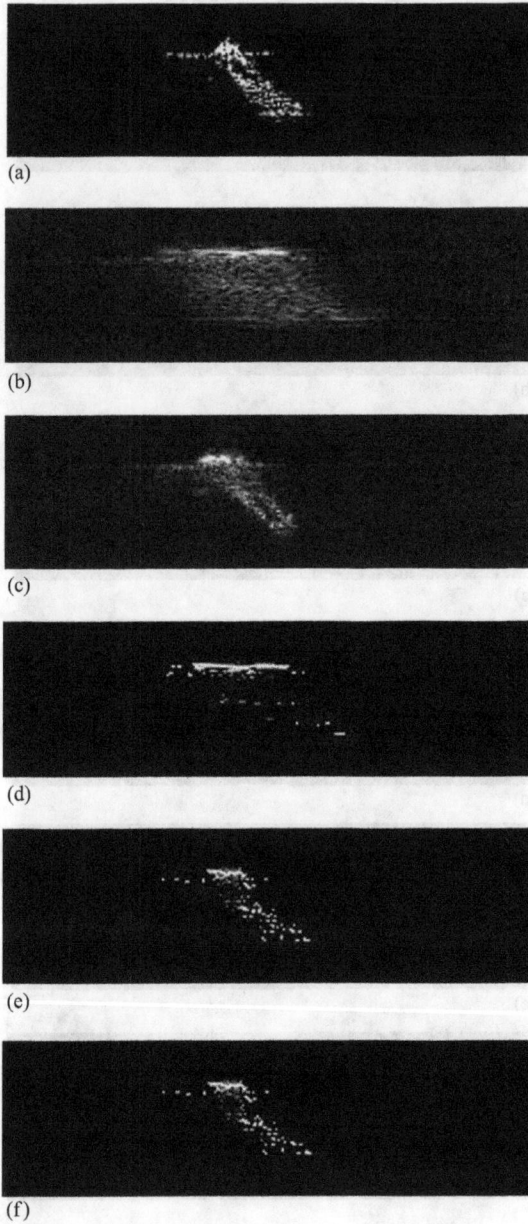

Figure 6.10   Comparison in terms of the imaging quality for moving ship P2:
(a) benchmark: PSR, 64-bit data; (b) defocused image directly
cropped from Figure 6.8; (c) PSR, 2-bit data; (d) QIHT, 2-bit data;
(e) PQIHT, 2-bit data; and (f) PQIHT, 1-bit data

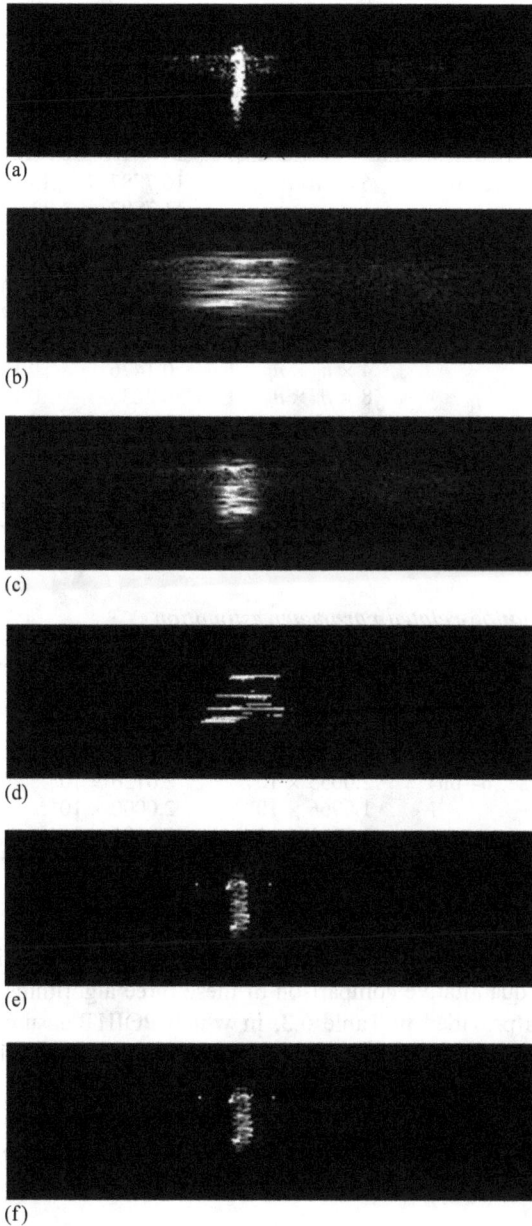

Figure 6.11   Comparison in terms of the imaging quality for moving ship P3: (a) benchmark: PSR, 64-bit data; (b) defocused image directly cropped from Figure 6.8; (c) PSR, 2-bit data; (d) QIHT, 2-bit data; (e) PQIHT, 2-bit data; and (f) PQIHT, 1-bit data

Table 6.3   The image contrast of the ROI data

Method	Data	Data volume	Image contrast		
			P1	P2	P3
PSR	Precise (64-bit)	$64 \times n_r \times n_a$	16.7787	13.6834	12.7505
PQIHT	1-bit	$1 \times n_r \times n_a$	11.7802	11.4633	11.5331
	2-bit	$2 \times n_r \times n_a$	11.9938	11.4684	11.6795
	4-bit	$4 \times n_r \times n_a$	12.1379	11.4926	11.7091
	8-bit	$8 \times n_r \times n_a$	12.1495	11.5052	11.7123
PRS	1-bit	$1 \times n_r \times n_a$	0.1913	0.1779	0.0964
	2-bit	$2 \times n_r \times n_a$	0.2151	0.1187	0.1626
	4-bit	$4 \times n_r \times n_a$	0.1476	0.1359	0.1636
	8-bit	$8 \times n_r \times n_a$	0.2235	0.1747	0.1624
QIHT	1-bit	$1 \times n_r \times n_a$	10.6864	10.4377	9.8382
	2-bit	$2 \times n_r \times n_a$	10.7481	10.4441	9.8374
	4-bit	$4 \times n_r \times n_a$	10.7361	10.4550	9.8361
	8-bit	$8 \times n_r \times n_a$	10.7157	10.4639	9.8374

Table 6.4   The motion-related parameter estimation

Method	Data	Estimation of α		
		P1	P2	P3
PSR	Precise (64-bit)	2.0055×10^{-8}	2.0126×10^{-8}	2.0042×10^{-8}
PQIHT	2-bit	1.9966×10^{-8}	2.0000×10^{-8}	1.9998×10^{-8}
QIHT	2-bit	1.9581×10^{-8}	1.9581×10^{-8}	1.9581×10^{-8}

and recognition. A similar phenomenon can be also found in Figure 6.10 and Figure 6.11. The quantitative comparison of these three algorithms in terms of the image contrast is provided in Table 6.3, in which PQIHT ensures higher image contrast even for 1-bit data. The estimation results of the motion-related parameter α are listed in Table 6.4. Still taking the result estimated by performing PSR on precise data as a benchmark, the relative estimation error of PQIHT is 0.43%, while the relative estimation error of QIHT is 2.46%, which is calculated by averaging the relative estimation errors for the three ships.

6.3   Enhanced 1-bit radar imaging by exploiting two-level block sparsity

In this section, the E-BIHT algorithm [20] is presented to enhance the quality of 1-bit radar imaging of stationary targets by combining the BIHT algorithm [12] with the two-level block sparsity model [21]. The two-level block sparsity includes both of the joint sparsity pattern and the clustering property of the sparse image, as described in Chapter 3.

6.3.1 Signal model

Consider the stepped frequency SAR as an example. The radar measurement data are acquired by M antenna positions at L frequency points $\{f_1, \ldots, f_l, \ldots, f_L\}$. The observed two-dimensional scene is discretized as P pixels. The radar echo received from the mth antenna position at the lth frequency point can be expressed as

$$y_{m,l} = \sum_i \theta_i \exp(-j2\pi f_l \tau_{i,m}) + w_{m,l} \tag{6.19}$$

where θ_i is the complex reflectivity of the ith pixel i, $\tau_{i,m}$ is the round-trip travel time of the signal from the mth antenna position to the location of ith pixel, and $w_{m,l}$ is the additive noise. The target reflectivity is assumed constant in the radar frequency band.

We can rewrite (6.19) as

$$\mathbf{y} = \mathbf{\Phi}\boldsymbol{\theta} + \mathbf{w} \tag{6.20}$$

where $\mathbf{y} \in \mathbb{C}^{ML \times 1}$ is the measurement vector, $\boldsymbol{\theta} \in \mathbb{C}^{P \times 1}$ is the complex reflectivity vector, that is, the complex image to be formed, $\mathbf{\Phi} \in \mathbb{C}^{ML \times P}$ denotes the dictionary matrix, and $\mathbf{w} \in \mathbb{C}^{ML \times 1}$ represents the noise vector. The element of the dictionary matrix $\mathbf{\Phi}$ is given by

$$\mathbf{\Phi}(l + (m-1)L, i) = \exp(-j2\pi f_l \tau_{i,m}) \tag{6.21}$$

where $l = 1, 2, \ldots, L$, $m = 1, 2, \ldots, M$, and $i = 1, 2, \ldots, P$. It is common that the number of measurements is smaller than the number of the discretized pixels, that is, $ML < P$, especially in high-resolution imaging applications. Consider the case that the scene of interest is inherently sparse, that is, only few coefficients in $\boldsymbol{\theta}$ are dominant and others are close to zero. The sparsity of $\boldsymbol{\theta}$ is assumed to be K, that is, the number of the dominant coefficients in $\boldsymbol{\theta}$ is not larger than K.

When 1-bit ADC is used for quantization, the vector of 1-bit quantization measurements can be written as

$$\bar{\mathbf{y}} = \text{sign} \begin{bmatrix} \text{Re}(\mathbf{y}) \\ \text{Im}(\mathbf{y}) \end{bmatrix} \tag{6.22}$$

where $\text{Re}(\cdot)$ and $\text{Im}(\cdot)$ represent the real part operator and the imaginary part operator, respectively. Then, from (6.20) and (6.22), we have

$$\bar{\mathbf{y}} = \text{sign}(\bar{\mathbf{\Phi}}\bar{\boldsymbol{\theta}} + \bar{\mathbf{w}}) \tag{6.23}$$

where

$$
\bar{\boldsymbol{\Phi}} = \begin{bmatrix} \mathrm{Re}(\boldsymbol{\Phi}) & -\mathrm{Im}(\boldsymbol{\Phi}) \\ \mathrm{Im}(\boldsymbol{\Phi}) & \mathrm{Re}(\boldsymbol{\Phi}) \end{bmatrix}
$$
$$
\bar{\boldsymbol{\theta}} = \begin{bmatrix} \mathrm{Re}(\boldsymbol{\theta}) \\ \mathrm{Im}(\boldsymbol{\theta}) \end{bmatrix} \tag{6.24}
$$
$$
\bar{\mathbf{w}} = \begin{bmatrix} \mathrm{Re}(\mathbf{w}) \\ \mathrm{Im}(\mathbf{w}) \end{bmatrix}
$$

The estimation of $\bar{\boldsymbol{\theta}}$ in (6.23) can be converted to 1-bit CS problem. The consistency between the original 1-bit measurement vector $\bar{\mathbf{y}}$ and the regenerated measurement vector $\bar{\boldsymbol{\Phi}}\bar{\boldsymbol{\theta}}$ can be measured by the quantization consistency function [12]:

$$
J(\bar{\boldsymbol{\theta}}) = \|[\bar{\mathbf{y}} \odot (\bar{\boldsymbol{\Phi}}\bar{\boldsymbol{\theta}})]_-\|_1 \tag{6.25}
$$

where $(\cdot)_-$ follows (6.16). Intuitively, when $\bar{\mathbf{y}}$ and $\bar{\boldsymbol{\Phi}}\bar{\boldsymbol{\theta}}$ have the same sign, $J(\bar{\boldsymbol{\theta}})$ achieves the minimum value. This suggests the following objective function for solving $\bar{\boldsymbol{\theta}}$:

$$
\widehat{\boldsymbol{\theta}} = \arg\min_{\bar{\boldsymbol{\theta}}} J(\bar{\boldsymbol{\theta}}) \quad \text{s.t.} \ \|\bar{\boldsymbol{\theta}}\|_0 \le 2K \tag{6.26}
$$

The sparsity level of $\bar{\boldsymbol{\theta}}$ is required to be smaller than $2K$ because both the real and imaginary parts of $\boldsymbol{\theta}$ have at most K dominant coefficients.

The problem in (6.26) can be solved by the BIHT algorithm [12], which is reviewed in Table 6.5. At each iteration, the temporary solution $\mathbf{a}$ is updated by gradient descent:

$$
\mathbf{a} = \bar{\boldsymbol{\theta}}^{t-1} - \mu \nabla J(\bar{\boldsymbol{\theta}}^{t-1}) \tag{6.27}
$$

Table 6.5   The BIHT algorithm

Input:
   $\bar{\mathbf{y}}, \bar{\boldsymbol{\Phi}}, K$, the descent step size μ, the maximum iteration count t_{max}, and the error tolerance parameter ε.
Initialization:
   $\bar{\boldsymbol{\theta}}^0 = \mathbf{0}^{2P \times 1}$; $t = 0$.
Iteration:
   (1) $t = t + 1$;
   (2) $\mathbf{a} = \bar{\boldsymbol{\theta}}^{t-1} - \frac{1}{2}\mu\bar{\boldsymbol{\Phi}}^T[\mathrm{sign}(\bar{\boldsymbol{\Phi}}\bar{\boldsymbol{\theta}}) - \bar{\mathbf{y}}]$;
   (3) $\bar{\boldsymbol{\theta}}^t = H_{2K}(\mathbf{a})$;
   Until $t \ge t_{max}$ or $\|\bar{\boldsymbol{\theta}}^t - \bar{\boldsymbol{\theta}}^{t-1}\|_2 / \|\bar{\boldsymbol{\theta}}^t\|_2 < \varepsilon$
Output:
   $\bar{\boldsymbol{\theta}}^t$

---

where $\mu > 0$ is the descent step size, and gradient of $J(\bar{\theta})$ is given by [12]

$$\nabla J(\bar{\theta}) = \frac{1}{2}\bar{\Phi}^T[\text{sign}(\bar{\Phi}\bar{\theta}) - \bar{y}] \tag{6.28}$$

Then a hard thresholding operator $H_{2K}(\cdot)$ is imposed on $\mathbf{a}$ to guarantee the sparsity level, as done in Step 3 in Table 6.5. $H_{2K}(\cdot)$ sets all the coefficients of a vector to zero but those having the $2K$ strongest amplitudes, which is similar with the operation in (6.17).

Although BIHT is capable of recovering the sparse vector $\bar{\theta}$ from 1-bit measurement data, the inherent structure in $\bar{\theta}$ is not fully considered. This motivates an extension by combining the BIHT framework and the structured sparsity model, as described in Section 6.3.2. Note that this extension can also be applied to other 1-bit CS algorithms. Here we specifically consider BIHT because of its computational simplicity.

6.3.2 Description of the E-BIHT algorithm

In this section, the E-BIHT algorithm is formulated in detail. The key idea of E-BIHT is to extend the original BIHT algorithm by exploiting the two-level block sparsity model, which is the combination of the joint sparsity and the clustered sparsity.

6.3.2.1 Joint sparsity

To a large extent, the real and imaginary parts of a sparse complex target reflectivity share a joint sparsity pattern [27]. Thus, the nonlinear sparse approximation of Step 3 of the original BIHT algorithm in Table 6.5 can be replaced with a jointly sparse approximation, as done in [28]:

$$\bar{\theta}^t = \arg\min_{\mathbf{b}} \|\mathbf{a} - \mathbf{b}\|_2^2 \quad \text{s.t.} \sum_{i=1}^{P} \|\sqrt{(b_i)^2 + (b_{i+P})^2}\|_0 \leq K \tag{6.29}$$

where $\mathbf{a}$ is calculated by (6.27), $\mathbf{b} \in \mathbb{R}^{2P\times 1}$ is a vector with joint sparsity constraint to be optimized, and b_i is the ith entry of $\mathbf{b}$. Compared to Step 3 of the original BIHT algorithm, the constraint of $\sum_{i=1}^{P} \|\sqrt{(b_i)^2 + (b_{i+P})^2}\|_0 \leq K$ in (6.29) enhances the joint sparsity pattern between the real and imaginary components of the target image to be reconstructed.

6.3.2.2 Clustered sparsity

A spatially extended target is not represented by an isolated point but rather by a clustered structure in a radar image. The Markov random field (MRF) defines a conditional probability distribution to capture the spatial dependence between a pixel and its neighbors. Here a one-order MRF model, the auto-logistic model [29], is used to cluster the nonzero coefficients in a sparse image.

For compact description, the support set Λ is defined as the set of indices corresponding to the nonzero coefficients in θ, that is, Λ represents the positions of "target area" in θ. The complementary set of Λ is denoted by $\bar{\Lambda}$, which corresponds to "clutter area" in θ. Define a support area s and let $s_\Lambda = 1$ and $s_{\bar{\Lambda}} = -1$. This interaction between the entries in s can be formulated by the following probability density function (PDF) [29]:

$$p(s_i|s_{N_i}) \propto \exp\left(a_i s_i + \sum_{i' \in N_i} \beta_{i,i'} s_i s_{i'} \right) \tag{6.30}$$

where N_i is the set of all neighbors of the ith pixel, $a_i \geq 0$ denotes the prior information about s_i and $\beta_{i,i'} \geq 0$ denote the interaction between s_i and its neighboring pixels. Here we set:

$$a_i = 0, \ \forall i \tag{6.31}$$

and

$$\beta_{i,i'} = \beta > 0, \ \forall i, i' \tag{6.32}$$

The condition in (6.32) means that each pair of neighboring pixels enforces the same influence on each other, while the condition in (6.31) indicates that no prior information in assumed about the pixel itself. A pseudo-likelihood function is given by

$$U(s) = -\log \left[\prod_{i=1}^{P} p(s_i|s_{N_i}) \right]$$
$$\propto -\sum_{i=1}^{P} \left(s_i \sum_{i' \in N_i} s_{i'} \right) \tag{6.33}$$

which is expected to be minimized to enforce the clustering structure. It can be deduced that when

$$s_i = s_{i'}, \ \forall i' \in N_i \tag{6.34}$$

the pseudo-likelihood function $U(s)$ in (6.33) is explicitly minimized. The parameter β is ignored in (6.33) since it is only a scaling factor.

Now we consider how to formulate the relationship between the image θ and the support area s. The following criterion is expected:

$$s_i = \begin{cases} 1 & |\theta_i| \neq 0 \\ -1 & |\theta_i| = 0 \end{cases} \tag{6.35}$$

Inspired by the smooth Gaussian indicator function [30], the following function is selected to approximately formulate the relationship between the image θ

and the support area **s**:

$$s_i = g(\theta_i) = 1 - 2 \exp\left(-\frac{\theta_i \theta_i^*}{2\sigma^2}\right) \tag{6.36}$$

where $\sigma > 0$ determines the quality of the approximation. Note that:

$$\lim_{\sigma \to 0} g(\theta_i) = \begin{cases} 1 & |\theta_i| \neq 0 \\ -1 & |\theta_i| = 0 \end{cases} \tag{6.37}$$

or approximately [30]:

$$g(\theta_i) \approx \begin{cases} 1 & |\theta_i| \gg \sigma \\ -1 & |\theta_i| \ll \sigma \end{cases} \tag{6.38}$$

Then, the pseudo-likelihood function in (6.33) can be rewritten by

$$U(\boldsymbol{\theta}) \propto -\sum_{i=1}^{P} \left\{ g(\theta_i) \left[\sum_{i' \in N_i} g(\theta_{i'}) \right] \right\} \tag{6.39}$$

Further, considering that the real and imaginary parts of θ_i are $\bar{\theta}_i$ and $\bar{\theta}_{i+P}$, respectively, the equivalent form of (6.39) is given by

$$\tilde{U}(\bar{\boldsymbol{\theta}}) \propto -\sum_{i=1}^{P} \left\{ g\left(\sqrt{(\bar{\theta}_i)^2 + (\bar{\theta}_{i+P})^2}\right) \left[\sum_{i' \in N_i} g(\sqrt{(\bar{\theta}_{i'})^2 + (\bar{\theta}_{i'+P})^2}) \right] \right\} \tag{6.40}$$

Note that the clustered sparsity model used here, which is suitable for the framework of iterative thresholding, is different from that introduced in Chapter 3, which is suitable for the matching pursuit strategy.

6.3.2.3 The E-BIHT algorithm

In what follows, we present an algorithm for 1-bit radar imaging, named E-BIHT, which attempts to impose the two-level block sparsity on the target image during the process of image formation from the 1-bit measurement data. Figure 6.12 presents a simple example of the two-level block sparsity. In real and imagery parts of the image, the colorized lattice indicates the target area, while the white lattice represents the nontarget area. Different colors are used to illustrate the difference between the real and imagery coefficients.

The E-BIHT algorithm is described in Table 6.6. At each iteration of E-BIHT, the temporary signal estimate **a** is calculated, as done in BIHT. Then, in Step 3 of E-BIHT, we extend (6.29) by adding a regularization term to further impose the clustering property:

$$\boldsymbol{\theta}^t = \arg \min_{\mathbf{b}} \|\mathbf{a} - \mathbf{b}\|_2^2 + \gamma \tilde{U}(\mathbf{b}) \quad \text{s.t.} \quad \sum_{i=1}^{P} \|\sqrt{(b_i)^2 + (b_{i+P})^2}\|_0 \leq K \tag{6.41}$$

where $\gamma > 0$ is the regularization parameter. The iterations of E-BIHT are terminated when the number of iterations exceeds the maximum iteration count $t_{\max}$ or the

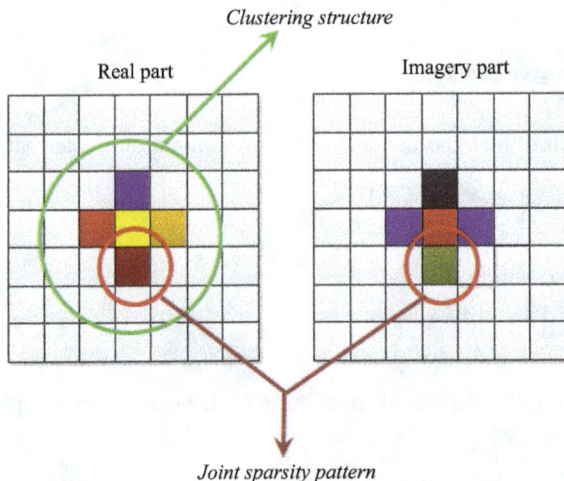

Figure 6.12   *A simple example of the two-level block sparsity*

Table 6.6   *The E-BIHT algorithm*

Input:
  $\bar{\mathbf{y}}, \bar{\mathbf{\Phi}}, K$, the descent step size μ, the maximum iteration count t_{max}, and the error tolerance ε.
Initialization:
  $\bar{\boldsymbol{\theta}}^0 = \mathbf{0}^{2P \times 1}$; $t = 0$.
Iteration:
  (1) $t = t + 1$;
  (2) $\mathbf{a} = \bar{\boldsymbol{\theta}}^{t-1} - \mu \nabla J(\bar{\boldsymbol{\theta}}^{t-1})$;
  (3) $\bar{\boldsymbol{\theta}}^t = \arg\min_{\mathbf{b}} \|\mathbf{a} - \mathbf{b}\|_2^2 + \lambda \tilde{U}(\mathbf{b})$ s.t. $\sum_{i=1}^{P} \|\sqrt{(b_i)^2 + (b_{i+P})^2}\|_0 \leq K$;
Until $t \geq t_{max}$ or $\|\bar{\boldsymbol{\theta}}^t - \bar{\boldsymbol{\theta}}^{t-1}\|_2 / \|\bar{\boldsymbol{\theta}}^t\|_2 < \varepsilon$
Output:
  $\bar{\boldsymbol{\theta}}^t$

---

difference between the solutions at two consecutive iterations $\|\bar{\boldsymbol{\theta}}^t - \bar{\boldsymbol{\theta}}^{t-1}\|_2 / \|\bar{\boldsymbol{\theta}}^t\|_2$ is smaller than the error tolerance parameter ε. The parameters t_{max} and ε are user-defined and application-dependent.

Now we discuss the implementation of Step 3 of the E-BIHT algorithm, that is, how to solve the problem in (6.41). This problem can be solved by using the iterative greedy block coordinate descent (GBCD) algorithm [31,32], which is summarized in Table 6.7. Let:

$$F(\mathbf{b}) = \|\mathbf{a} - \mathbf{b}\|_2^2 + \lambda \tilde{U}(\mathbf{b}) \qquad (6.42)$$

Table 6.7   GBCD for solving the problem in Step 3 of E-BIHT

Input:
     $\mathbf{a}, K$, the maximum iteration count $\tilde{\imath}_{\max}$, and the error tolerance $\tilde{\varepsilon}$.
   Initialization:
     $\mathbf{b}^0 = \mathbf{a}; \tilde{\imath} = 0$.
   Iteration:
(1) $\mathbf{g} = \mathbf{b}^{\tilde{\imath}}$;
(2) $\tilde{\imath} = \tilde{\imath} + 1$;
     For $j = 1, 2, \ldots, 2P$
(3) $b_j^{\tilde{\imath}} = \underset{b_j^{\tilde{\imath}-1}}{\arg \min} \, F(g_1, \ldots, g_{j-1}, b_j^{\tilde{\imath}-1}, g_{j+1}, \ldots, g_{2P})$;

(4) $g_j = b_j^{\tilde{\imath}}$;
     End For
(5) $\mathbf{b}^{\tilde{\imath}} = \Gamma(\mathbf{b}^{\tilde{\imath}}, K)$;
     Until $\tilde{\imath} \geq \tilde{\imath}_{\max}$ or $\|\mathbf{b}^{\tilde{\imath}} - \mathbf{b}^{\tilde{\imath}-1}\|_2 / \|\mathbf{b}^{\tilde{\imath}}\|_2 < \tilde{\varepsilon}$
   Output:
     $\bar{\theta}^l = \mathbf{b}^{\tilde{\imath}}$.

In Steps 1–4 of GBCD, the objective function in (6.42) is minimized by using the block coordinate descent. In Step 3 of GBCD, only the single variable $b_j^{\tilde{\imath}}$ is optimized with other variables fixed, which can be solved by the gradient descent method [33] through a small number of iterations. The gradient descent proceeds by computing the following recursion until the predefined stopping tolerance or the maximum iteration count is reached:

$$b_j^{\tilde{\imath}} = b_j^{\tilde{\imath}} - \tilde{\mu} \frac{\partial F}{\partial b_j^{\tilde{\imath}}} \tag{6.43}$$

where $\tilde{\mu}$ is the descent step size and $j \in \{1, 2, \ldots, 2P\}$. When $j \leq P$,

$$\frac{\partial F}{\partial b_j^{\tilde{\imath}}} = -2\left(a_j - b_j^{\tilde{\imath}}\right)$$

$$- \frac{4\lambda b_j^{\tilde{\imath}}}{\sigma^2} \exp\left(-\frac{\left(b_j^{\tilde{\imath}}\right)^2 + \left(b_{j+P}^{\tilde{\imath}}\right)^2}{2\sigma^2}\right) \left[\sum_{j' \in N_j} g\left(\sqrt{\left(b_{j'}^{\tilde{\imath}}\right)^2 + \left(b_{j'+P}^{\tilde{\imath}}\right)^2}\right)\right]$$

$$\tag{6.44}$$

otherwise

$$\frac{\partial F}{\partial b_j^{\tilde{i}}} = -2\left(a_j - b_j^{\tilde{i}}\right)$$

$$- \frac{4\lambda b_j^{\tilde{i}}}{\sigma^2}\exp\left(-\frac{\left(b_j^{\tilde{i}}\right)^2 + \left(b_{j-P}^{\tilde{i}}\right)^2}{2\sigma^2}\right)\left[\sum_{j' \in N_j} g\left(\sqrt{\left(b_{j'}^{\tilde{i}}\right)^2 + \left(b_{j'-P}^{\tilde{i}}\right)^2}\right)\right]$$

$$(6.45)$$

In Step 5 of Table 6.7, the greedy selection rule is imposed to guarantee the joint sparsity pattern, where $\Gamma(\mathbf{b}^{\tilde{i}}, K)$ is an elementwise operation: when $j \leq P$,

$$\Gamma\left(b_j^{\tilde{i}}, K\right) = \begin{cases} b_j^{\tilde{i}} & \sqrt{\left(b_j^{\tilde{i}}\right)^2 + \left(b_{j+P}^{\tilde{i}}\right)^2} \geq \rho \\ 0 & \sqrt{\left(b_j^{\tilde{i}}\right)^2 + \left(b_{j+P}^{\tilde{i}}\right)^2} < \rho \end{cases} \qquad (6.46)$$

otherwise

$$\Gamma\left(b_j^{\tilde{i}}, K\right) = \begin{cases} b_j^{\tilde{i}} & \sqrt{\left(b_j^{\tilde{i}}\right)^2 + \left(b_{j-P}^{\tilde{i}}\right)^2} \geq \rho \\ 0 & \sqrt{\left(b_j^{\tilde{i}}\right)^2 + \left(b_{j-P}^{\tilde{i}}\right)^2} < \rho \end{cases} \qquad (6.47)$$

In (6.46)–(6.47), ρ is the Kth largest element in $\left\{\sqrt{\left(b_i^{\tilde{i}}\right)^2 + \left(b_{i+P}^{\tilde{i}}\right)^2} | i = 1, 2, \ldots, P\right\}$

Similar to BIHT, the E-BIHT requires the sparsity K as a part of its input variables. As suggested in [34], the following condition can guarantee the successful recovery of sparse signals from 1-bit measurements:

$$M \times L = O(K \log(P/K)) \qquad (6.48)$$

Note that (6.48) also provides a strategy for the selection of K. An alternative approach of choosing the value of K is to run the algorithm using a range of sparsity levels such as $K = 1, 2, 4, \ldots, ML$ and then select the appropriate sparse recovery result, at the cost of increasing the computational complexity by $O(\log(ML))$ [35]. During the iterative process of E-BIHT, another parameter to be determined is σ in (6.36), which separates large and small signal coefficients according to (6.38). The value of σ can be set as the Kth largest element in $\left\{\sqrt{(a_i)^2 + (a_{i+P})^2} | i = 1, 2, \ldots, P\right\}$.

6.3.3   *Experimental results*

In this section, three 1-bit CS algorithms, that is, BIHT [12], 1-Bit-MAP [7] and E-BIHT presented in this chapter, are compared by using simulated and measured stepped-frequency SAR data. For E-BIHT, the maximum iteration count, the error tolerance, and the descent step size are set as $t_{max} = \tilde{t}_{max} = 100$, $\varepsilon = \tilde{\varepsilon} = 10^{-2}$, and $\mu = \tilde{\mu} = 10^{-1}$, respectively. The regularization parameter γ in (6.41) is application-dependent and needs to be tuned appropriately. Here we suggest $\gamma = 0.4\sigma^2$ empirically by considering both of the convergence speed and the recovery accuracy. In all the reconstructed images, the maximum intensity value is normalized to 0 dB. The reconstructed images are quantitatively evaluated by the target-clutter-ratio (TCR), which is defined as

$$
\text{TCR} = 10 \log_{10} \frac{(1/P_T) \sum_{(x,y) \in T} |G(x,y)|^2}{(1/P_C) \sum_{(x,y) \in C} |G(x,y)|^2}
\tag{6.49}
$$

where $G(x,y)$ is the magnitude of the pixel (x,y) in the image, T and C denote the target and clutter areas, respectively, and P_T and P_C denote the numbers of pixels in target and clutter areas, respectively. A large value of TCR means that the dominant coefficients are clustering in the target area and the artifacts outside this area are effectively suppressed.

6.3.3.1 Simulations

The simulated radar system consists of 61-antenna array with an aperture of 0.6 m. The stepped-frequency signal covers a 2 GHz bandwidth, ranging from 2 to 4 GHz, with 201 frequency points available. There are three point-like targets in the simulated scene, as shown in Figure 6.13. The distance from these targets to the

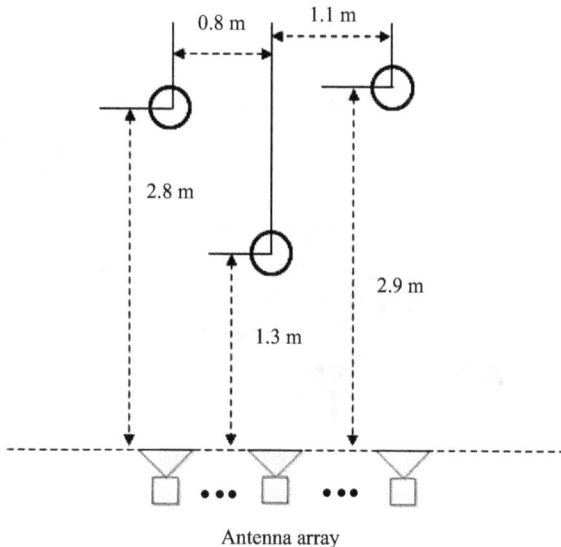

Figure 6.13 Layout of the simulated scene

antenna array is 2.8 m, 1.3 m, and 2.9 m, respectively. The size of the image to be reconstructed is set to $P = 41 \times 41$. The sparsity level is set to $K = 30$.

Select $M \times L = 1,500$ spatial and frequency samples through uniform sub-sampling. Before the 1-bit quantization, the Gaussian white noise is added to the measurements with SNR = 15 dB. The radar images reconstructed by BIHT, 1-Bit-MAP, and E-BIHT are provided in Figure 6.14. As shown in Figure 6.14(a) and (b), BIHT and 1-Bit-MAP locate all the targets correctly but exhibit some isolated artifacts in the nontarget area. In the imaging result of E-BIHT, given in Figure 6.14(c), the dominant pixels are well clustering and the artifacts are greatly alleviated. The TCR values of Figure 6.14(a)–(c) are 28.2003 dB, 28.7149 dB, and

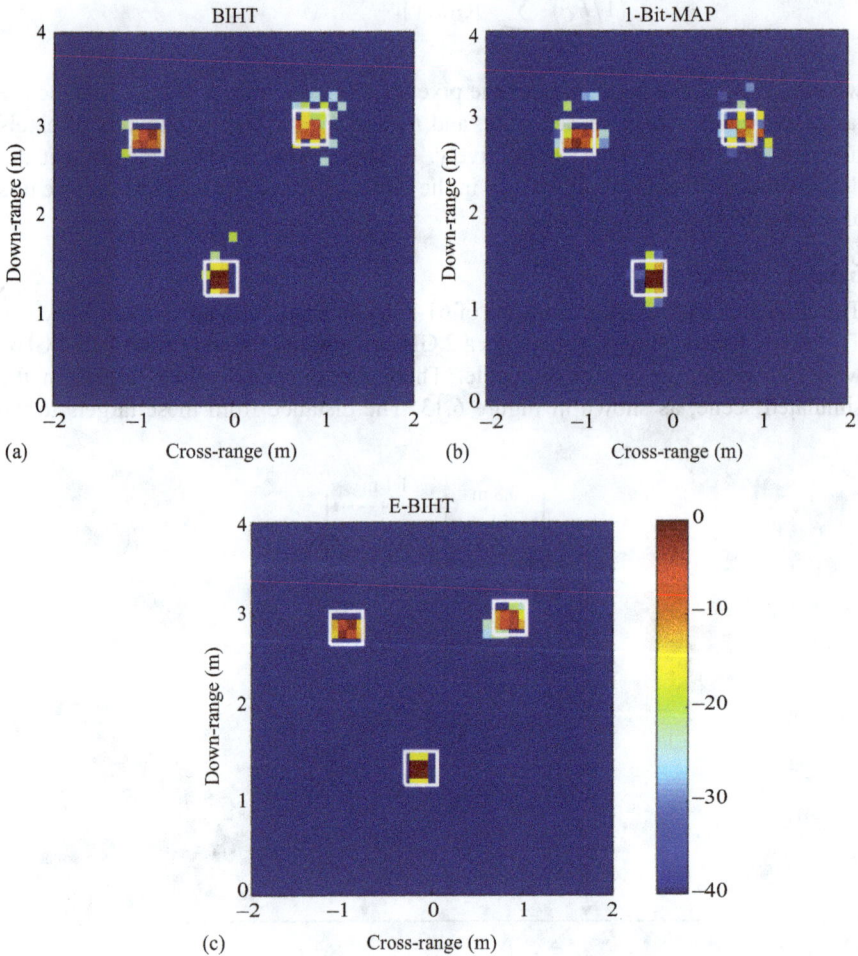

Figure 6.14   Imaging results of the simulated scene yielded by (a) BIHT, (b) 1-Bit-MAP, and (c) E-BIHT

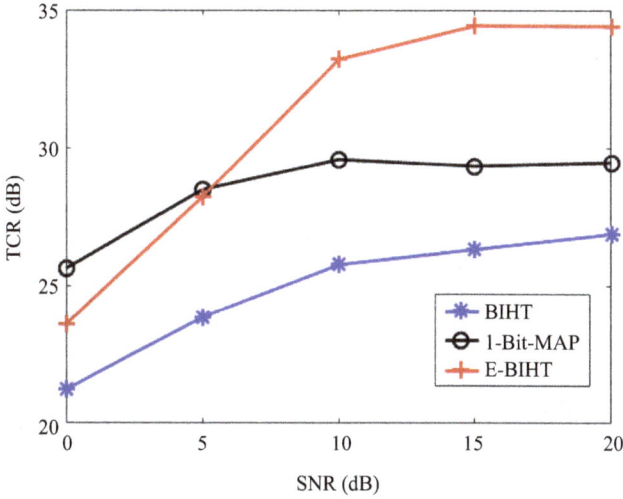

Figure 6.15   TCR versus SNR when the bit budget is 2ML = 1,500

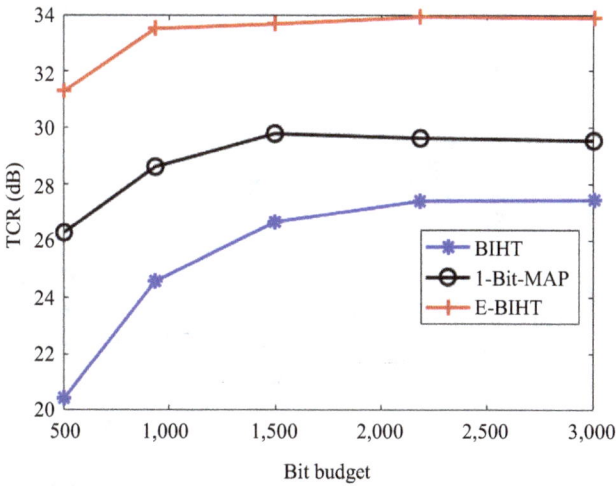

Figure 6.16   TCR versus bit budget when SNR = 15 dB

32.1669 dB, respectively. This quantitative comparison also indicates the superiority of E-BIHT for radar imaging with 1-bit quantized data.

Next, the performances of these 1-bit CS algorithms with varying SNR and bit budget are evaluated. Since the measurements are collected from I/Q channels independently, the total bit budget is $2ML$. The spatial samples and frequency

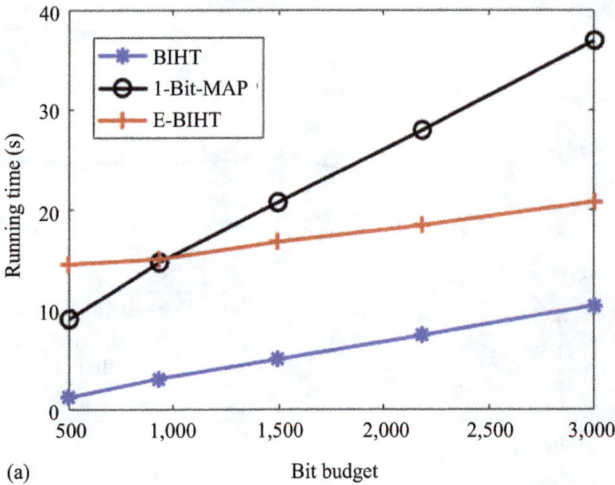

(a)

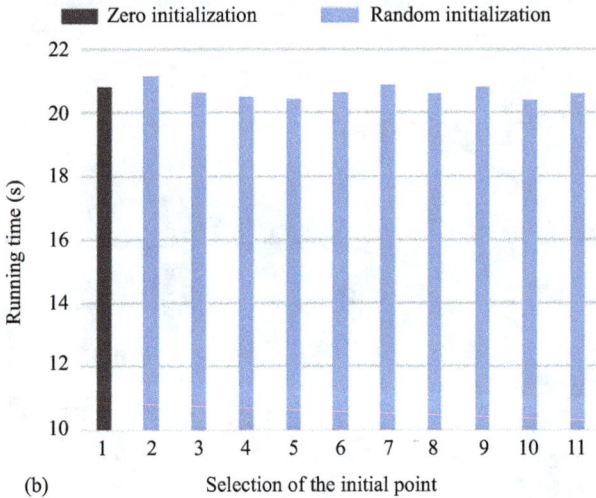

(b)

*Figure 6.17   (a) Average running time versus bit budget when SNR = 15 dB and
(b) average running time of E-BIHT with different initializations
when SNR =15 dB and the bit budget is 2ML = 3,000*

samples are randomly selected in each trial. The TCR values of three algorithms versus SNR are plotted in Figure 6.15, in which each point is the result of averaging over 50 trials. One can see that the imaging quality of E-BIHT is better than that of BIHT and 1-Bit-MAP when SNR $\geq$ 5 dB. The TCR values of these algorithms versus bit budget are depicted in Figure 6.16, where E-BIHT consistently outperforms the other two algorithms.

In Figure 6.17 and Figure 6.18, the complexity and the convergence of E-BIHT are investigated, respectively. Compared to BIHT, exploiting the two-level block sparsity in E-BIHT increases the computational complexity. It is difficult to quantify the increase of the computational load due to the undeterminable number of iterations. In Figure 6.17(a), the running time of BIHT, 1-Bit-MAP, and the E-BIHT are compared by averaging over 50 trials. It can be seen that the computational complexity of E-BIHT is larger than that of BIHT but less than that of 1-Bit-MAP as the bit budget increases. In the above simulations, a zero vector is selected as the initialization point of E-BIHT. In Figure 6.17(b), we further evaluate the effect of different initialization points, including zero vector and 10 random vectors, on the computational complexity of E-BIHT. One can see that the change of running time caused by different initialization points is negligible. In Figure 6.18, the trends of TCR of the images produced by E-BIHT versus the number of iterations are plotted with different SNRs and bit budgets, which show the convergence of E-BIHT as the iterations proceed.

6.3.3.2 Experiments on real radar data

In this section, we compare the above three 1-bit CS algorithms, that is, BIHT, 1-Bit-MAP, and E-BIHT based on the experiments on real radar data. The original measured data were collected by the Radar Imaging Lab of the Center for Advanced

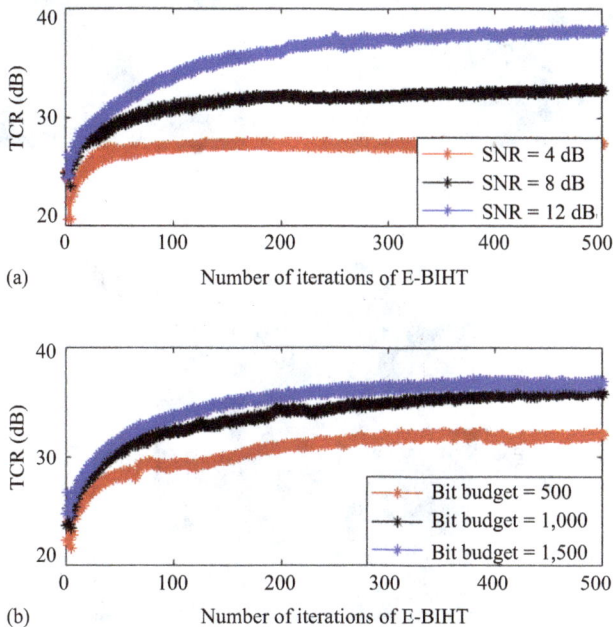

Figure 6.18 TCR versus the number of iterations of E-BIHT: (a) for different SNRs with fixed bit budget of 1,500 and (b) for different bit budgets with fixed SNR of 10 dB

Communications at Villanova University [36], which is the same as that used in Section 2.4.4 of Chapter 2. The measurements from the HH channel are used here.

Total $M \times L = 1,750$ spatial and frequency samples are selected from the original measurement data and requantized by 1-bit quantizer at I/Q channels. The size of the image to be imaged is set as $P = 66 \times 61$. The sparsity level K is set to 70, which is sufficient for the scene of interest. The imaging results of three 1-bit CS algorithms are given in Figure 6.19, where the target areas are marked by white squares. As shown in Figure 6.19(a), BIHT suffers from severe artifacts outside the target areas. The reason is that any underlying signal structure is not considered except for sparsity. The artifacts are slightly mitigated in the imaging result of 1-Bit-MAP given in Figure 6.19(b) because the joint sparsity pattern of the real and imagery parts of the complex target reflectivity is imposed by 1-Bit-MAP. By

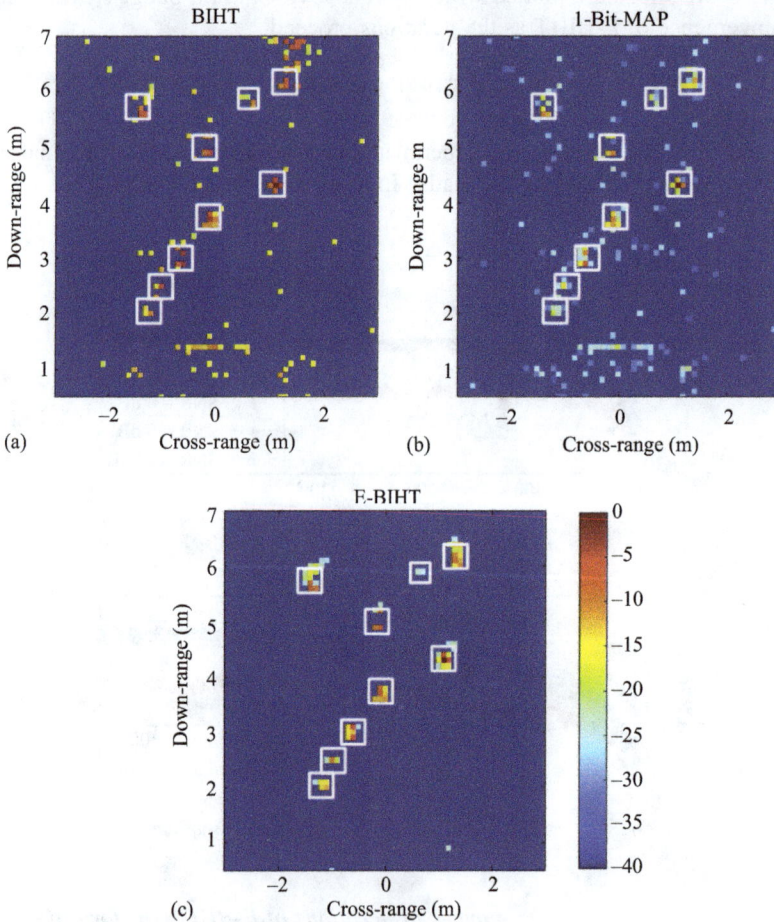

Figure 6.19   Imaging results of measured 1-bit data produced by (a) BIHT, (b) 1-Bit-MAP, and (c) E-BIHT

imposing the two-level block sparsity constraint, E-BIHT enhances both the clustering sparsity and the joint sparsity of the imaging result. As shown in Figure 6.19(c), E-BIHT outperforms the other two algorithms in the following two aspects: (1) the dominant coefficients are more concentrated around the positions of true scatterers; and (2) the clutter and artifacts are significantly suppressed. The TCR values of Figure 6.19(a), (b), and (c) are 15.6962 dB, 23.0237 dB, and 27.2799 dB, respectively, which quantitatively demonstrate the superiority of E-BIHT.

6.4   Conclusion

In this chapter, we presented two algorithms, that is, the PQIHT algorithm and the E-BIHT algorithm, for enhancing the radar imaging quality with coarsely quantized data. The first algorithm, PQIHT, is basically the combination of the original QIHT algorithm and the PSR framework. It aims to eliminate the negative effect of the model uncertainty caused by the target motion and the quantization error on the SAR imaging quality. Experimental results demonstrate that the PQIHT algorithm can achieve moving target refocusing for coarsely quantized data and even 1-bit data, with little sacrifice of the image quality compared to that generated from precise data. The second algorithm, E-BIHT, is based on the combination of the original BIHT algorithm and the two-level block sparsity model. Experimental results demonstrate that the E-BIHT algorithm can enhance the quality of radar imaging of stationary targets with 1-bit data, by effectively removing the isolated artifacts and clustering the dominant pixels, at the cost of the increase in computational complexity. It is worth emphasizing that similar extensions can be also applied to other CS algorithms based on coarsely quantized data. The combination of the advance sparse signal models with 1-bit CS algorithms has great potential in simplifying hardware implementation without severe degradation in radar imaging quality.

References

[1]   Zymnis A., Boyd S., and Candes E. "Compressed sensing with quantized measurements." *IEEE Signal Processing Letters*. 2010; 17(2): 149–152.

[2]   Jacques L., Hammond D. K., and Fadili J. M. "Dequantizing compressed sensing: When oversampling and non-Gaussian constraints combine." *IEEE Transactions on Information Theory*; 2011; 57(1): 559–571.

[3]   Boufounos P. and Baraniuk R. "1-bit compressive sensing." *Proceeding of 42nd IEEE Annual Conference on Information Sciences and Systems (CISS)*. Princeton, NJ, USA: IEEE; 2008, pp. 16–21.

[4]   Le B., Rondeau T. W., Reed J. H., and Bostian C. W. "Analog-to-digital converters." *IEEE Signal Processing Magazine*. 2005; 22(6): 69–77.

[5]   Laska J. and Baraniuk R. "Regime change: Bit-depth versus measurement-rate in compressive sensing." *IEEE Transactions on Signal Processing*. 2012; 60(7): 3496–3505.

[6]   Plan Y. and Vershynin R. "Robust 1-bit compressed sensing and sparse logistic regression: A convex programming approach." *IEEE Transactions on Information Theory*. 2013; 59(1): 482–494.

[7]   Dong X. and Zhang Y. "A MAP approach for 1-bit compressive sensing in synthetic aperture radar imaging." *IEEE Geoscience and Remote Sensing Letters*. 2015; 12(6): 1237–1241.

[8]   Yang Z., Xie L., and Zhang C. "Variational Bayesian algorithm for quantized compressed sensing." *IEEE Transactions on Signal Processing*. 2013; 61(11): 2815–2824.

[9]   Boufounos P. "Greedy sparse signal reconstruction from sign measurements." *Proceedings of 43rd Asilomar Conference on Signals, Systems and Computers*. Pacific Grove, CA, USA: IEEE. 2009, pp. 1305–1309.

[10] Laska J., Wen Z., Yin W., and Baraniuk R. "Trust, but verify: Fast and accurate signal recovery from 1-bit compressive measurements." *IEEE Transactions on Signal Processing*. 2011; 59(11): 5289–5301.

[11] Jacques L., Degraux K., and Vleeschouwer C. D. "Quantized iterative hard thresholding: Bridging 1-bit and high-resolution quantized compressed sensing." *Mathematics*. 2013.

[12] Jacques L., Laska J. N., Boufounos P. T., and Baraniuk R. G. "Robust 1-bit compressive sensing via binary stable embeddings of sparse vectors." *IEEE Transactions on Information Theory*. 2013; 59(4): 2082–2102.

[13] Fornaro G., Pascazio V., and Schirinzi G. "Synthetic aperture radar interferometry using one bit coded raw and reference signals." *IEEE Transactions on Geoscience and Remote Sensing*. 1997; 35(5): 1245–1253.

[14] Franceschetti G., Merolla S., and Tesauro M. "Phase quantized SAR signal processing: Theory and experiments." *IEEE Transactions on Aerospace Electronic Systems*. 1999; 35(1): 201–214.

[15] Pascazio V. and Schirinzi G. "Synthetic aperture radar imaging by one bit coded signals." *Electronics & Communication Engineering Journal*. 1998; 10(1): 17–28.

[16] Zhou C., Liu F., Li B., Hu J., and Lv Y. "A 1-bit compressive sensing approach for SAR imaging based on approximated observation." *Proceedings of Eighth International Conference on Digital Image Processing (ICDIP)*. Chengdu, China: SPIE; 2016, pp. 100333J–100333J-7.

[17] Li J., Naghsh M. M., Zahabi S. J., and Hashemi M. M. "Compressive radar sensing via one-bit sampling with time-varying thresholds." *Proceedings of IEEE 50th Asilomar Conference on Signals, Systems and Computers*. Pacific Grove, CA, USA: IEEE; 2016, pp. 1164–1168.

[18] Han J., Li G., and Zhang X.-P. "Refocusing of moving targets based on low-bit quantized SAR data via parametric quantized iterative hard thresholding." *IEEE Transactions on Aerospace and Electronic Systems*. 2020; 56(3): 2198–2211.

[19] Chen Y., Li G., Zhang Q., and Sun J. "Refocusing of moving targets in SAR images via parametric sparse representation." *Remote Sensing*. 2017; 9(8): doi:10.3390/rs9080795.

[20] Wang X., Li G., Liu Y., and Amin M. G. "Enhanced 1-bit radar imaging by exploiting two-level block sparsity." *IEEE Transactions on Geoscience and Remote Sensing*. 2019; 57(2): 1131–1141.

[21] Wang X., Li G., Liu Y., and Amin M. G. "Two-level block matching pursuit for polarimetric through-wall radar imaging." *IEEE Transactions on Geoscience and Remote Sensing*. 2018; 56(3): 1533–1545.

[22] Shao Y. F., Wang R., Deng Y., *et al.* "Fast backprojection algorithm for bistatic SAR imaging." *IEEE Geoscience Remote Sensing Letters*. 2013; 10(5): 1080–1084.

[23] Bamler R. "A comparison of range-Doppler and wavenumber domain SAR focusing algorithms." *IEEE Transactions on Geoscience Remote Sensing*. 1992; 30(4): 706–713.

[24] Raney R. K., Runge H., Bamler R., Cumming I. G., and Wong F. H. "Precision SAR processing using chirp scaling." *IEEE Transactions on Geoscience Remote Sensing*. 1994; 32(4): 786–799.

[25] Cafforio C., Prati C., and Rocca F. "SAR data focusing using seismic migration and techniques." *IEEE Transactions on Aerospace Electronic Systems*. 1991; 27(2): 194–207.

[26] Zhang Y., Sun J. P., Lei P., Li G., and Hong W. "High-resolution SAR-based ground moving target imaging with defocused ROI data." *IEEE Transactions on Geoscience Remote Sensing*. 2016; 54(2): 1062–1073.

[27] Wu Q., Zhang Y. D., Amin M. G., and Himed B. "Complex multitask Bayesian compressive sensing." *Proceedings of the IEEE International Conference on Acoustics, Speech and Signal Processing (ICASSP)*. Florence, Italy: IEEE; 2014, pp. 3375–3379.

[28] Baraniuk R. G., Cevher V., Duarte M. F., and Hegde C. "Model-based compressive sensing." *IEEE Transactions on Information Theory*. 2010; 56(4): 1982–2001.

[29] Li S. Z. *Markov random field modeling in image analysis*. 2nd ed. New York, NY: Springer-Verlag, 2001.

[30] Mohimani H., Babaie-Zadeh M., and Jutten C. "A fast approach for over-complete sparse decomposition based on smoothed l_0 norm." *IEEE Transactions on Signal Processing*. 2009; 57(1): 289–301.

[31] Wei X., Yuan Y., and Ling Q. "DOA estimation using a greedy block coordinate descent algorithm." *IEEE Transactions on Signal Processing*. 2012; 60(12): 6382–6394.

[32] Li H., Fu Y., Hu R., and Rong R. "Perturbation analysis of greedy block coordinate descent under RIP." *IEEE Signal Processing Letters*. 2014; 21(5): 518–522.

[33] Boyd S. and Vandenberghe L. *Convex optimization*. Cambridge, UK: Cambridge University Press, 2004.

[34] Baraniuk R. G., Foucart S., Needell D., Planb Y., and Woottersset M. "Exponential decay of reconstruction error from binary measurements of sparse signals." *IEEE Transactions on Information Theory*. 2017; 63(6): 3368–3385.

[35]   Needell D. and Tropp J. A. "CoSaMP: Iterative signal recovery from incomplete and inaccurate samples." *Applied and Computational Harmonic Analysis*. 2009; 26: 301–321.

[36]   Dilsavor R., Ailes W., Rush P., *et al*. "Experiments on wideband through the wall imaging." *Proceedings of SPIE Symposium on Defense and Security, Algorithms for Synthetic Aperture Radar Imagery XII Conference. Vol. 5808*. Orlando, Florida, USA; SPIE; 2005, pp. 196–209.

Chapter 7

Sparsity aware micro-Doppler analysis for radar target classification

7.1 Introduction

The Doppler frequency of the radar echo is proportional to the radial velocity of a target. The rotation, vibration, or coning motion of a target or its parts may produce additional Doppler modulations of the received signal, which is called the micro-Doppler effect. Micro-Doppler parameters such as the Doppler repetition period, the Doppler amplitude, and the initial phase can directly indicate the characteristics of the target and, therefore, aid target classification and recognition [1–3]. Take the micro-Doppler signal reflected from the rotating wind turbine as an example. The Doppler repetition frequency is related to the number of blades and the rotational speed of the turbine, the maximal Doppler frequency is proportional to the length of each blade given the rotational speed of the turbine, and the difference among the initial phases of multiple signal components reflects the geometry of the scatterers on the turbine. Especially, the micro-Doppler analysis is an essential tool for target classification and recognition with narrow-band radar systems, in which the high-resolution one-dimensional (1D) image (i.e., the range profile) and two-dimensional (2D) image are not available.

Time–frequency analysis is necessary to visualize the time-varying Doppler behavior of the radar signal and to retrieve the micro-Doppler characteristics of the target. Typical algorithms of time–frequency analysis include the short-time Fourier transform (STFT), the Wigner–Ville distribution (WVD), and the Cohen's class with different kernel functions [1–3]. STFT is easy to be implemented by using the windowed Fourier transform with varying window center, but high time resolution and high frequency resolution cannot be obtained simultaneously. WVD can be expressed as

$$\text{WVD}(t,f) = \iint A(\tau, v) \, \exp\{j2\pi(vt - f\tau)\} d\tau dv \tag{7.1}$$

where $A(\tau, v)$ is the ambiguity function defined by $A(\tau, v) = \int y(t + \tau/2)y^*(t - \tau/2) \exp\{-j2\pi vt\} dt$, $y(t)$ is the received signal. Due to the inherent autocorrelation, WVD suffers from the problem of cross-term interference when more than one signal components exist. To attenuate the cross-term interference, the Cohen's class has been proposed by embedding a low-pass filter into WVD:

$$C(t,f) = \iint A(\tau, v)\gamma(\tau, v)\exp\{j2\pi(vt - f\tau)\}d\tau dv \tag{7.2}$$

where $\gamma(\tau, v)$ denotes a low-pass filter kernel. The cost of the Cohen's class is the loss of resolution.

Another popular method for time–frequency analysis is the Gabor decomposition [4,5]:

$$y(t) = \sum_p x_p h_p(t) \tag{7.3}$$

where x_p denotes the corresponding coefficient and the pth Gabor function is basically a single-frequency signal with a Gaussian window, expressed as

$$h_p(t) = \left(\pi\sigma_p^2\right)^{-0.25}\exp\left(-\frac{(t-t_p)^2}{2\sigma_p^2}\right)\exp\left(j2\pi f_p t\right) \tag{7.4}$$

where t_p and f_p denote the time center and the frequency center, σ_p denotes the width of the Gaussian window. The Gabor decomposition aims to decompose the received signal into a group of Gabor functions that best fit the time–frequency behavior of the received signal.

The compressed sensing (CS) has been applied to characterize the time–frequency domain [6], where the sparse WVD coefficients are retrieved from a small number of measurements of the ambiguity function. The benefits resulting from CS include reduction of the number of the measurements, enhancement of the time–frequency resolution, and capability of dealing with the missing data case. The Gabor decomposition can also be regarded as a CS-based method with the Gabor dictionary. In [7], some state-of-the-art algorithms of time–frequency analysis based on sparse signal recovery are summarized.

After the energy distribution in the time–frequency domain is characterized by the above time–frequency analysis tools, the micro-Doppler parameters can be estimated by mapping the time–frequency distribution onto the parameter space by a pattern recognition tool, such as the Hough transform [8]. For example, the Hough transform for parameter estimation of the sinusoidal frequency modulated signals is defined as

$$H(\alpha, \omega, \theta) = \int \mathrm{TF}(t, \alpha\cos(\omega t + \theta))dt \tag{7.5}$$

where $\mathrm{TF}(\cdot)$ denotes the absolute value of the time–frequency distribution obtained by the time–frequency analysis tools. The Hough transform is basically equivalent to energy accumulation along every path determined by a set of parameter candidates in the time–frequency domain. Therefore, the micro-Doppler parameters can be estimated by finding the accumulation peaks in the parameter space [9]. The Hough transform has been combined with the pseudo-Wigner-Ville-distribution

(PWVD) and the reassigned smoothed pseudo-Wigner-Ville-distribution (RSPWVD) in the pseudo-Wigner-Hough transform (PWHT) algorithm [10] and in the Hough-RSPWVD algorithm [9], respectively. The key idea of these algorithms is that the time–frequency distribution is first localized by PWVD or RSPWVD and then the correct micro-Doppler parameters are estimated from the candidate sets by finding the maximal Hough integration peak. However, the Hough-based methods are not suitable for the case when the time–frequency behavior is difficult to be formulated as an analytical function, which usually happens for nonrigid-body targets. This is the reason why the Hough-based methods are commonly used for analyzing the radar signals reflected from rigid-body targets.

In this chapter, two sparsity-aware algorithms for radar micro-Doppler analysis will be presented in order to classify rigid-body and nonrigid-body targets, respectively.

1. The first algorithm aims to decompose multiple micro-Doppler signal components reflected from a rotating, coning, or vibrating rigid-body target and estimate their parameters such as the Doppler repetition period, the Doppler amplitude and the initial phase [11]. The key idea of this algorithm is to formulate the radar echo reflected from a rigid-body via the parametric sparse representation (PSR) method. That is, the common parameter of the multiple micro-Doppler components such as the Doppler repetition period is embed into the parametric dictionary, while the sparse solution to be solved consists of the individual parameters such as the Doppler amplitude and the initial phase. Instead of putting all of the common parameters and the individual parameters into the solution space, the PSR model significantly reduces the problem size. To efficiently estimate the micro-Doppler parameter with the PSR model, a pruned orthogonal matching pursuit (POMP) algorithm is presented. Simulation results demonstrate that, compared with the traditional Hough-based methods, the combination of the PSR model and the POMP algorithm is capable of yielding better time–frequency resolution and more accurate micro-Doppler parameter estimation. The micro-Doppler parameter estimates directly represent the physical characteristics of the rigid-body target and therefore help target classification and recognition.

2. The second algorithm is presented for recognition of nonrigid-body objects such as hand gestures [12]. The radar signals reflected from hand gestures are generally nonperiodical and difficult to be formulated as an analytical form, so estimating some specific feature parameters that have physical meanings is not a good choice for gesture recognition. The time–frequency locations and the corresponding reflectivity coefficients of the dominant components are extracted from the radar signal via the Gabor decomposition and regarded as the micro-Doppler features. Then, these micro-Doppler features are fed into the modified-Hausdorff-distance-based nearest neighbor (NN) classifier for gesture recognition. This algorithm is referred to as the Gabor–Hausdorff algorithm. The experiments with real radar data show that the Gabor–Hausdorff algorithm outperforms the state-of-the-art methods such as the

principal component analysis (PCA) and the deep convolutional neural network (DCNN) in the condition of small training dataset.

7.2 Micro-Doppler parameter estimation via PSR

In this section, we introduce how to estimate the micro-Doppler parameters of a rigid body by using the PSR method.

7.2.1 Signal model

The radar echo from a coning rigid-body target can be expressed as [1,2]

$$y(t) = \sum_{k=1}^{K} a_k \exp\left\{j\frac{4\pi}{\lambda} d_k \sin(\omega t + \theta_k)\right\}, t = t_1, t_2, \ldots, t_M \tag{7.6}$$

where ω is the angular speed of the target and is assumed as constant if the observation duration is not too long, a_k is the complex reflectivity of the kth scatterer, λ is the radar wavelength, d_k is dependent on the spatial position of the kth scatterer and proportional to the maximal Doppler amplitude, θ_k is the initial phase and it is related to the relative geometrical structure between the kth scatterer and the rotation center, and K is the number of dominant scatterers. The micro-Doppler frequency corresponding to the kth scatterer can be directly obtained by taking the time derivative of the phase term in (7.6):

$$f_{MD,k} = \frac{2}{\lambda} d_k \omega \cos(\omega t + \theta_k) \tag{7.7}$$

One can see that the micro-Doppler frequencies of different scatterers on a coning target vary periodically with the same period but different peak values and different starting phase. Our goal is to estimate the micro-Doppler parameters ω and $\{d_k, \theta_k\}$ for $k = 1, 2, \ldots, K$. It is worth emphasizing that the model in (7.6) and (7.7) can also be used to formulate other kinds of micro-Doppler signals reflected from a rigid body that is vibrating, rotating or tumbling [2].

If the Doppler amplitude domain that d_k belongs to and the initial phase domain that θ_k belongs to are uniformly discretized into $P \times Q$ discrete values, that is, $d_k \in \{d_1, \ldots, d_p, \ldots, d_P\}$ and $\theta_k \in \{\theta_1, \ldots, \theta_q, \ldots, \theta_Q\}$, the received signal in (7.6) can be rewritten as

$$\mathbf{Y} = \mathbf{\Phi}(\omega) \cdot \mathbf{X} \tag{7.8}$$

where $\mathbf{Y} = [y(t_1), y(t_2), \ldots, y(t_M)]^T$ is an $M \times 1$ measurement vector, $\mathbf{\Phi}(\omega) \in C^{M \times (PQ)}$ and its element is

$$\mathbf{\Phi}(\omega)_{m,p+(q-1)P} = \exp\left\{j\frac{4\pi}{\lambda} d_p \sin(\omega t_m + \theta_q)\right\} \tag{7.9}$$

$\mathbf{X} \in C^{(PQ) \times 1}$ is a K-sparse signal and its nonzero element $\mathbf{X}_{p+(q-1)P} = a_k$ if and only if $d_p = d_k$ and $\theta_q = \theta_k$. The value of K is assumed unknown. Without loss of generality, we assume $K \ll M < PQ$, and accordingly, the micro-Doppler parameter estimation problem is converted into a problem of sparse signal recovery. Note that the dictionary matrix $\mathbf{\Phi}(\omega)$ is related to the unknown angular velocity of the target, which follows the definition of the PSR model in [13–15].

What advantages does the PSR model offer over the traditional CS model for micro-Doppler analysis? The traditional CS model was used for micro-Doppler analysis in [16]. The main difference between the PSR model and the traditional CS model in [16] is that in the former the dictionary matrix is adjustable during the solution process while in the latter the dictionary matrix is predesigned and fixed during the solution process. If one follows the strategy in [16], that is, further discretizing the ω domain into L values and synthesizing the micro-Doppler basis-signals according to all possible candidates $\{\omega_l, d_p, \theta_q\}$ for $l = 1, 2, \ldots, L$, $p = 1, 2, \ldots, P, q = 1, 2, \ldots, Q$, the sizes of the dictionary matrix and the sparse solution will be enlarged to $M \times (PQL)$ and $(PQL) \times 1$, respectively, which may result in an unreliable sparse recovery due to the limited number of measurements [17,18]. Note that ω is the same for all the scatterers on a coning rigid-body target and $\{d_k, \theta_k\}$ vary with the scatterer index. It should also be pointed out that, this property holds not only for the micro-Doppler signals induced by coning motion but also for those induced by rotation, vibration, and tumbling with minor expression differences [2]. Therefore, it is reasonable to separate the common parameter ω from individual parameters $\{d_k, \theta_k\}$ in the signal decomposition process. The PSR model offers a feasible way to do so, and as a result, the problem size is reduced to $M \times (PQ)$ and the reliability of sparse signal recovery is definitely improved.

It is desirable that the solution of (7.8) ensures the sparsity of $\mathbf{X}$ and makes the recovery error as small as possible, that is,

$$\{\widehat{\omega}, \widehat{\mathbf{X}}\} = \arg\min_{\{\omega, \mathbf{X}\}} \|\mathbf{X}\|_0 \quad \text{s.t.} \quad \|\mathbf{Y} - \mathbf{\Phi}(\omega) \cdot \mathbf{X}\|_2^2 \leq \varepsilon \tag{7.10}$$

We define the recovery error as

$$E(\omega, d_1, \theta_1, \ldots, d_K, \theta_K) \triangleq \|\mathbf{Y} - \mathbf{\Phi}(\omega) \cdot \mathbf{X}\|_2^2$$

$$= \sum_{m=1}^{M} \left| y(t_m) - \sum_{k=1}^{K} a_k \exp\left\{ j\frac{4\pi}{\lambda} d_k \sin(\omega t_m + \theta_k) \right\} \right|^2 \tag{7.11}$$

Note that $E(\omega, d_1, \theta_1, \ldots, d_K, \theta_K)$ is more sensitive to the relative change of the value of ω than to the relative change of values of $\{d_k, \theta_k\}$, as analyzed below. Consider the case of two micro-Doppler components as an example and focus on the error of the mth moment in (7.11):

$$
\begin{aligned}
&E_m(\omega, d_1, \theta_1, d_2, \theta_2) \\
&= \left| y(t_m) - a_1 \exp\left\{ j\frac{4\pi}{\lambda} d_1 \sin(\omega t_m + \theta_1) \right\} - a_2 \exp\left\{ j\frac{4\pi}{\lambda} d_2 \sin(\omega t_m + \theta_2) \right\} \right|^2 \\
&= |y(t_m)|^2 + |a_1|^2 + |a_2|^2 \\
&\quad -2\left[\mathrm{Re}(y(t_m)a_1)\cos\left(\frac{4\pi}{\lambda} d_1 \sin(\omega t_m + \theta_1)\right) - \mathrm{Im}(y(t_m)a_1)\sin\left(\frac{4\pi}{\lambda} d_1 \sin(\omega t_m + \theta_1)\right) \right] \\
&\quad -2\left[\mathrm{Re}(y(t_m)a_2)\cos\left(\frac{4\pi}{\lambda} d_2 \sin(\omega t_m + \theta_2)\right) - \mathrm{Im}(y(t_m)a_2)\sin\left(\frac{4\pi}{\lambda} d_2 \sin(\omega t_m + \theta_2)\right) \right] \\
&\quad +2\left[\mathrm{Re}(a_1 a_2)\cos\left(\frac{4\pi}{\lambda}(d_1 \sin(\omega t_m + \theta_1) - d_2 \sin(\omega t_m + \theta_2))\right) \right. \\
&\quad \left. - \mathrm{Im}(a_1 a_2)\sin\left(\frac{4\pi}{\lambda}(d_1 \sin(\omega t_m + \theta_1) - d_2 \sin(\omega t_m + \theta_2))\right) \right]
\end{aligned}
$$

(7.12)

where $\mathrm{Re}(\cdot)$ and $\mathrm{Im}(\cdot)$ stand for the real and imaginary parts, respectively. Then we have the following derivatives:

$$
\begin{aligned}
\frac{\partial E_m}{\partial \omega} &= \frac{8\pi}{\lambda} \left\{ d_1 t_m \cos(\omega t_m + \theta_1) \cdot \left[\mathrm{Re}(y(t_m)a_1)\sin\left(\frac{4\pi}{\lambda} d_1 \sin(\omega t_m + \theta_1)\right) \right. \right. \\
&\quad \left. + \mathrm{Im}(y(t_m)a_1)\cos\left(\frac{4\pi}{\lambda} d_1 \sin(\omega t_m + \theta_1)\right) \right] \\
&\quad + d_2 t_m \cos(\omega t_m + \theta_2) \cdot \left[\mathrm{Re}(y(t_m)a_2)\sin\left(\frac{4\pi}{\lambda} d_2 \sin(\omega t_m + \theta_2)\right) \right. \\
&\quad \left. + \mathrm{Im}(y(t_m)a_2)\cos\left(\frac{4\pi}{\lambda} d_2 \sin(\omega t_m + \theta_2)\right) \right] \\
&\quad - t_m(d_1 \cos(\omega t_m + \theta_1) - d_2 \cos(\omega t_m + \theta_2)) \\
&\quad \cdot \left[\mathrm{Re}(a_1 a_2)\sin\left(\frac{4\pi}{\lambda}(d_1 \sin(\omega t_m + \theta_1) - d_2 \sin(\omega t_m + \theta_2))\right) \right. \\
&\quad \left. \left. + \mathrm{Im}(a_1 a_2)\cos\left(\frac{4\pi}{\lambda}(d_1 \sin(\omega t_m + \theta_1) - d_2 \sin(\omega t_m + \theta_2))\right) \right] \right\}
\end{aligned}
$$

(7.13)

and

$$
\begin{aligned}
\frac{\partial E_m}{\partial d_1} &= \frac{8\pi}{\lambda} \sin(\omega t_m + \theta_1) \left[\mathrm{Re}(y(t_m)a_1)\sin\left(\frac{4\pi}{\lambda} d_1 \sin(\omega t_m + \theta_1)\right) \right. \\
&\quad + \mathrm{Im}(y(t_m)a_1)\cos\left(\frac{4\pi}{\lambda} d_1 \sin(\omega t_m + \theta_1)\right) \\
&\quad - \mathrm{Re}(a_1 a_2)\sin\left(\frac{4\pi}{\lambda}(d_1 \sin(\omega t_m + \theta_1) - d_2 \sin(\omega t_m + \theta_2))\right) \\
&\quad \left. - \mathrm{Im}(a_1 a_2)\cos\left(\frac{4\pi}{\lambda}(d_1 \sin(\omega t_m + \theta_1) - d_2 \sin(\omega t_m + \theta_2))\right) \right]
\end{aligned}
$$

(7.14)

and

$$\frac{\partial E_m}{\partial \theta_1} = \frac{8\pi}{\lambda} d_1 \cos\left(\omega t_m + \theta_1\right) \left[\operatorname{Re}(y(t_m)a_1)\sin\left(\frac{4\pi}{\lambda} d_1 \sin\left(\omega t_m + \theta_1\right)\right) \right.$$

$$+ \operatorname{Im}(y(t_m)a_1)\cos\left(\frac{4\pi}{\lambda} d_1 \sin\left(\omega t_m + \theta_1\right)\right)$$

$$- \operatorname{Re}(a_1 a_2)\sin\left(\frac{4\pi}{\lambda}(d_1 \sin\left(\omega t_m + \theta_1\right) - d_2 \sin\left(\omega t_m + \theta_2\right))\right)$$

$$\left. - \operatorname{Im}(a_1 a_2)\cos\left(\frac{4\pi}{\lambda}(d_1 \sin\left(\omega t_m + \theta_1\right) - d_2 \sin\left(\omega t_m + \theta_2\right))\right) \right]$$

$$(7.15)$$

Assume that the values of d_1 and d_2 are comparable and the values of a_1 and a_2 are comparable, then $\partial E_m/\partial d_2$ and $\partial E_m/\partial \theta_2$ have the same order magnitude with (7.14) and (7.15), respectively. The change of E_m is denoted as ΔE_m and it can be approximated as

$$\Delta E_m \approx \frac{\partial E_m}{\partial \omega} \Delta \omega = \omega \frac{\partial E_m}{\partial \omega} \cdot \frac{\Delta \omega}{\omega}$$

$$\Delta E_m \approx \frac{\partial E_m}{\partial d_1} \Delta d_1 = d_1 \frac{\partial E_m}{\partial d_1} \cdot \frac{\Delta d_1}{d_1}$$

$$(7.16)$$

$$\Delta E_m \approx \frac{\partial E_m}{\partial \theta_1} \Delta \theta_1 = \theta_1 \frac{\partial E_m}{\partial \theta_1} \cdot \frac{\Delta \theta_1}{\theta_1}$$

where $\frac{\Delta \chi}{\chi}$ stands for the relative change of the variable χ, and the $\chi \frac{\partial \chi}{\partial \chi}$ denotes the rate of the change of E_m in terms of the relative change of the variable χ. From (7.13)–(7.16), one can see that the absolute value of $\omega \frac{\partial E_m}{\partial \omega}$ is approximately ωt_m times larger than the absolute value of $d_k \frac{\partial E_m}{\partial d_k}$ and $\frac{\omega t_m}{\theta_k}$ times larger than the absolute value of $\theta_k \frac{\partial E_m}{\partial \theta_k}$, respectively. The value of ωt_m increases as the observation time duration increases. In general, to effectively observe the periodic rotation, vibration, or coning motion of a target and to accurately estimate the period of the micro-Doppler modulation, the observation time duration is required to be large enough such that more than one-period data are recorded, that is, $\omega t_m > 2\pi$. Accordingly, we have $\frac{\omega t_m}{\theta_k} > 1$ since $0 \le \theta_k < 2\pi$. Therefore, we can conclude that the rate of the change of E_m in terms of the relative change of ω is larger than in terms of the relative change of $\{\theta_k, d_k\}$. In other words, the recovery error in (7.11) is more sensitive to the value change of ω.

Taking the signal containing a single micro-Doppler component as an example, we also demonstrate this fact by simulations as shown in Figure 7.1. Assume that the radar wavelength is $\lambda = 8$ mm and the micro-Doppler parameters are $\omega = 2.2\pi$ rad/s, $d = 2$ mm, and $\theta = 0.6\pi$. Figure 7.1 plots $E(\omega, d, \theta)$ versus varying relative change of the micro-Doppler parameters. Here the "relative change" means the change relative to the true value, for example, relative changes of 0.8 and 1.1 provide parameter candidate values that are equal to 80% and 110% of the true value, respectively. This simulation result is consistent with the above

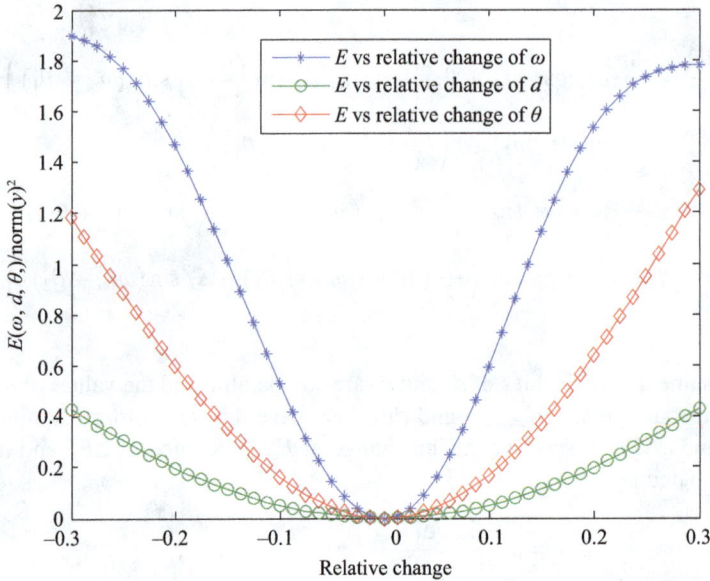

Figure 7.1   Recovery errors versus relative changes of micro-Doppler parameters

theoretical analysis, that is, the recovery error in (7.11) is more sensitive to the change of the parameter ω.

Both of the theoretical analysis and the simulation result in Figure 7.1 imply that the wrong candidate values of ω result in serious recovery error. This allows us to estimate ω as follows: first generate multiple dictionaries with multiple candidate values of ω according to (7.9), next obtain multiple solutions via the orthogonal matching pursuit (OMP) algorithm [17,19] with multiple dictionaries, and then search for the correct candidate values of ω corresponding to the minimum recovery error. This process can be expressed as

$$
\begin{aligned}
\widehat{\omega} &= \arg \min_{\omega} \left\{ \|\mathbf{Y} - \mathbf{\Phi}(\omega) \cdot \mathbf{X}(\omega)\|_2^2 \right\} \\
\mathbf{X}(\omega) &= \mathrm{OMP}(\mathbf{Y}, \mathbf{\Phi}(\omega), \varepsilon)
\end{aligned}
\tag{7.17}
$$

where $\mathbf{X}(\omega)$ is the recovery result obtained by the OMP algorithm with the measurement $\mathbf{Y}$ and the dictionary $\mathbf{\Phi}(\omega)$, ε is the error tolerance threshold. The details of the OMP algorithm are provided in Chapter 1. Here the iterative process of the OMP algorithm is stopped when the residual error is below a threshold ε, as suggested in [19]. Once a good estimate $\widehat{\omega}$ is found, the corresponding sparse solution $\mathbf{X}(\widehat{\omega})$ is also obtained, in which the indices of dominant coefficients indicate the accurate estimates of $\{d_k, \theta_k\}$ for $k = 1, 2, \dots, K$. The process expressed in (7.17) is computationally expensive since the whole procedure of OMP needs to be repeated for a number of candidate values of ω. In what follows, we present an algorithm to improve computational efficiency without sacrificing the parameter estimation accuracy.

7.2.2 Description of the POMP algorithm

To efficiently solve the problem in (7.17), we try to avoid unnecessary computations with wrong candidate values of ω in the process of sparse recovery as far as possible. Note that OMP operates in an iterative manner and extracts the largest component from the residual by finding the maximal correlation coefficient of the residual and the basis-signals at each iteration. By combining a pruning process with OMP, the POMP algorithm is summarized in Table 7.1, where $\Omega^{(\alpha)} = \{\omega_1, \omega_2 \ldots, \omega_n, \ldots\}$ is a set consisting of candidate values of ω and its cardinality is $N^{(\alpha)} = \text{card}(\Omega^{(\alpha)})$, the superscript (α) denotes iteration index, $\mathbf{r}(\widetilde{\omega})$ is the recovery residual with the dictionary $\mathbf{\Phi}(\widetilde{\omega})$ for $\widetilde{\omega} \in \Omega^{(\alpha)}$, $\Lambda(\widetilde{\omega})$ is the support set indicating the indices of nonzero coefficients in the solution $\mathbf{X}(\widetilde{\omega})$, $\mathbf{\Phi}_\Lambda(\widetilde{\omega})$ is composed of those columns of $\mathbf{\Phi}(\widetilde{\omega})$ whose column indices belong to $\Lambda(\widetilde{\omega})$, $\mathbf{X}_\Lambda(\widetilde{\omega})$ is a subvector of $\mathbf{X}(\widetilde{\omega})$ generated by selecting the entries whose indices belong to $\Lambda(\widetilde{\omega})$.

Now we give more explanations about the POMP algorithm. Consider the first iteration, in which all the candidate values of ω are included. Using a dictionary $\mathbf{\Phi}(\omega)$ with a candidate value of ω equal or close to the true value, the most significant signal component will be correctly localized by finding $\max_i |\varphi_i^H(\omega)\mathbf{Y}|$. In

Table 7.1 The POMP algorithm

Input:
 The candidate set $\Omega^{(1)}$ and the received signal $\mathbf{Y}$.
Initialization:
 Let $\alpha = 1$; for every $\widetilde{\omega} \in \Omega^{(1)}$, initialize the dictionary $\mathbf{\Phi}(\widetilde{\omega})$ according to (7.9), the residuals $\mathbf{r}(\widetilde{\omega}) = \mathbf{Y}$, and the support sets $\Lambda(\widetilde{\omega}) = \varnothing$.
Iteration:
 (1) For every $\widetilde{\omega} \in \Omega^{(\alpha)}$, go through the steps from (1.1) to (1.2).
  (1.1) Find the index $\widetilde{i} = \arg\max_i |\varphi_i^H(\widetilde{\omega})\mathbf{r}(\widetilde{\omega})|$, where $\varphi_i(\widetilde{\omega})$ denotes the ith column
   of $\mathbf{\Phi}(\widetilde{\omega})$, and let $\Lambda(\widetilde{\omega}) = \Lambda(\widetilde{\omega}) \cup \{\widetilde{i}\}$;
  (1.2) Update the sparse solution $\mathbf{X}(\widetilde{\omega})$, whose nonzero entries are located at the
   indices indicated by $\Lambda(\widetilde{\omega})$ with the coefficients $\mathbf{X}_\Lambda(\widetilde{\omega}) = (\mathbf{\Phi}_\Lambda(\widetilde{\omega}))^\dagger \mathbf{Y}$, and the
   residual $\mathbf{r}(\widetilde{\omega}) = \mathbf{Y} - \mathbf{\Phi}_\Lambda(\widetilde{\omega})\mathbf{X}_\Lambda(\widetilde{\omega})$;
 (2) Remove $\lceil N^{(\alpha)}/2 \rceil$ candidate values of ω that correspond to $\lceil N^{(\alpha)}/2 \rceil$ largest residual
  errors from $\Omega^{(\alpha)}$, where $\lceil \cdot \rceil$ denotes the ceiling function, and re-denote the new
  candidate set as $\Omega^{(\alpha+1)}$ with the cardinality $N^{(\alpha+1)} = N^{(\alpha)} - \lceil N^{(\alpha)}/2 \rceil$.
 (3) If $\Omega^{(\alpha+1)}$ contains more than one element, increment the iteration counter α by one and
  return to Step (1). Otherwise, that is, if $\Omega^{(\alpha+1)}$ contains only one element $\widetilde{\omega}$, repeat
  the steps from (1.1) to (1.2) until $\|\mathbf{r}(\widetilde{\omega})\|_2^2 \le \varepsilon$.
Output:
 The correct candidate value $\widehat{\omega} = \widetilde{\omega}$ and the sparse solution $\widehat{\mathbf{X}} = \mathbf{X}(\widehat{\omega})$.

---

contrast, using a dictionary $\mathbf{\Phi}(\omega)$ with a wrong/biased candidate value of ω, any column in $\mathbf{\Phi}(\omega)$ will not fit well with any one of the actual signal components. In other words, the energy of any actual signal component will spread over many coefficients of $\left|\mathbf{\Phi}^H(\omega)\mathbf{Y}\right|$, so the largest magnitude $\max\left|\varphi_i^H(\omega)\mathbf{Y}\right|$ only corresponds to a part of the energy of most significant signal component. This means that the recovery error after the first iteration will more likely become larger as the candidate value of ω goes away from the true value. After α iterations, $N^{(\alpha)}$ candidate values of ω will still be active, denoted as $\widetilde{\omega}_1, \widetilde{\omega}_2, \ldots, \widetilde{\omega}_{N^{(\alpha)}}$. All active support sets $\Lambda(\widetilde{\omega}_1), \Lambda(\widetilde{\omega}_2), \ldots, \Lambda(\widetilde{\omega}_{N^{(\alpha)}})$ have the same cardinality, that is, $\mathrm{card}(\Lambda(\widetilde{\omega}_i)) = \alpha$ for $i = 1, 2, \ldots, N^{(\alpha)}$. The rule for keeping viable candidate values of ω is: if the selected columns in $\mathbf{\Phi}_\Lambda(\widetilde{\omega}_i)$ correspond to small least squares residues, they are most likely regarded to span the correct subspace that the received signal $\mathbf{Y}$ lies in. Therefore, some infeasible candidate values of ω can be excluded by error comparison after each iteration, that is, candidate values of ω that yield large residual errors are removed from the candidate set. This pruning process is performed during each iteration until only one candidate value of ω is left. The computational efficiency is due to the fact that the whole procedure of OMP is performed only for the true value of ω, and the number of actual iterations goes down as the candidate value gets away from the true value.

7.2.3   Discussions
7.2.3.1   Connection with dictionary learning
The POMP algorithm can be seen as a special case of dictionary learning since the estimate of ω and, therefore, the dictionary $\mathbf{\Phi}(\omega)$ becomes more accurate as the iterative process continues. Typical dictionary learning algorithms [20–22] generally include two-stage operations: (1) sparse coding determines a sparse solution with a fixed dictionary matrix, and (2) dictionary updating updates the dictionary matrix with the previously determined sparse solution. Different from such two-stage operations, the POMP algorithm achieves sparse coding and dictionary updating simultaneously by embedding the pruning process into OMP, based on the fact that a parametric form is enforced on the dictionary. Compared to the existing dictionary learning algorithms [20–22], the advantage of the POMP algorithm is that one does not need to worry about the local minimum issue because it basically performs the search on the candidate values of ω. The computational burden caused by the large range of candidate values of ω can be mitigated by a smart search strategy, that is, the search grid on ω may be rough at the beginning and then can be refined as the procedure approaches convergence.

7.2.3.2   Computational complexity
Compared with the full-search based method in [23] that solves multiple sparse solutions with multiple dictionaries generated by multiple candidate values of Doppler repetition period, the POMP algorithm is computationally efficient because unnecessary computations corresponding to wrong dictionary candidates are most likely avoided by the pruning process. At the αth iteration, the

computational load of method in [23] comes from the following steps: find the maximal correlation coefficient, next calculate the least squares projection, and then update the residual. The correlation costs one matrix-vector multiplication with complexity MPQ, and the search for the largest coefficient in the correlation results can be done in about PQ floating point operation. Assume that the least squares projection is implemented by QR factorization, then $(2M\alpha + 3M)$ floating point operations are required to update the QR factorization at the αth iteration. Updating the residual requires M floating point operations. Therefore, the total complexity of performing the entire procedure of OMP on all the candidates of ω is given by

$$C_{\text{full-serach-OMP}} = \sum_{\widetilde{\omega} \in \Omega^{(1)}} \sum_{\alpha=1}^{K(\widetilde{\omega})} [MPQ + 2M\alpha + 4M] \tag{7.18}$$

where $\Omega^{(1)}$ is the set consisting of all the candidate values of ω and its cardinality is $N^{(1)} = \text{card}(\Omega^{(1)})$, and $K(\widetilde{\omega})$ is the required number of iterations with the dictionary $\Phi(\widetilde{\omega})$ to keep the residual error small enough. As discussed in Sections 7.2.1 and 7.2.2, a dictionary $\Phi(\widetilde{\omega})$ with an incorrect candidate value $\widetilde{\omega}$ requires larger support set (and therefore larger number of iterations) to suppress the residual error below the threshold. Therefore, it is more likely that $K(\omega) \leq K(\widetilde{\omega})$ when the candidate values $\widetilde{\omega}$ goes away from the true value ω, and then we have:

$$C_{\text{full-serach-OMP}} \geq N^{(1)} \cdot \sum_{\alpha=1}^{K(\omega)} [MPQ + 2M\alpha + 4M] \tag{7.19}$$

In contrast, in the POMP algorithm, the number of reliable candidate values of ω decreases by half as the iteration index increases. Therefore, the computational complexity of the POMP algorithm is given by

$$C_{\text{POMP}} = \sum_{\alpha=1}^{K(\omega)} \left(\frac{N^{(1)}}{2^{\alpha-1}} (MPQ + 2M\alpha + 4M) + O\left(\frac{N^{(1)}}{2^{\alpha-1}} \log \frac{N^{(1)}}{2^{\alpha-1}} \right) \right) \tag{7.20}$$

where the additional floating point operations $O\left(\frac{N^{(1)}}{2^{\alpha-1}} \log \frac{N^{(1)}}{2^{\alpha-1}} \right)$ at each iteration are required for sorting multiple residual errors corresponding to multiple candidate values of ω. The comparison between (7.19) and (7.20) implies that the computational complexity of the POMP algorithm is much less than that of performing the entire procedure of OMP on all the candidates of ω. Since only one candidate value of ω is regarded as reliable when the iterative process of the POMP algorithm terminates, we have $\frac{N^{(1)}}{2^{\alpha-1}} = 1$ when $\alpha = K(\omega)$, that is, $N^{(1)} = 2^{K(\omega)-1}$. For example, when $P = Q = 50$, $M = 200$, and $N^{(1)} = 64$, the computational complexity of performing the entire procedure of OMP on all the candidates of ω and the POMP algorithm are about 2.3×10^8 and 6.4×10^7, respectively. The superiority of the POMP algorithm in terms of the complexity will become more pronounced as the number of candidate values of ω increases.

7.2.3.3 Comparison with Hough-kind algorithms

It is expected that the POMP algorithm will outperform Hough-kind algorithms such as PWHT [10] and Hough-RSPWVD [9] in terms of the parameter estimation accuracy and the time–frequency resolution. In the PWHT and the Hough-RSPWVD algorithms, PWVD and RSPWVD are respectively performed on the windowed data to characterize the energy distribution in the time–frequency domain before the Hough accumulation. The sliding window (or the embedded low-pass filter) that divides the whole observation period into several sub time-intervals guarantees that the assumption of locally linear frequency modulation holds within every subtime-interval and allows one to estimate the time–frequency distribution of nonlinear frequency modulated signals by using PWVD or RSPWVD. However, the time–frequency resolution is limited by the sliding window, and as a result, the accuracy of the subsequent Hough accumulation degenerates when the signal components are closely located in the time–frequency domain or when strong noise exists. In contrast, the POMP algorithm carries out the parameter estimation by the matching pursuit strategy, in which the basis-signal selection at each iteration is implemented by first evaluating the correlation coefficients between the received signal and the basis-signals and then performing the least square projection. The correlation between the received signal and the basis-signals ensures coherent processing on the whole observation period while the least square projection eliminates the cross-interference from other signal components, so the time–frequency resolution and the parameter estimation accuracy are better than that of the PWHT and the Hough-RSPWVD algorithms. This superiority of the POMP algorithm is also shown by experiments in Section 7.2.4.

7.2.4 Simulation results

Some simulation experiments are conducted and performance results are provided to validate the accuracy and efficiency of the POMP algorithm. We consider a millimeter wave radar and assume that the radar wavelength $\lambda = 8$ mm (i.e., the carrier frequency $= 37.5$ GHz), the sampling rate (i.e., the pulse repetition frequency) $f_s = 200$ Hz, and $M = 200$ samples are collected.

7.2.4.1 Analysis of parameter estimation accuracy in the noise-free case

We first consider the noise-free case and compare the results obtained by PWHT, Hough-RSPWVD, and POMP. Assume that there are two scatterers on a coning target with the angular speed $\omega = 4\pi$ rad/s, and the received signal contains two micro-Doppler components with parameters $\{d_1, d_2\} = \{4.24, 8.91\}$ mm, $\{\theta_1, \theta_2\} = \{0.28\pi, 1.48\pi\}$, and $\{a_1, a_2\} = \{1, 0.9\}$. It can be calculated that the maximal Doppler frequencies of these signal components are 13.3 Hz and 28 Hz, respectively. There are 40 candidate values of ω and the $\{\theta, d\}$ domain is discretized into 50×50 coordinates. The results obtained by the PWHT, the Hough-RSPWVD, and the POMP algorithms are provided in Figure 7.2, Figure 7.3, and Figure 7.4, respectively. Figure 7.2(a) shows the time–frequency characteristics of

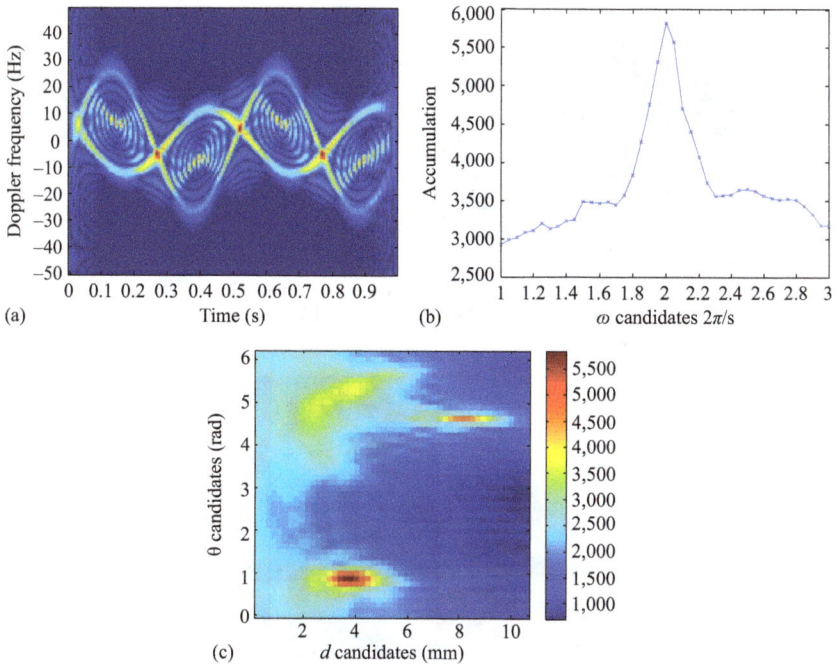

Figure 7.2   *Results obtained by the PWHT algorithm: (a) time–frequency domain obtained by PWVD, (b) Hough accumulation versus ω candidate, and (c) {θ, d} domain obtained by PWHT*

the received signal obtained by the PWVD algorithm, where the two micro-Doppler components can be seen but the cross-term interference is quite considerable. Then micro-Doppler parameter estimation is carried out by performing the Hough transform with different micro-Doppler parameter sets on the time–frequency plane shown in Figure 7.2(a). Figure 7.2(b) plots the Hough accumulation results versus varying candidate values of ω, where the correct angular speed can be obtained by finding the maximal value of this curve. The (θ, d) domain obtained by the PWHT algorithm is shown in Figure 7.2(c), where the peak positions indicate the correct values of parameters $\{\theta_1, d_1\}$ and $\{\theta_2, d_2\}$. The results obtained by the Hough-RSPWVD algorithm are shown in Figure 7.3. As mentioned earlier, the difference between the PWHT and the Hough-RSPWVD algorithms is that the time–frequency characteristics are obtained by PWVD in the former and by RSPWVD in the latter. In Figure 7.3(a), we can see the cross-term interference in the time–frequency domain is well suppressed, thanks to the use of a 2D low-pass filter (Gaussian modulated exponential function) in RSPWVD. Performing the Hough transform with different micro-Doppler parameter sets on the time–frequency plane shown in Figure 7.3(a) yields Figure 7.3(b) and (c). In Figure 7.3(b), the correct angular speed can also be found by finding the maximal Hough accumulation value, but the curve is not as sharp as that in Figure 7.2(b). The reason is

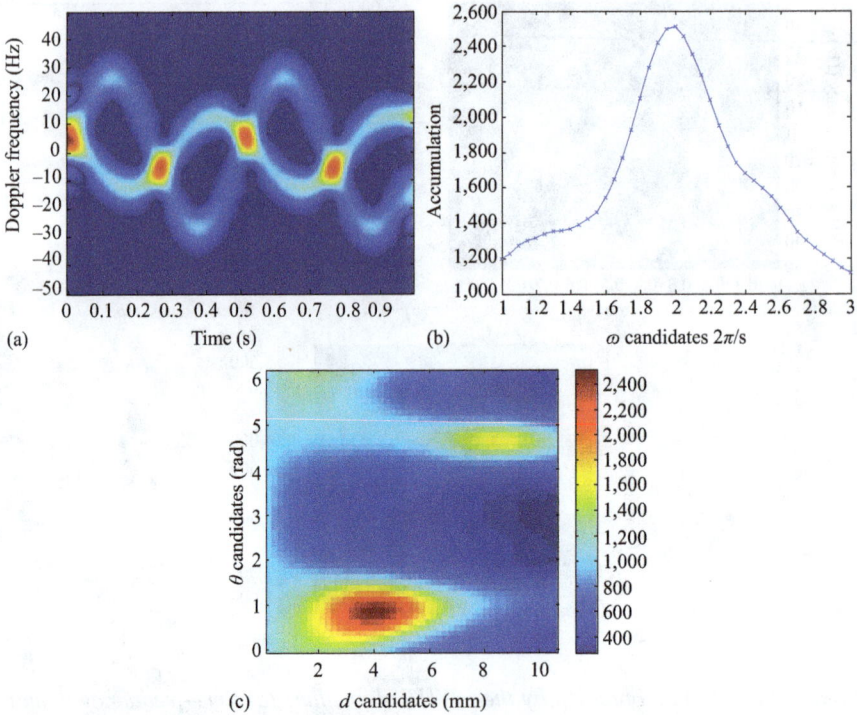

Figure 7.3   Results obtained by the Hough-RSPWVD algorithm: (a) time–frequency domain obtained by RSPWVD, (b) Hough accumulation versus ω candidate, and (c) {θ, d} domain obtained by Hough-RSPWVD

that the 2D low-pass filter in RSPWVD slightly flattens the energy distribution in the time–frequency domain. The (θ, d) domain obtained by the Hough-RSPWVD algorithm is shown in Figure 7.3(c). The energy in Figure 7.3(c) is more concentrated in the coordinates of the correct values of $\{\theta_1, d_1\}$ and $\{\theta_2, d_2\}$, in comparison with that in Figure 7.2(c). The results obtained by the POMP algorithm are provided in Figure 7.4, where the error threshold is set as $\varepsilon = 0.05\|\mathbf{Y}\|_2^2$, that is, the iteration process is stopped when the residual energy is equal to or smaller than 5% of the energy of the received signal. Different from the PWVD and the Hough-RSPWVD algorithms that have to first estimate the time–frequency distribution and then map to micro-Doppler parameter domain, the POMP algorithm is capable of directly obtaining the parameter domain from the received signal. The pruning process of the POMP algorithm is demonstrated in Figure 7.4(b), where "star" denotes active candidate values of ω in every iteration. One can see that the cardinality of the candidate set of ω is continually reduced as the iterations proceed, and the correct candidate value of ω is successfully selected when the pruning process terminates. The (θ, d) domain recovered by the POMP algorithm is shown

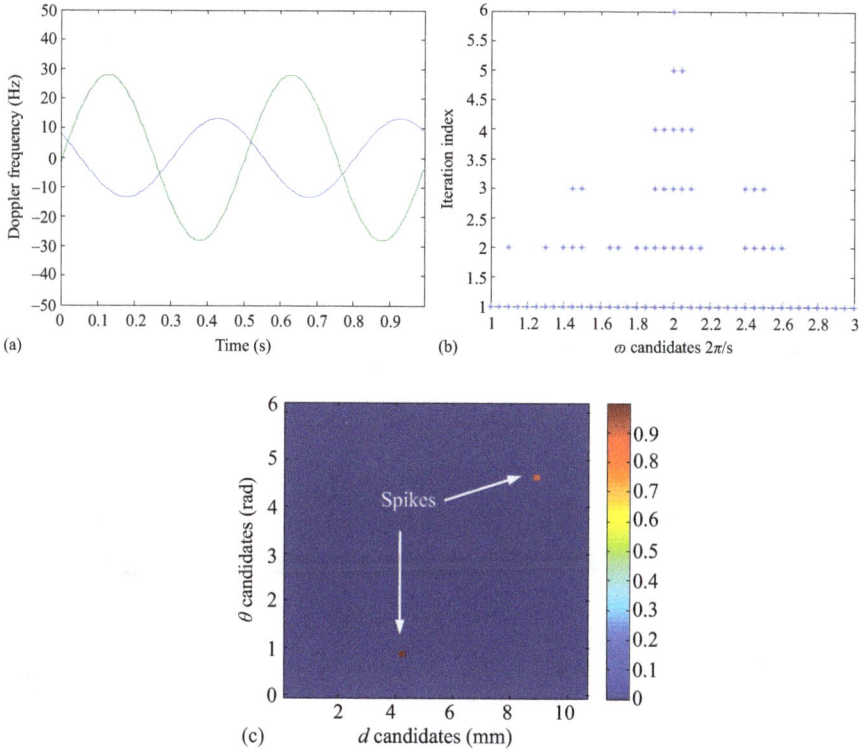

Figure 7.4   Results obtained by the POMP algorithm: (a) time–frequency domain obtained by POMP, (b) pruning process for estimation of ω, and (c) {θ, d} domain obtained by POMP

in Figure 7.4(c), where the spike-like peaks indicating the correct micro-Doppler parameters $\{\theta_1, d_1\}$ and $\{\theta_2, d_2\}$ are obviously sharper than those in Figure 7.2(c) and Figure 7.3(c). Using the estimated micro-Doppler parameters, the time–frequency characteristics are retrieved by substituting the parameter estimates into (7.7). As shown in Figure 7.4(a), the time–frequency characteristics retrieved by the POMP algorithm clearly follow the micro-Doppler signal model and do not suffer from the cross-term problem. The above simulation results demonstrate that all of the PWHT, the Hough-RSPWVD and the POMP algorithms are effective for micro-Doppler parameter estimation and imply that the POMP algorithm may outperform the other two in noisy environments.

7.2.4.2 Analysis of parameter estimation accuracy in noisy environment

In this experiment, we evaluate the accuracy of micro-Doppler parameter estimation of the PWHT, the Hough-RSPWVD, and the POMP algorithm in noisy environments. We consider the signal composed of two micro-Doppler components

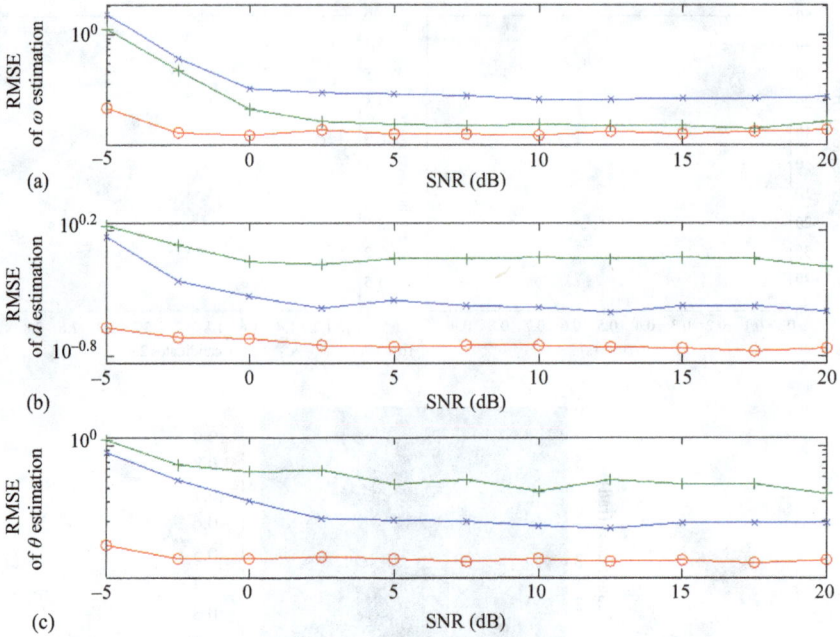

Figure 7.5   *RMSE versus SNR: (a) for estimation of ω; (b) for estimation of d; and (c) for estimation of θ.* ——+——: *PWHT;* ——×——: *Hough-RSPWVD; and* ——○——: *POMP*

with the same angular speed and different Doppler amplitudes and initial phases. Assume that the micro-Doppler parameters randomly vary in specific ranges without any Doppler ambiguity, that is, $2\pi \le \omega \le 6\pi, 0 < d_k \le 10.6$ mm, $0 \le \theta_k < 2\pi$, and $k = 1, 2$. The received signal is assumed to be corrupted by an additive Gaussian noise, and the signal to noise ratio (SNR) is defined as SNR $= \|\mathbf{Y}\|_2^2/(M\sigma^2)$, where σ^2 is the variance of the noise. Figure 7.5 represents the root-mean-squared error (RMSE) of micro-Doppler parameter estimation versus varying SNR, where each point is obtained by averaging over 100 Monte Carlo trials. We can see that the Hough-RSPWVD algorithm is more accurate than the PMHT algorithm in terms of the estimation of $\{\theta, d\}$. The reason is that the 2D low-pass filter in the Hough-RSPWVD algorithm suppresses the cross-term interference well and, therefore, yields more concentrated Hough accumulation result, which exactly agrees with the comparison between Figure 7.2(c) and Figure 7.3(c). However, the 2D low-pass filter in the Hough-RSPWVD algorithm also flattens the energy distribution of the actual signal components in the time–frequency domain, as shown in Figure 7.2(a) and Figure 7.3(a). This is the reason why the PMHT algorithm is better than the Hough-RSPWVD algorithm in terms of the estimation of ω, which agrees with the comparison between Figure 7.2(b) and Figure 7.3(b). The estimation accuracy of the POMP algorithm with the error threshold $\varepsilon = 0.05\|\mathbf{Y}\|_2^2$ is better than that of the PWHT and the Hough-RSPWVD algorithms, thanks to the following two factors. (1) The time–frequency characteristics are represented by a specific kind of analytic functions, so any WVD-like operations for time–frequency analysis are not required. (2) The parametric dictionary makes it possible to map the received signal onto the micro-Doppler domain, so the Hough-like accumulation for pattern recognition is not required. It should also be pointed out that performances of the PWHT, the Hough-RSPWVD, and the POMP algorithms are related to the search grid since both the Hough transform and the pruning operation in POMP are basically based on search in the micro-Doppler parameter space. Generally speaking, when the true value of the parameter of interest is not located on the search grids, the performance of a search based algorithm may degenerate, similar to the sidelobe effect. Here the values of micro-Doppler parameters are randomly selected in some ranges. This corresponds to the general case that the values of micro-Doppler parameters are most likely not located on the discrete search grids. From Figure 7.5, we can see that the accuracy loss caused by off-grid problem is significantly mitigated by the POMP algorithm thanks to the assumption of sparsity of the received signal.

7.2.4.3   Analysis of resolution

In this experiment, we consider two closely spaced micro-Doppler components in the $\{\theta, d\}$ parameter domain and empirically investigate the probability of successful separation. The angular speed of the target is set as $\omega = 4\pi$ rad/s, and one micro-Doppler component is fixed with the parameters $d_1 = 5.3$ mm and $\theta_1 = 0.76\pi$. The second micro-Doppler component is set to be close to the first one with the parameters $d_2 = d_1 + \Delta d$ and $\theta_2 = \theta_1 + \Delta\theta$, where Δd and $\Delta\theta$ denote the gap between the two components along d-axis and θ-axis, respectively. The SNR is set equal to 20 dB. Giving different values of Δd and $\Delta\theta$, we try to separate these two micro-Doppler components with the PWHT, the Hough-RSPWVD, and the POMP algorithms. The separation is regarded as successful if two significant peaks are detected in the neighborhood of (θ_1, d_1). The successful separation rates of these algorithms versus the component gap are plotted in Figure 7.6, where each point is obtained by averaging over 100 Monte Carlo trials. In Figure 7.6(a), $\Delta\theta$ is set equal to zero and Δd is varying; in Figure 7.6(b), Δd is set equal to zero and $\Delta\theta$ is varying. When the two components are closely located with a small difference in the value of θ, the PWHT algorithm has higher successful separation rate than the Hough-RSPWVD algorithm; while when the two components are closely located with a small difference in the value of d, the situation is reversed. The POMP algorithm with the error threshold $\varepsilon = 0.05\|\mathbf{Y}\|_2^2$ has the best successful separation rate among these algorithms for both varying $\Delta\theta$ and varying Δd, and, therefore, better resolution in the micro-Doppler parameter domain. According to (7.7), the time–frequency characteristics are dependent on the micro-Doppler parameters, so Figure 7.6 also implies that the POMP algorithm has higher resolution in the time–frequency domain. As mentioned in previous experiments, the performance superiority of the POMP algorithm comes from the fact that the PSR provides a way to directly estimate the micro-Doppler parameters without any WVD-like and Hough-like operations.

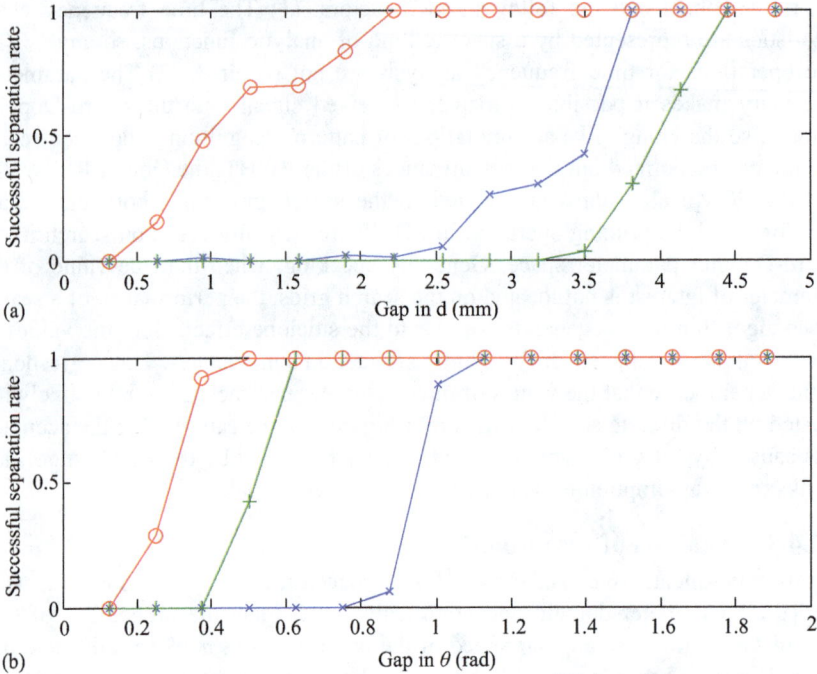

Figure 7.6   Successful separation rate versus component gap: (a) with $\Delta\theta = 0$ and varying Δd; and (b) with $\Delta d = 0$ and varying $\Delta\theta$. ——+——: PWHT; ——×——: Hough-RSPWVD; and ——○——: POMP

7.3   Dynamic hand gesture recognition via Gabor–Hausdorff algorithm

In this section, we introduce how to classify the micro-Doppler signals reflected from dynamic hand gestures. This method can also be applied to radar recognition of other nonrigid body targets.

7.3.1   Measurements of dynamic hand gestures

The measured data are collected by a K-band continuous wave (CW) radar system. The carrier frequency and the base-band sampling frequency are 25 GHz and 1 kHz, respectively. The antenna is oriented directly to the human hand at a distance of 0.3 m. The following four dynamic hand gestures are considered: (a) hand rotation, (b) beckoning, (c) snapping fingers, and (d) flipping fingers. The illustrations and descriptions of the four dynamic hand gestures are shown in Figure 7.7 and Table 7.2, respectively. The data are collected from three personnel targets: two males and one female. Each person repeats a particular dynamic hand gesture for 20 times. Each 0.6 s time interval containing a complete dynamic hand

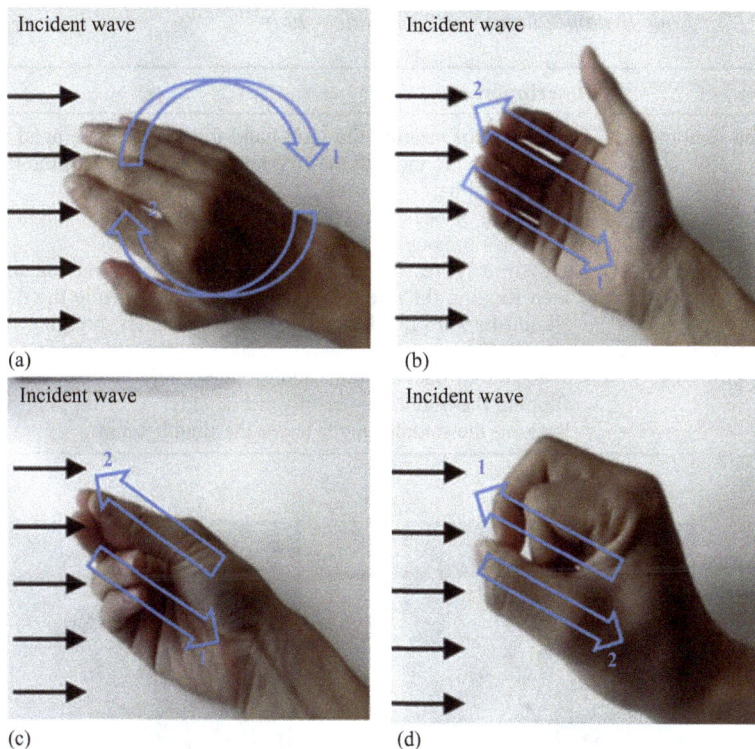

Figure 7.7   Illustrations of four dynamic hand gestures: (a) hand rotation,
(b) beckoning, (c) snapping fingers, and (d) flipping fingers

gesture is recorded as a signal segment. The total number of the signal segments is (4 gestures) $\times$ (3 personnel targets) $\times$ (20 repeats) = 240.

To visualize the time-varying characteristics of the dynamic hand gestures, the STFT with a Kaiser window is applied to the received signals to obtain the corresponding spectrograms. The resulting spectrograms of the four dynamic hand gestures from one personnel target are shown in Figure 7.8. It is clear that the time–frequency trajectories of these dynamic hand gestures are different from each other. The Doppler shifts corresponding to the gesture "hand rotation" continuously change along the time-axis, because the velocity of the hand continuously changes during the rotation process. The radar echo of the gesture "beckoning" contains a negative Doppler shift and a positive Doppler shift, which are corresponding to the back and forth movements of the fingers, respectively. The negative Doppler shift of the gesture "snapping fingers" is larger than its positive Doppler shift since the velocity corresponding to the retreating movement of the fingers is much larger than the velocity corresponding to the returning movement. The time–frequency trajectory of the gesture "flipping fingers" starts with a positive Doppler shift that corresponds to the middle finger flipping towards the radar. The differences among

Table 7.2   Four dynamic hand gestures under study

Gesture	Description
(a) Hand rotation	The gesture of rotating the right hand for a cycle. The hand moves away from the radar in the first half cycle and toward the radar in the second half
(b) Beckoning	The gesture of beckoning someone with the fingers swinging back and forth for one time
(c) Snapping fingers	The gesture of pressing the middle finger and the thumb together and then flinging the middle finger onto the palm while the thumb sliding forward quickly. After snapping fingers, pressing the middle finger and the thumb together again
(d) Flipping fingers	The gesture of bucking the middle finger under the thumb and then flipping the middle finger forward quickly. After flipping fingers, bucking the middle finger under the thumb again

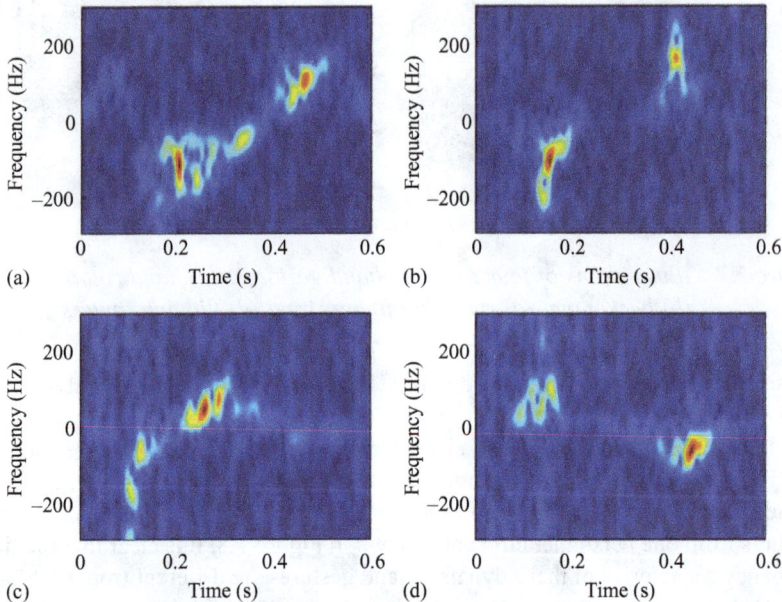

Figure 7.8   Spectrograms of received signals corresponding to four dynamic hand gestures from one personnel target: (a) hand rotation, (b) beckoning, (c) snapping fingers, and (d) flipping fingers

the time–frequency trajectories imply the potential to distinguish different dynamic hand gestures. From Figure 7.8, we can also see that most of the power of the dynamic hand gesture signals are distributed in limited areas in the time–frequency domain. This allows us to use sparse signal processing techniques to extract micro-Doppler features of dynamic hand gestures. Figure 7.9 shows the spectrograms of received signals corresponding to dynamic hand gesture "hand rotation" from three personnel targets. It can be seen that the time–frequency spectrograms of the same gesture from different personnel targets have similar patterns.

7.3.2   Sparsity-driven recognition of hand gestures

The scheme of the sparsity-driven algorithm for hand gesture recognition is illustrated in Figure 7.10. This method contains two subprocesses, that is, the training process and the testing process. The training process is composed of two steps. First, the time–frequency trajectory of each training signal is extracted using the OMP algorithm [17,19]. Second, the K-means algorithm [24,25] is employed to cluster the time–frequency trajectories of all training signals and generate the central trajectory corresponding to each dynamic hand gesture. In the testing process, the modified Hausdorff distances [26,27] between the time–frequency trajectory of the testing signal and the central trajectories of all the hand gesture patterns are computed and inputted into the nearest neighbor classifier to determine the type of the hand gesture under test.

7.3.2.1   Extraction of the time–frequency trajectory

As discussed in Section 7.3.1, the time–frequency distributions of the dynamic hand gesture signals are generally sparse. Denoting the received signal as an $N \times 1$ vector $\mathbf{y}$, the model of the sparse representation of $\mathbf{y}$ in time–frequency domain can be expressed as [26]

$$\mathbf{y} = \mathbf{\Phi}\mathbf{x} + \boldsymbol{\eta} \tag{7.21}$$

where $\mathbf{\Phi}$ is an $N \times M$ time–frequency dictionary, $\mathbf{x}$ is an $M \times 1$ sparse vector, and $\boldsymbol{\eta}$ is an $N \times 1$ noise vector. Assume that $\mathbf{x}$ is a P-sparse signal, that is, there are at most P nonzero entries in $\mathbf{x}$. Here we use the Gabor function in (7.4) to generate the dictionary $\mathbf{\Phi}$. The mth column of dictionary $\mathbf{\Phi}$ can be expressed as

$$\mathbf{\Phi}[:, m] = [\phi_m(1), \phi_m(2), \ldots, \phi_m(N)]^T \tag{7.22}$$

where

$$\phi_m(n) = \phi(n|t_m, f_m) = \left(\pi\sigma^2\right)^{-0.25} \exp\left(-\frac{(n - t_m)^2}{2\sigma^2}\right) \exp(j2\pi f_m n) \tag{7.23}$$

where t_m and f_m represent the time shift and the frequency shift of the Gabor basissignal, respectively, σ denotes the width of the Gaussian window, $n = 1, 2, \ldots, N$.

When $P \ll M < N$, the sparse vector $\mathbf{x}$ in (7.21) can be solved by the OMP algorithm [17,19], and the sparse solution is denoted as

$$\hat{\mathbf{x}} = \text{OMP}(\mathbf{y}, \mathbf{\Phi}, P) = [0, \ldots, \hat{x}_{i_1}, 0, \ldots, \hat{x}_{i_2}, 0, \ldots, \hat{x}_{i_P}, \ldots]^T \tag{7.24}$$

where $\hat{\mathbf{x}}$ is the P-sparse vector and the nonzero element at the time–frequency position (t_{i_p}, f_{i_p}) is $\hat{x}_{i_p}$ for $p = 1, 2, \ldots, P$. This implies that the time–frequency characteristics of $\mathbf{y}$ can be described by a group of basis signals with time–frequency parameters $(t_{i_p}, f_{i_p}, \hat{x}_{i_p})$, $p = 1, 2, \ldots, P$. Based on this observation, we

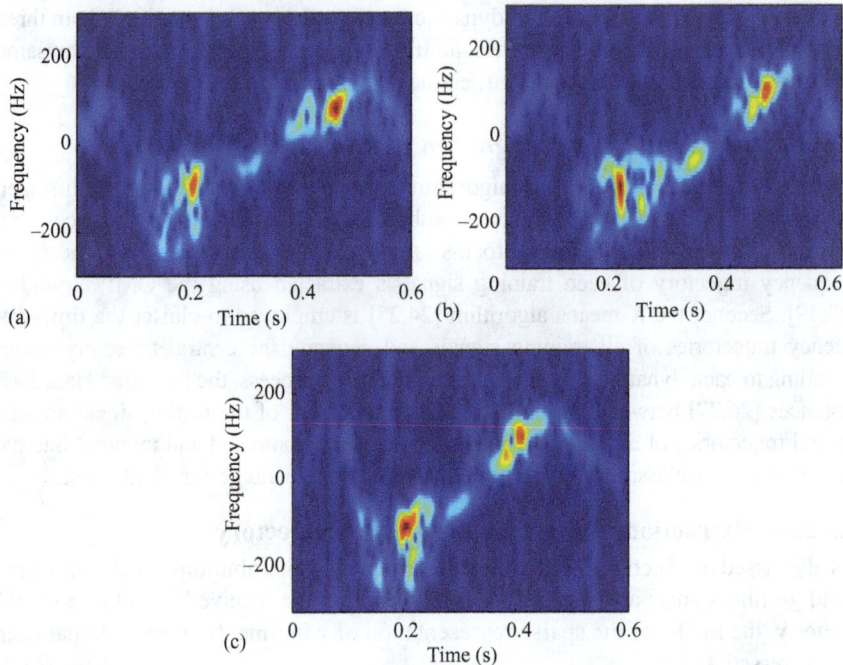

Figure 7.9   *Spectrograms of the received signals corresponding to dynamic hand gesture "hand rotation" from three personnel targets: (a) Target 1, (b) Target 2, and (c) Target 3*

define the time–frequency trajectory of $\mathbf{y}$ as

$$\mathbf{T}(\mathbf{y}) = \{(t_{i_p}, f_{i_p}, A_{i_p}), p = 1, 2, \ldots, P\} \tag{7.25}$$

where $A_{i_p} \triangleq |\widehat{x}_{i_p}|$ indicates the intensity at the time–frequency position (t_{i_p}, f_{i_p}), for $p = 1, 2, \ldots, P$.

To explain the sparse signal representation clearer, the OMP algorithm is applied to analyze the measured signals presented in Figure 7.8. The length of each dynamic hand gesture signal is 0.6 s and the sampling frequency is 1 kHz, which means that the value of N is 600. The sparsity P is set to be 10. The window length parameter σ is set to be 32. The dictionary Φ is designed as discussed above and its size is $600 \times 2,400$. The OMP algorithm is used to solve sparse vector $\widehat{\mathbf{x}}$, and then the reconstructed signal is obtained by $\mathbf{y}_{rec} = \Phi \widehat{\mathbf{x}}$. The time–frequency spectrograms of the reconstructed signals $\mathbf{y}_{rec}$ are plotted in Figure 7.11. By comparing Figure 7.8 and Figure 7.11, we can find that the reconstructed signals contain the majority part of the original time–frequency features. In addition, it is clear that the noise energy has been significantly suppressed in the reconstructed signals, which is helpful for dynamic hand gesture recognition. The locations of time–frequency trajectory, that is, (t_{i_p}, f_{i_p}), $p = 1, 2, \ldots, P$, extracted by OMP are plotted in

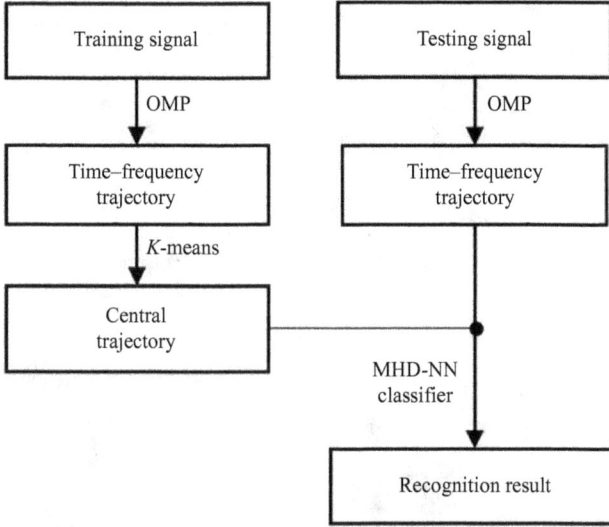

*Figure 7.10 The flowchart the sparsity-driven algorithm for hand
          gesture recognition*

Figure 7.12. By comparing Figure 7.8 and Figure 7.12, we can see that the extracted
time–frequency trajectories are capable of representing the time–frequency patterns
of corresponding dynamic hand gestures.

7.3.2.2   Clustering for central time–frequency trajectory

In the training process, a central time–frequency trajectory is clustered for each
dynamic hand gesture by using the K-means algorithm [24,25] based on the time–
frequency trajectories of the training signals. The details of clustering process are
introduced below.

Assume there are S segments of training signals for each dynamic hand ges-
ture, denoted as $\mathbf{y}_g^{(s)}$ ($s = 1, 2, \ldots, S$), where s and g denote the indices of training
segments and dynamic hand gestures, respectively. The time–frequency trajectory
of $\mathbf{y}_g^{(s)}$ is denoted as $\mathrm{T}\left(\mathbf{y}_g^{(s)}\right)$, which is composed of P time–frequency positions as
presented in (7.25). In an ideal case, different realizations of a certain dynamic
hand gesture are expected to have the same time–frequency trajectory. However, in
realistic scenarios, a human can hardly repeat one dynamic hand gesture in completely the same way. Therefore, there are minor differences among the time–
frequency trajectories extracted from different realizations of one dynamic hand
gesture. In order to explain this phenomenon more clearly, we plot the time–frequency
trajectories extracted from eight segments of signals corresponding to the gesture
"snapping fingers" in Figure 7.13(a) for an example. It is clear that the time–
frequency trajectories of different signal segments are distributed closely to each other
with slight differences. In order to extract the main pattern from the time–frequency
trajectories of training data, the K-means algorithm [24,25], which is a clustering

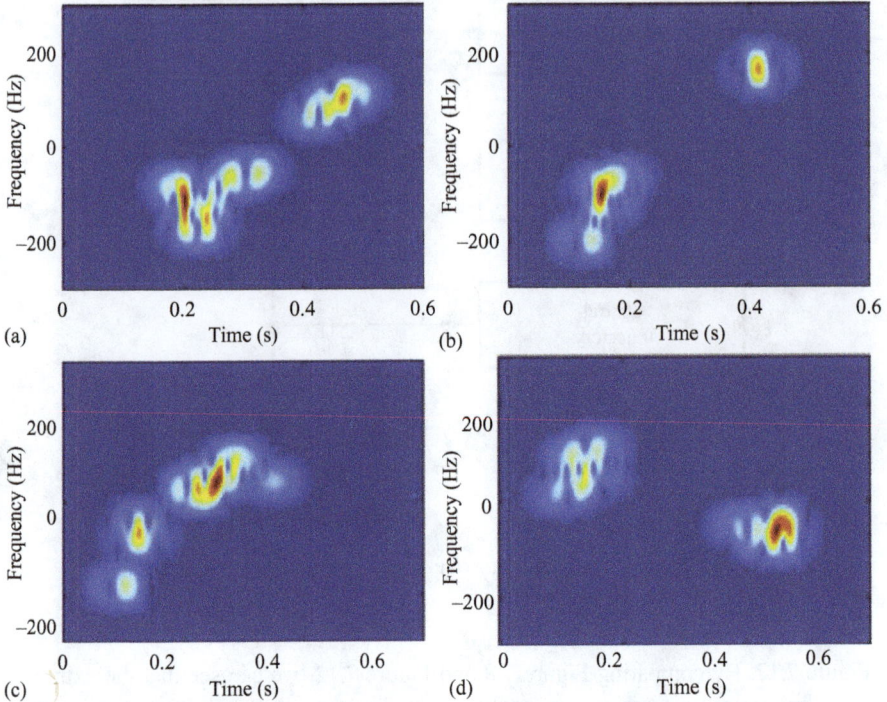

Figure 7.11   Spectrograms of the signals reconstructed by OMP with P = 10:
(a) hand rotation, (b) beckoning, (c) snapping fingers, and
(d) flipping fingers

technique widely used in pattern recognition, is employed to generate the central
time–frequency trajectory of each dynamic hand gesture. The inputs of the K-means
algorithm are the time–frequency trajectories of S training signals, and accordingly,
the total number of input time–frequency positions is $P \times S$. With the K-means
algorithm, P central time–frequency positions are produced to minimize the mean
squared distance from each input time–frequency position to its nearest central
position. We denote the central time–frequency trajectory of dynamic hand gesture
g generated by the K-means algorithm as

$$\begin{aligned}
\mathrm{T}_{c,g} &= K\text{-means}\left(\mathrm{T}\left(\mathbf{y}_g^{(1)}\right), \mathrm{T}\left(\mathbf{y}_g^{(2)}\right), \dots, \mathrm{T}\left(\mathbf{y}_g^{(S)}\right)\right) \\
&= \left\{\left(t_{c,g}^{(p)}, f_{c,g}^{(p)}, A_{c,g}^{(p)}\right), p = 1, 2, \dots, P\right\}
\end{aligned} \tag{7.26}$$

where $t_{c,g}^{(p)}, f_{c,g}^{(p)}$, and $A_{c,g}^{(p)}$ denote the time shift, the frequency shift, and the magnitude of the pth time–frequency position on the central time–frequency trajectory,
respectively, and the superscript g is the dynamic hand gesture index. Figure 7.13(b)
shows the location of the central time–frequency trajectory generated by the K-means

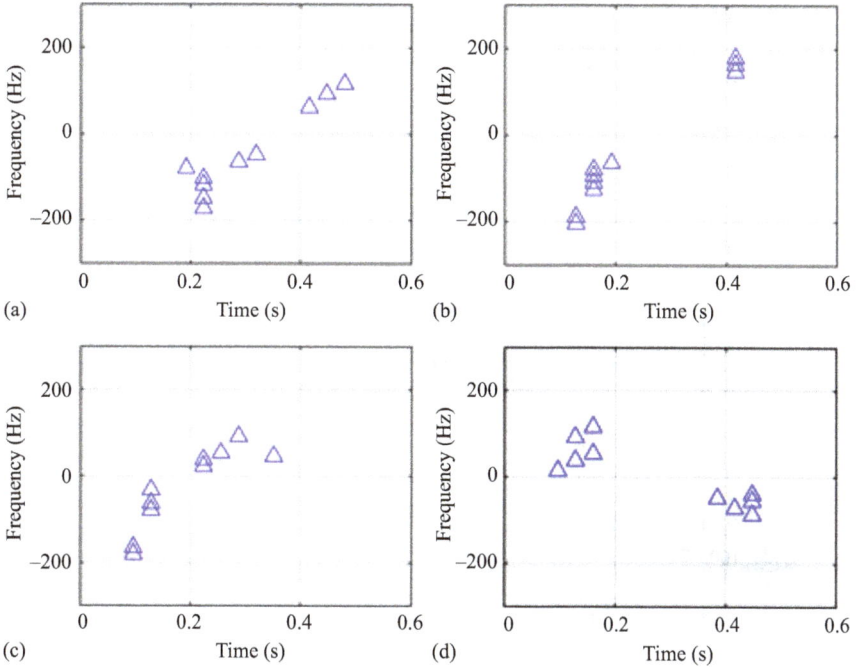

Figure 7.12 *The time–frequency trajectories extracted by OMP with P = 10:*
(a) hand rotation, (b) beckoning, (c) snapping fingers, and
(d) flipping fingers

algorithm. It is clear that the majority of time–frequency positions in Figure 7.13(a) are located around the central time–frequency trajectory in Figure 7.13(b), which implies that the central time–frequency trajectory is capable of representing the time–frequency pattern of a dynamic hand gesture.

7.3.2.3 The nearest neighbor classifier based on modified Hausdorff distance

In the testing process, the type of dynamic hand gesture corresponding to a given testing signal is determined by the NN classifier. The modified Hausdorff distance [26,27] is used to measure the similarity between the time–frequency trajectory of the testing signal and the central time–frequency trajectory of each dynamic hand gesture.

For a testing signal $\mathbf{y}_{test}$, the classification process can be divided into three steps.

1. The time–frequency trajectory $T(\mathbf{y}_{test})$ is extracted by using the OMP algorithm as described in Section 7.3.2.1.

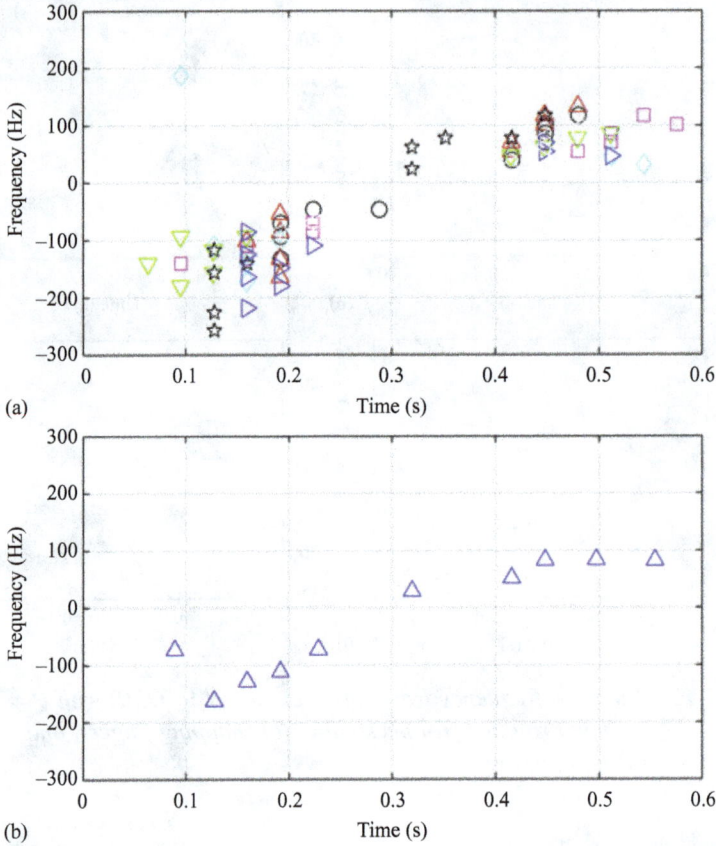

(a)

(b)

Figure 7.13   (a) Time–frequency trajectories extracted from eight segments of signals corresponding to the gesture "snapping fingers" with P = 10, where different markers indicate different signal segments. (b) The clustered time–frequency trajectory generated by the K-means algorithm

2. The modified Hausdorff distances between $T(\mathbf{y}_{\text{test}})$ and the central time–frequency trajectories $T_{c,g}$ are computed as [26]

$$\text{MHD}\big(T(\mathbf{y}_{\text{test}}), T_{c,g}\big) = \sum_{\tau \in T(\mathbf{y}_{\text{test}})} d_H\big(\tau, T_{c,g}\big) \qquad (7.27)$$

where τ is an element in set $T(\mathbf{y}_{\text{test}})$, that is, τ is a parameter set composed of the time shift, the frequency shift and the amplitude as described in (7.25), and $d_H(\cdot, \cdot)$ represents the Hausdorff distance, which is defined as [27]

$$d_H\big(\tau, T_{c,g}\big) = \min_{z \in T_{c,g}} \|\tau - z\|_2 \qquad (7.28)$$

where z denotes the element in set $T_{c,g}$.

3. The type of the dynamic hand gesture corresponding to $\mathbf{y}_{\text{test}}$ is determined by the following NN classifier:

$$\widehat{g} = \underset{g \in \{1,2,...,G\}}{\arg \min} \ \text{MHD}\left(\text{T}(\mathbf{y}_{\text{test}}), \text{T}_{c,g}\right) \tag{7.29}$$

where G represents the total number of dynamic hand gestures and $\widehat{g}$ is the recognition result, that is, the index of the determined gesture.

7.3.3 Experimental results

In this section, the real data measured with the K-band radar are used to validate the sparsity-driven algorithm of hand gesture recognition. The recognition accuracy is defined as the proportion of correctly recognized dynamic hand gesture signals among all the testing signals.

7.3.3.1 Recognition results with varying sparsity

In this experiment, we evaluate the recognition accuracy of the sparsity-driven algorithm of hand gesture recognition with different values of sparsity P. The Gabor–Hausdorff algorithm is compared with the Sparse-SVM method [28] in terms of the recognition accuracy. With the Sparse-SVM method, the time–frequency trajectories of the dynamic hand gestures are extracted by the OMP algorithm and inputted into support vector machine (SVM) for recognition. The sparsity P is varied from 7 to 21 with a step size of 2, and the recognition accuracy is calculated using cross-validation. For each value of sparsity P, a certain proportion of measured data are randomly selected from the training data set, and the remaining data are used for testing. The recognition accuracy is averaged over 50 trials with randomly selected training data. The window width σ in the Gobar function is set to be 32. The recognition performances yielded by the Gabor–Hausdorff algorithm and the Sparse-SVM method are illustrated in Figure 7.14, and the confusion matrix yielded by the Gabor–Hausdorff algorithm with $P = 17$ with 30% training data are presented in Table 7.3.

It is clear from Figure 7.14 that the recognition accuracies of the Gabor–Hausdorff algorithm increases as the sparsity P increases when $P \leq 15$. This is because more features of the dynamic hand gestures are extracted as the sparsity P increases. The recognition accuracies change slightly as the sparsity P increases when $P \geq 17$. This is because no more useful features can be extracted in this condition. Therefore, the sparsity P is selected to be larger than 15 for the experimental dataset to achieve satisfying recognition accuracy. If the Gabor–Hausdorff algorithm is applied to other datasets, the sparsity P should be selected large enough to extract the micro-Doppler features sufficiently. However, a too large value of sparsity P results in more computational burden. In order to determine the proper value of sparsity P, the Gabor–Hausdorff algorithm for dynamic hand gesture recognition can be evaluated under different values of sparsity P by conducting multifold validation within the training dataset off-line. The value of sparsity P

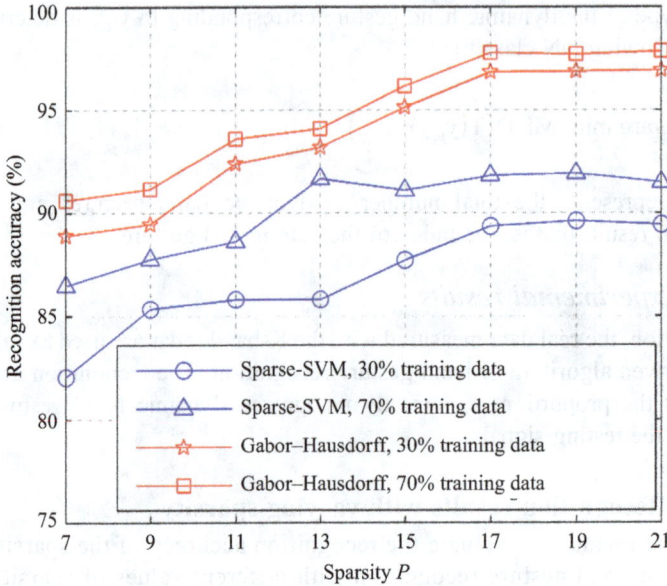

Figure 7.14   Recognition accuracies of the Gabor–Hausdorff algorithm and the Sparse-SVM method versus the value of sparsity P

Table 7.3   Confusion matrix yielded by the Gabor–Hausdorff algorithm with $p = 17$ and 30% of data for training

	Hand rotation (%)	Beckoning (%)	Snapping fingers (%)	Flipping fingers (%)
Hand rotation	96.67	2.32	0.72	0
Beckoning	3.21	95.42	3.45	0
Snapping fingers	0.12	2.26	95.71	0
Flipping fingers	0	0	0.12	100

corresponding to the highest recognition accuracy can be selected in the final recognition system. In addition, it can be seen from Figure 7.14 that the Gabor–Hausdorff algorithm outperforms the Sparse-SVM method under each sparsity value.

7.3.3.2   Recognition results with varying time–frequency resolution

In this experiment, the performance of the Gabor–Hausdorff algorithm versus the window length of the Gabor function is evaluated under different proportions of training data and different values of sparsity. The change of the window length in the Gabor function results in varying time–frequency resolution. The window length σ in the Gabor function varies from 8 to 48 with a step size of 8. The sparsity P is chosen from $\{17, 19, 21\}$. The recognition accuracies yielded by the Gabor–Hausdorff algorithm using 30% and 70% of data for training are illustrated in Figure 7.15(a) and (b), respectively. It can be seen that the Gabor–Hausdorff algorithm is pretty robust against the change of the time–frequency resolution and achieves the best performance when $16 \leq \sigma \leq 40$, which implies that the Gabor–Hausdorff algorithm is quite robust to the window length of the time–frequency dictionary. The performance of the Gabor–Hausdorff algorithm declines when σ is less than 16 or larger than 40. The reason is that the frequency resolution or the time resolution of the Gabor function is poor when the window length is too small or too large, which leads to the quality reduction of micro-Doppler feature extraction.

7.3.3.3 Recognition results with the varying size of training data

In this experiment, the performance of the Gabor–Hausdorff algorithm is analyzed with different sizes of training dataset. We compare the recognition accuracies yielded by the Gabor–Hausdorff algorithm with that yielded by the Sparse-SVM method, the PCA-based methods, and the DCNN-based method. With the PCA-based methods, the micro-Doppler features of dynamic hand gestures are obtained by extracting the principal components of the received signals and inputted into SVM for recognition. Two kinds of PCA-based methods are considered here: (1) PCA in the time-domain, which extracts the features from the time-domain data [29]; and (2) PCA in the time–frequency domain, which extracts the features in the time–frequency domain as presented in [30,31]. As for the DCNN-based method, the time–frequency spectrograms are fed into a deep convolutional neural network, where the micro-Doppler features are extracted using convolutional filters and the recognition is performed through fully connected perceptron functions. The structure of the DCNN used here is similar to that used in [32]. The proportions of training data are set to be varied from 10% to 90% with a step size of 10%, the sparsity P is set to be 17, and the window width σ of the Gabor function is set to be 32. The recognition accuracies of these algorithms are shown in Figure 7.16, where the Gabor–Hausdorff algorithm obtains the highest recognition accuracies under different sizes of training set. In addition, the advantages of Gabor–Hausdorff algorithm over the PCA-based and the DCNN-based methods are remarkable especially when the proportion of the training data is less than 50%. This implies the Gabor–Hausdorff algorithm is more applicable than the PCA-based and the DCNN-based methods when the training set is small.

7.3.3.4 Recognition results for unknown personnel targets

During the data collection, the dynamic hand gesture signals are measured from three personnel targets, denoted as Targets 1, 2, and 3, respectively. In the above experiments, the data measured from all the three personnel targets are mixed together, and a part of the data is used for training and the remaining data are used for testing. In this experiment, the data measured from one of Targets 1, 2, and 3 are used for training, and the data measured from the other two personnel targets are used for testing. This experiment aims to validate the Gabor–Hausdorff algorithm

(a)

(b)

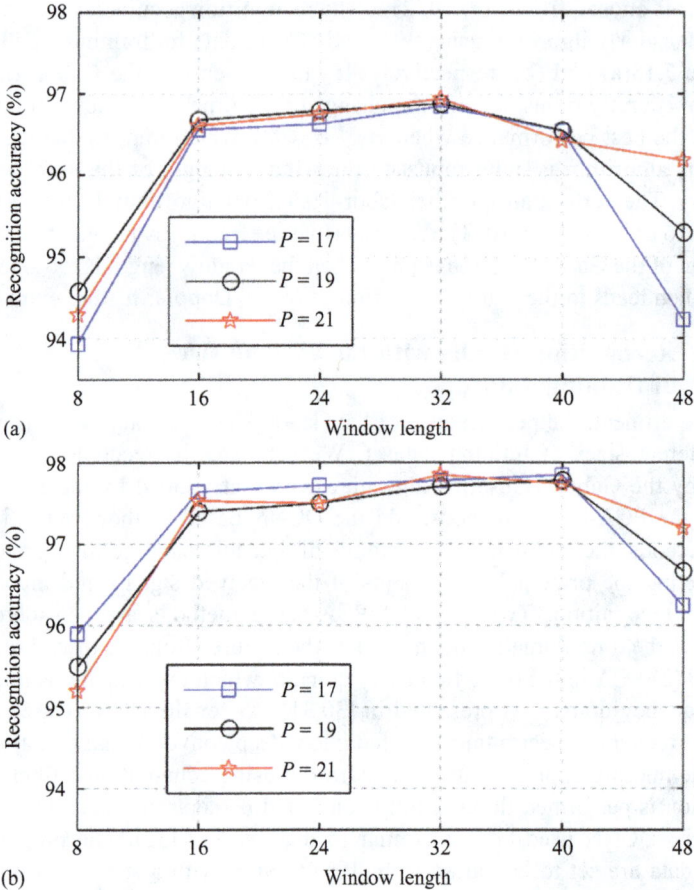

Figure 7.15   Recognition accuracy yielded by the Gabor–Hausdorff algorithm versus the window width of the Gobar function: (a) using 30% of data for training; and (b) using 70% of data for training

in the condition of recognizing the dynamic hand gestures of unknown personnel targets. Cross-validation is employed in this experiment. We randomly select 70% of the data measured from one of Targets 1, 2, and 3 for training and 70% of the data measured from the other two personnel targets for testing. The recognition accuracy is calculated by averaging over 50 trials with randomly selected training data and testing data. The sparsity P is set to be 17, and the window length σ of the Gabor function is set to be 32. The resulting recognition accuracies of five algorithms are listed in Table 7.4. It can be seen that the Gabor–Hausdorff algorithm produces the highest recognition accuracies under all conditions. This implies that the Gabor–Hausdorff algorithm is superior to the Sparse-SVM method, the PCA-based methods, and the DCNN-based method in terms of recognizing dynamic hand gestures of unknown personnel targets.

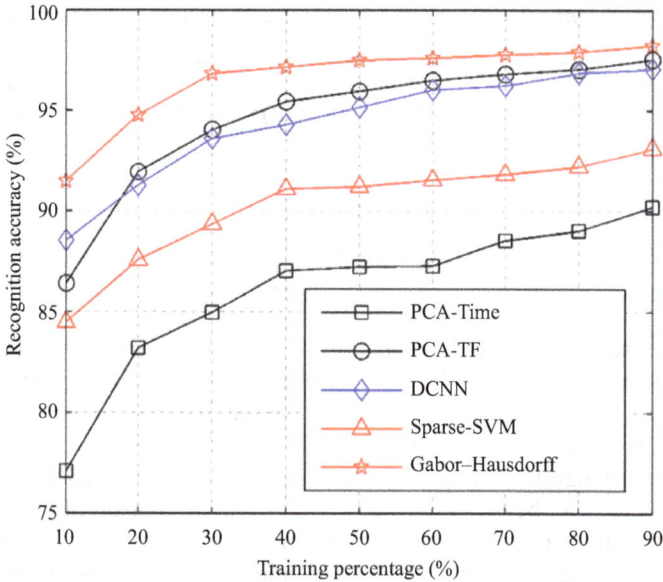

*Figure 7.16   Recognition accuracies yielded by the Gabor–Hausdorff algorithm,
the Sparse-SVM method, the time-domain PCA-based method
(denoted as PCA-Time), the time–frequency domain PCA-based
method (denoted as PCA-TF), and the DCNN-based method (denoted
as DCNN) under different sizes of training dataset*

Table 7.4   Recognition accuracies for unknown personnel targets

Algorithms	Training data from Target 1 (%)	Training data from Target 2 (%)	Training data from Target 3 (%)
Gabor–Hausdorff	96.96	96.88	95.48
Sparse-SVM	90.54	90.54	88.21
DCNN	94.38	95.25	91.87
PCA-TF	92.14	91.80	91.14
PCA-Time	84.68	84.59	81.07

7.3.3.5   Analysis of the computational complexity

The computational time consumed by the dynamic hand gesture recognition methods is measured in this experiment. The hardware platform is a laptop with an Intel(R) Core(TM) i5-4200M CPU inside, and the CPU clock frequency and the memory size are 2.5 GHz and 3.7 GB, respectively. The software platform is MATLAB® 2014a and the operating system is Windows 10. For each method, the running time for training and classifying is measured by averaging over 100

Table 7.5   Time consumption of the dynamic hand gesture recognition methods

	Training time for one sample (ms)	Testing time for one sample (ms)
Gabor–Hausdorff	650	220
Sparse-SVM	130	154
DCNN	850	18
PCA-TF	4	11
PCA-Time	0.1	0.5

trials. The results of running time are presented in Table 7.5. It can be seen that the Gabor–Hausdorff algorithm and the DCNN-based method consume the longest time for classifying and training dynamic hand gesture signals, respectively, among all the tested approaches. In realistic applications, the dynamic hand gesture recognition needs to be real-time processing. Considering that the training process can be accomplished off-line, the bottleneck of real-time processing is the time consumption of the classifying phase. Since the running time for classifying one hand gesture by the Gabor–Hausdorff algorithm with the nonoptimized MATLAB code is only 0.22 s, it is promising to achieve real-time processing with optimized code on embedded computing platforms in practical applications.

7.4 Conclusion

In this chapter, two sparsity-driven algorithms of micro-Doppler analysis were presented for radar classification of rigid-body and nonrigid body targets, respectively. The first algorithm aimed to accurately estimate the micro-Doppler parameters of a rigid-body target. A parametric dictionary, which is dependent on the unknown angular speed of the target, was designed to decompose the radar echo into several dominant micro-Doppler components. By doing so, the problem of micro-Doppler parameter estimation was converted into the problem of sparse signal recovery with a parametric dictionary. To avoid the time-consuming full search, the POMP algorithm was presented by embedding the pruning process into the OMP procedure. Simulation results have demonstrated that the POMP algorithm can yield more accurate micro-Doppler parameter estimates and better time–frequency resolution in comparison with some well-recognized algorithms based on WVD and Hough transform. The second algorithm, referred to as the Gabor–Hausdorff algorithm, was presented for micro-Doppler feature extraction and applied to radar recognition of nonrigid body targets such as hand gestures. Taking advantage of the sparse properties of radar echoes reflected from dynamic hand gestures, the Gabor decomposition was used to extract the time–frequency locations and corresponding coefficients of the dominant signal components. The extracted micro-Doppler features were inputted into modified-Hausdorff-distance-based NN classifier to determine the type of dynamic hand gestures. Experimental results based on real radar data have shown that the Gabor–Hausdorff algorithm outperforms the PCA-based and the DCNN-based methods in conditions of small training dataset.

References

[1] Chen V. C., Li F., Ho S.-S., and Wechsler H. "Micro-Doppler effect in radar: Phenomenon, model, and simulation study." *IEEE Transactions on Aerospace and Electronic Systems*. 2006; 42(1): 2–21.

[2] Chen V. C. *The micro-Doppler effect in radar*. London: Artech House, 2011.

[3] Clemente C., Balleri A., Woodbridge K., and Soraghan J. "Developments in target micro-Doppler signatures analysis: Radar imaging, ultrasound and through-the-wall radar." *EURASIP Journal on Advances in Signal Processing*. 2013; 47.

[4] Wang Y., Ling H., and Chen V. C. "ISAR motion compensation via adaptive joint time–frequency technique." *IEEE Transactions on Aerospace and Electronic Systems*. 1998; 34(2): 670–677.

[5] Trintinalia L. C. and Ling H. "Joint time–frequency ISAR using adaptive processing." *IEEE Transactions on Antennas and Propagation*. 1997; 45(2): 221–227.

[6] Orović I., Stanković S., and Amin M. G. "Compressive sensing for sparse time–frequency representation of nonstationary signals in the presence of impulsive noise." *Proceedings of SPIE 8717, Compressive Sensing II, 87170A*. Baltimore, Maryland, USA. SPIE Defense, Security and Sensing, 2013, doi:10.1117/12.2015916.

[7] Amin M. G., Jokanovic B., Zhang Y. D., and Ahmad F. "A sparsity-perspective to quadratic time–frequency distributions." *Digital Signal Processing*. 2015; 46: 175–190.

[8] Hough P. V. C. "Methods and means for recognizing complex patterns." U.S. Patent 069 654. U.S. Patent Office, Washington, DC, 1962.

[9] Barbarossa S. and Lemoine O. "Analysis of nonlinear FM signals by pattern recognition of their time–frequency representation." *IEEE Signal Processing Letters*. 1996; 3(4): 112–115.

[10] Cirillo L., Zoubir A., and Amin M. G. "Parameter estimation for locally linear FM signals using a time–frequency Hough transform." *IEEE Transactions on Signal Processing*. 2008; 56(9): 4162–4175.

[11] Li G. and Varshney P. K. "Micro-Doppler parameter estimation via parametric sparse representation and pruned orthogonal matching pursuit." *IEEE Journal of Selected Topics in Applied Earth Observations and Remote Sensing*. 2014; 7(12): 4937–4948.

[12] Li G., Zhang R., Ritchie M., and Griffiths H. "Sparsity-driven micro-Doppler feature extraction for dynamic hand gesture recognition." *IEEE Transactions on Aerospace and Electronic Systems*. 2018; 54(2): 655–665.

[13] Li G., Zhang H., Wang X., and Xia X.-G. "ISAR 2-D imaging of uniformly rotating targets via matching pursuit." *IEEE Transactions on Aerospace and Electronic Systems*. 2012; 48(2): 1838–1846.

[14] Rao W., Li G., Wang X., and Xia X.-G. "Adaptive sparse recovery by parametric weighted L1 minimization for ISAR imaging of uniformly rotating targets." *IEEE Journal of Selected Topics in Applied Earth Observations and Remote Sensing*. 2013; 6(2): 942–952.

[15] Chen Y., Li G., Zhang Q., and Sun J. "Refocusing of moving targets in SAR images via parametric sparse representation." *Remote Sensing*. 2017; 9: 795.

[16] Luo Y., Zhang Q., Qiu C., Li S., and Yeo T.-S. "Micro-Doppler feature extraction for wideband imaging radar based on complex image orthogonal matching pursuit decomposition." *IET Radar, Sonar & Navigation*. 2013; 7(8): 914–924.

[17] Tropp J. A. and Gilbert A. C. "Signal recovery from random measurements via orthogonal matching pursuit." *IEEE Transactions on Information Theory*. 2007; 53(12): 4655–4666.

[18] Candes E., Romberg J., and Tao T. "Stable signal recovery from incomplete and inaccurate measurements." *Communications on Pure and Applied Mathematics*. 2006; 59(8): 1207–1223.

[19] Tropp J. A., Gilbert A. C., and Strauss M. J. "Algorithms for simultaneous sparse approximation. Part I: Greedy pursuit." *Signal Processing*. 2006; 86(3): 572–588.

[20] Engan K. and Husøy S. A. H. "Method of optimal directions for frame design." *Proceedings of IEEE International Conference on Acoustics, Speech, Signal Processing*. Phoenix, AZ, USA: IEEE; 1999; vol. 5, pp. 2443–2446.

[21] Bryta O. and Elad M. "Compression of facial images using the K-SVD algorithm." *Journal of Visual Communication and Image Representation*. 2008; 19(4): 270–282.

[22] Dai W., Xu T., and Wang W. "Simultaneous codeword optimization (SimCO) for dictionary update and learning." *IEEE Transactions on Signal Processing*. 2012; 60(12): 6340–6353.

[23] Li G., Zhang R., Rao W., and Wang X. "Separation of multiple micro-Doppler components via parametric sparse recovery." *Proceedings of IEEE Geoscience and Remote Sensing Symposium (IGARSS)*. Melbourne, VIC, Australia: IEEE; 2013, pp. 2978–2981.

[24] Hastie T., Tibshirani R., and Friedman J. *The elements of statistical learning: Data mining, inference, and prediction*. Berlin, Germany: Springer Series in Statistics, 2009.

[25] Kanungo T., Mount D. M., Netanyahu N. S., Piatko C. D., Silverman R., and Wu A. Y. "An efficient k-means clustering algorithm: Analysis and implementation." *IEEE Transactions on Pattern Analysis and Machine Intelligence*. 2002; 24(7): 881–892.

[26] Dubuisson M. P. and Jain A. K. "A modified Hausdorff distance for object matching." *Proceedings of the International Conference on Pattern Recognition (ICPR'94)*. Jerusalem, Israel: IEEE; 1994; pp. 566–568.

[27] Huttenlocher D. P., Klanderman G. A., and Rucklidge W. J. "Comparing images using the Hausdorff distance." *IEEE Transactions on Pattern Analysis and Machine Intelligence*. 1993; 15(9): 850–863.

[28] Li G., Zhang R., Ritchie M., and Griffiths H. "Sparsity-based dynamic hand gesture recognition using micro-Doppler signatures." *Proceeding of 2017 IEEE Radar Conference*. Seattle, WA, USA: IEEE; 2017, pp. 0928–0931.

[29] Balleri A., Chetty K., and Woodbridge K. "Classification of personnel targets by acoustic micro-Doppler signatures." *IET Radar, Sonar & Navigation*. 2011; 5(9): 943–951.

[30] Wu Q., Zhang Y. D., Tao W., and Amin M. G. "Radar-based fall detection based on Doppler time–frequency signatures for assisted living." *IET Radar, Sonar & Navigation*. 2015; 9(2): 164–172.

[31] Jokanovic B., Amin M. G., Ahmad F., and Boashash B. "Radar fall detection using principal component analysis." SPIE Defense+ Security. International Society for Optics and Photonics, Baltimore, Maryland, USA: SPIE; May 2016; pp. 982919.

[32] Kim Y. and Toomajian B. "Hand gesture recognition using micro-Doppler signatures with convolutional neural network." *IEEE Access*. 2016: 7125–7130.

Chapter 8
Distributed detection of sparse signals in radar networks via locally most powerful test

8.1 Introduction

Radar networks are composed of a number of spatially distributed radar transmitters and receivers that observe the same region and operate in a collaborative way [1–6]. The radar networks can offer not only the spatial diversity but also waveform diversity, frequency diversity, and polarization diversity. The concepts of multistatic radar, bistatic radar, and multi-input-multi-output (MIMO) radar can be viewed as the special cases of radar networks. The data or the features acquired at all the local radar sensors are fused to improve the performance of target localization, detection, and classification.

The radar networks are called "active" if the signals are originally designed for radar tasks and the region of interest is illuminated by the emission from the radar transmitters. The radar networks can also work in a "passive" fashion, in which the emission from other radio frequency systems is used to illuminate the region of interest. The typical signals used in passive radar networks include the Wi-Fi signals, the digital video broadcasting signals, the frequency-modulation radio signals, the mobile communication signals, and the Global Navigation Satellite System (GNSS) signals [7].

An essential task of radar sensor networks is distributed signal detection. In practice, it is very common that the signals to be detected are sparsely distributed in time domain, spatial domain, or frequency domain. This inspires us to achieve sparse signal detection in radar sensor networks via compressed sensing (CS) approaches. The CS theories and methods were originally proposed for sparse signal recovery with a small number of measurements instead of signal detection. Besides the CS-based signal recovery, another emerging need is the extraction of test statistics from the compressed measurements [8,9] or from partly/completely reconstructed sparse signals [10–14] for sparse signal detection. In [11], it is also shown that sparse signal detection requires much fewer measurements than sparse signal reconstruction.

CS-based methods have resulted in an efficient scheme of sparse signal detection with sensor networks. Instead of high dimensional data, compressed measurements can be transmitted within the sensor network and further processed to save resource usage. In [10,12–14], the problem of detection of sparse signals with sensor networks has been investigated by using compressed measurements. In [10], a generalized likelihood ratio test (GLRT) is introduced for the detection of sparse signals in sensor networks, where the sparse signal recovery is integrated into the detection framework. Two variants of distributed orthogonal matching pursuit (DOMP) are proposed in [12], where the test statistic is derived based on partial estimation of the support set. In [13,14], the distributed subspace pursuit (DSP) algorithm is proposed for sparse signal detection by exploiting the joint sparsity pattern. All the above methods of detection of sparse signals assume that the observed sparse signals are deterministic.

Here we consider the detection of sparse stochastic signals with radar sensor networks. The sparse signals are assumed to follow the Bernoulli-Gaussian (BG) distribution, which has been widely used for sparsity modeling in existing algorithms of signal recovery and detection [8,15–18]. In the BG model, the sparsity degree is zero if the signals are absent, while it is nonzero but unknown in the presence of sparse signals. From this observation, the problem of detection of sparse signals in radar sensor networks can be formulated as the problem of determining whether the parameter sparsity degree is zero. This problem is equivalent to one-sided hypothesis testing since the sparsity degree is non-negative. It is also equivalent to close hypothesis testing problem [19,20] since the sparsity degree is close to zero when sparse signals are present.

In this chapter, we present a detector based on the locally most powerful test (LMPT) to determine the presence or absence of sparse signals with radar sensor networks [21]. LMPT has been applied to solve the problems of one-sided and close hypothesis testing [19,22]. Different from those methods of sparse signal detection that require partial or complete reconstruction of sparse signals, the LMPT detector operates without any requirement of signal reconstruction. This ensures that the LMPT detector is computationally efficient. The asymptotic detection performance of the LMPT detector is theoretically analyzed. Simulation results corroborate our theoretical analysis and show that the LMPT detector can produce almost the same detection performance as the DOMP-based detector [12] that requires partial reconstruction of sparse signals, but with much less computational burden. This LMPT detector is derived based on the assumption that the measurements collected at all the local radar sensors are high-precision data. In this case, the distributed detector based on LMPT is called "original LMPT detector" through this chapter.

In practice, stringent bandwidth constraints necessitate the transmission of quantized data within the radar sensor networks. The extreme case is 1-bit quantization, where only the signs of the measurements are maintained at all the local sensors. Distributed detection using coarsely quantized data significantly reduces the hardware cost [23,24] and alleviates the data transmission overhead [10,25]. However, there is a notable performance gap between the detector based on coarsely quantized measurements and the original detector based on high-precision measurements, due to the considerable information loss caused by the quantization [26,27]. In this chapter, the LMPT detector is further extended to the case where only coarsely quantized data are available in the radar sensor network. The corresponding distributed detector is called "quantized LMPT detector" [28] throughout this chapter. The quantization thresholds

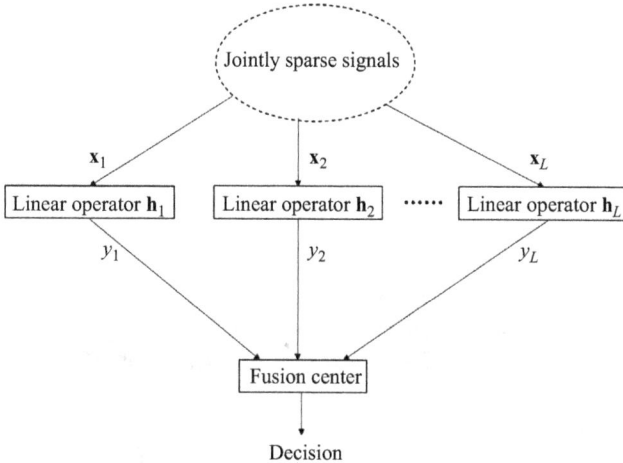

Figure 8.1 Distributed detection system

at the local radar sensors are specifically designed to ensure the near-optimal detection performance of the quantized LMPT detector. It will be shown that (1) the quantized LMPT detector with $3.3L$ 1-bit measurements achieves approximately the same detection performance as the original LMPT detector with L measurements, and (2) the quantized LMPT detector with 3-bit measurements can achieve almost the same detection performance as the original LMPT detector.

8.2 The original LMPT detector

In this section, the distributed detector based on LMPT is formulated for the detection of the sparse signals from high-precision measurements.

8.2.1 Problem formulation

Consider the problem of detecting sparse stochastic signals in the presence of noise based on high-precision measurements collected by the distributed radar sensors. The detection system is illustrated in Figure 8.1. There are L independent measurement vectors that share the joint sparsity pattern, which holds for multiple radar nodes observing the same region of interest. Assume that there are G independent radar sensors and the gth sensor acquires L_g independent measurements. Then the total number of independent measurements is $L = \sum_{g=1}^{G} L_g$. The observation model under hypotheses H_1 (the signal is present) and H_0 (the signal is absent) is given by as

$$\begin{cases} H_0 : y_l = w_l & l = 1, 2, \dots, L \\ H_1 : y_l = \mathbf{h}_l^T \mathbf{x}_l + w_l & l = 1, 2, \dots, L \end{cases} \tag{8.1}$$

where l denotes the index of the measurement, $y_l \in \mathbb{R}$ is the compressed measurement, $\mathbf{x}_l \in \mathbb{R}^{N \times 1}$ is the sparse signal containing only a few nonzero coefficients, $\mathbf{h}_l \in \mathbb{R}^{N \times 1}$ is the known linear operator, and $\{w_l, l = 1, 2, \ldots, L\}$ are the independent and identically distributed (i.i.d.) Gaussian noise variables with $w_l \sim \mathcal{N}(0, \sigma_w^2)$ for $l = 1, 2, \ldots, L$. Upon receiving the compressed measurements $\{y_l, l = 1, 2, \ldots, L\}$, the fusion center produces a global decision about the absence or presence of the sparse signals.

Note that here all the measurements are assumed to be real numbers. This holds true when the radar waveform is the frequency modulation with rapid chirps [29]. In such a radar system, the radar receiver only contains a single receiving channel, that is, the in-phase channel. For other radar systems in which the in-phase and quadrature receiving channels exist in parallel, it is not difficult to extend the signal model in (8.1) to fit the complex measurements.

The sparse signals can be modeled by the BG distribution as done in [8,15–17]. By taking the same discretization rule of the observed region at all the radar sensors, it is reasonable to assume the joint sparsity pattern of the sparse signals across all the radar sensors. Under the constraint of the joint sparsity pattern, the locations of nonzero coefficients in $\mathbf{x}_l$ are the same across all the radar sensors. Define a $N \times 1$ binary vector $\mathbf{s}$ to describe the joint sparsity pattern of $\{\mathbf{x}_l, l = 1, 2, \ldots, L\}$, where

$$\begin{cases} s_n = 1, & \text{for } \{x_{l,n} \neq 0, l = 1, 2, \ldots, L\} \\ s_n = 0, & \text{for } \{x_{l,n} = 0, l = 1, 2, \ldots, L\} \end{cases} \tag{8.2}$$

for $n = 1, 2, \ldots, N$. The elements $\{s_n, n = 1, 2, \ldots, N\}$ are assumed to be independent and identically distributed (i.i.d.) Bernoulli random variables with a common parameter p $(0 \leq p \leq 1)$ [19], where

$$\begin{cases} s_n = 1, & \text{with probability } p \\ s_n = 0, & \text{with probability } 1 - p \end{cases} \tag{8.3}$$

for $n = 1, 2, \ldots, N$. According to (8.3), each coefficient in the sparse vector $\mathbf{x}_l$ is nonzero with probability p, for $l = 1, 2, \ldots, L$. Moreover, each nonzero coefficient is assumed to follow the same Gaussian distribution $\mathcal{N}(0, \sigma_x^2)$. Therefore, the distribution of $x_{l,n}$ can be modeled by the following BG distribution [8,15–17]:

$$x_{l,n} \sim p\mathcal{N}(0, \sigma_x^2) + (1 - p)\delta(x_{l,n}) \tag{8.4}$$

for $n = 1, 2, \ldots, N$ and $l = 1, 2, \ldots, L$, where $\delta(\cdot)$ is the Dirac delta function. In (8.4), p is called as the sparsity degree, which is close to zero under H_1 [8]. The sparsity degree p is unknown *a priori*. The values of σ_w^2 and σ_x^2 are measurable, for example, the noise variance σ_w^2 can be estimated in the absence of the signal and the signal variance σ_x^2 can be obtained from statistics [8].

8.2.2 Formulation of the original LMPT detector

The problem in (8.1) is equivalent to

$$\begin{cases} H_0 : p = 0 \\ H_1 : 0 < p \le 1 \end{cases} \tag{8.5}$$

which is a problem of one-sided hypothesis testing. In addition, (8.5) is also a close hypothesis testing problem [19,20] since the sparsity degree p is close to zero under H_1.

Theoretically speaking, the uniformly most powerful test (UMPT) has the optimal detection performance. However, usually no UMPT exists because its test statistic contains unknown parameters [22]. When no UMPT exists, a suitable approach to deal with the unknown parameters is to estimate them under both hypotheses and then use the estimates in the likelihood ratio (LR). If the maximum likelihood estimation is taken, the corresponding method is the so-called GLRT [8]. The GLRT has no optimality property in general, but it asymptotically approaches the UMPT. However, it is difficult to select the decision threshold for GLRT in the case of one-sided hypothesis testing [22]. Moreover, the performance of GLRT may degrade for the problem of close hypothesis testing [22].

A powerful tool for solving the problem of one-sided and close hypothesis testing is LMPT [19,22]. Different from the GLRT detector, the decision threshold of the LMPT detector can be unambiguously determined, as described below.

The probability density function (PDF) of $y_l|H_0$ follows:

$$y_l|H_0 \sim \mathcal{N}\left(0, \sigma_w^2\right) \tag{8.6}$$

for $l = 1, 2, \ldots, L$. When $\mathbf{x}_l$ is a high dimensional signal, the distribution of $y_l|H_1$ can be approximately expressed by

$$y_l|H_1 \overset{a}{\sim} \mathcal{N}\left(0, p\kappa_l^2 + \sigma_w^2\right) \tag{8.7}$$

for $l = 1, 2, \ldots, L$, where "a" denotes an asymptotic PDF and

$$\kappa_l^2 \overset{\Delta}{=} \sigma_x^2 \|\mathbf{h}_l\|_2^2 \tag{8.8}$$

The proof of (8.7) was provided in [8], based on the Lyapunov Central Limit Theorem [30]. All the measurements received by the fusion center from all the local radar sensors are denoted by a vector:

$$\mathbf{y} \overset{\Delta}{=} [y_1, y_2, \ldots, y_L] \tag{8.9}$$

where all the elements in $\mathbf{y}$ are assumed to be conditionally independent under each hypothesis. Accordingly, the PDFs of $\mathbf{y}$ under H_0 and H_1 can be expressed as

$$P(\mathbf{y}|H_0) = \prod_{l=1}^{L} P(y_l|H_0) \tag{8.10}$$

and

$$P(\mathbf{y}|H_1;p) = \prod_{l=1}^{L} P(y_l|H_1;p) \tag{8.11}$$

respectively. The detection can be achieved by the likelihood ratio test (LRT) [22]:

$$\frac{P(\mathbf{y}|H_1;p)}{P(\mathbf{y}|H_0)} \underset{H_0}{\overset{H_1}{\underset{<}{>}}} \eta \tag{8.12}$$

where η is the decision threshold. The logarithmic form of (8.12) is

$$\ln P(\mathbf{y}|H_1;p) - \ln P(\mathbf{y}|H_0) \underset{H_0}{\overset{H_1}{\underset{<}{>}}} \ln \eta \tag{8.13}$$

Since the value of the sparsity level p is small, the first-order Taylor's series expansion of $\ln P(\mathbf{y}|H_1;p)$ about $p=0$ is given by

$$\ln P(\mathbf{y}|H_1;p) \approx \ln P(\mathbf{y}|H_1;p=0) + p \cdot \left(\frac{\partial \ln P(\mathbf{y}|H_1;p)}{\partial p}\right)_{p=0} \tag{8.14}$$

where

$$\frac{\partial \ln P(\mathbf{y}|H_1;p)}{\partial p} = \frac{1}{2}\sum_{l=1}^{L}\left[\frac{y_l^2\kappa_l^2}{\left(\sigma_w^2+p\kappa_l^2\right)^2} - \frac{\kappa_l^2}{\sigma_w^2+p\kappa_l^2}\right] \tag{8.15}$$

The Fisher information is given by

$$\begin{aligned}\mathrm{FI}(p) &= E\left[\left(\frac{\partial \ln P(\mathbf{y}|H_1;p)}{\partial p}\right)^2\right]\\ &\overset{\langle 1 \rangle}{=} \sum_{l=1}^{L} E\left[\left(\frac{\partial \ln P(y_l|H_1;p)}{\partial p}\right)^2\right]\\ &= \sum_{l=1}^{L}\mathrm{FI}_l(p)\end{aligned} \tag{8.16}$$

and

$$\mathrm{FI}_l(p) = \frac{\kappa_l^4}{2\left(\sigma_w^2+p\kappa_l^2\right)^2} \tag{8.17}$$

for $l=1,2,\ldots,L$. The equation $\langle 1 \rangle$ in (8.16) can be derived from the independence of $\{y_1,y_2,\ldots,y_L\}$ and the regularity condition $E\left(\frac{\partial \ln P(y_l|H_1;p)}{\partial p}\right) = 0$ (see [31, Appendix 3A]).

From (8.6) and (8.7), we have

$$\ln P(\mathbf{y}|H_1; p = 0) = \ln P(\mathbf{y}|H_0) \tag{8.18}$$

By substituting (8.14) and (8.18) into (8.13), we can rewrite (8.13) as

$$\left. \frac{\partial \ln P(\mathbf{y}|H_1; p)}{\partial p} \right|_{p=0} \begin{array}{c} H_1 \\ \gtrless \\ H_0 \end{array} \begin{array}{c} \ln \eta \\ p \end{array} \tag{8.19}$$

Multiplying both sides of (8.19) by a scale factor $1/\sqrt{\mathrm{FI}(0)}$, the resulting LMPT detector is given by

$$T(\mathbf{y}) \triangleq \frac{\left. \frac{\partial \ln P(\mathbf{y}|H_1; p)}{\partial p} \right|_{p=0}}{\sqrt{\mathrm{FI}(0)}} \begin{array}{c} H_1 \\ \gtrless \\ H_0 \end{array} \eta' \tag{8.20}$$

where $\eta' = \frac{\ln \eta}{p\sqrt{\mathrm{FI}(0)}}$ denotes the decision threshold. Substituting (8.15) into (8.20), we have

$$T(\mathbf{y}) = \frac{1}{2\sqrt{\mathrm{FI}(0)}} \sum_{l=1}^{L} \left(\frac{y_l^2 \kappa_l^2}{\sigma_w^4} - \frac{\kappa_l^2}{\sigma_w^2} \right) \begin{array}{c} H_1 \\ \gtrless \\ H_0 \end{array} \eta' \tag{8.21}$$

When the value of L is large, which generally holds in applications of radar sensor networks, the test statistic $T(\mathbf{y})$ in (8.20) asymptotically follows the Gaussian distribution [22]:

$$T(\mathbf{y}) \overset{a}{\sim} \begin{cases} \mathcal{N}(0, 1) \ H_0 \\ \mathcal{N}(\mu_c, 1) \ H_1 \end{cases} \tag{8.22}$$

where the mean value μ_c is given by

$$\mu_c = p\sqrt{\mathrm{FI}(0)} = p\sqrt{\sum_{l=1}^{L} \frac{\kappa_l^4}{2\sigma_w^4}} \tag{8.23}$$

From (8.22), the relationship among the probability of false alarm $P_{FA} = P(T(\mathbf{y}) > \eta'|H_0)$, the probability of detection $P_D = P(T(\mathbf{y}) > \eta'|H_1)$, and the threshold η' can be derived by

$$P_{FA} = 1 - F(\eta') \tag{8.24}$$

$$P_D = 1 - F_{\mu_c}(\eta') \tag{8.25}$$

and

$$\eta' = F^{-1}(1 - P_{FA}) \tag{8.26}$$

where $F(\cdot)$ is the cumulative distribution function (CDF) of the standard normal distribution, that is,

$$F(\beta) = \int_{-\infty}^{\beta} \frac{1}{\sqrt{2\pi}} \exp\left(-\frac{\alpha^2}{2}\right) d\alpha \tag{8.27}$$

and $F_\omega(\cdot)$ is the CDF of the normal distribution with mean value of ω and variance of 1, that is,

$$F_\omega(\beta) = \int_{-\infty}^{\beta} \frac{1}{\sqrt{2\pi}} \exp\left[-\frac{(\alpha - \omega)^2}{2}\right] d\alpha \tag{8.28}$$

and $F^{-1}(\cdot)$ denotes the inverse function of $F(\cdot)$. From (8.26), it is clear that the decision threshold η' can be determined with the desired value of P_{FA}, without any requirement of the knowledge of the sparsity degree p.

It is worth emphasizing that sparse signal recovery is not required for the LMPT detector in (8.20). This means that the LMPT detector is more computationally efficient than the detectors based on partial or complete signal reconstruction reported in [10,12–14]. In particular, the complexity of the detector in [12] based on partial signal recovery by DOMP is $O(tNL)$, where t is the number of iterations of DOMP. In contrast, the computational complexity of the LMPT detector in (8.21) is only $O(L)$.

8.2.3 Simulation results

In this section, simulation results are presented to validate the LMPT detector. The signal-to-noise ratios (SNR) of the lth measurement is defined by

$$\text{SNR} = \frac{1}{N} \cdot \frac{E(\|\mathbf{x}_l\|_2^2)}{E(w_l^2)} = \frac{p\sigma_x^2}{\sigma_w^2} \tag{8.29}$$

Consider the homogeneous environment where the SNRs of all the measurements are assumed to be the same. The length of sparse signals to be detected is set to $N = 1,000$. The elements of linear operator $\mathbf{h}_l$ are taken from a standard normal distribution, and all the linear operators are normalized so that $\|\mathbf{h}_l\|_2^2 = 1$, for $l = 1, 2, \ldots, L$.

The Monte Carlo simulations (10,000 trials) are carried out to validate the theoretical analysis in (8.22)–(8.26). In Figure 8.2, the receiver operating characteristic (ROC) curves of the LMPT detector with different sparsity degrees are plotted. The solid lines represent the theoretical performance, while the star marks denote the performance of the Monte Carlo simulations. As observed in Figure 8.2, the theoretical analysis and simulations are quite consistent. Given $P_{FA} = 0.05$ and $L = 100$, the LMPT detector can successfully detect the signals with $P_D \geq 0.95$

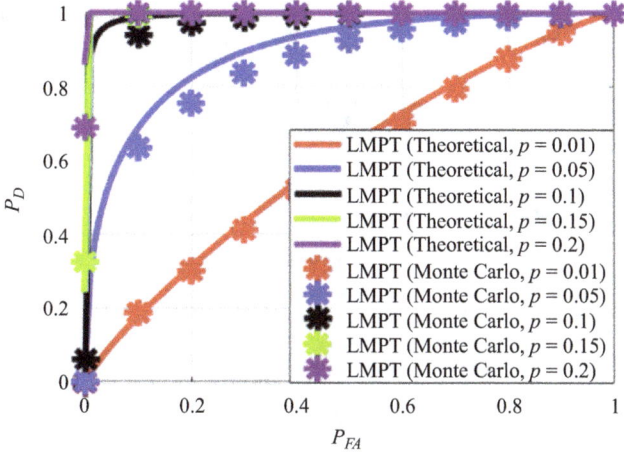

*Figure 8.2   ROC curves of the LMPT detector with different sparsity degrees,
where $\sigma_x^2 = 5$, $\sigma_w^2 = 1$, and $L = 100$*

when $p \geq 0.1$, that is, SNR ≥ -3 dB according to (8.29). In Figure 8.3, the ROC curves of the LMPT detector with different number of measurements are plotted, which validate the consistency between the simulations and the theoretical analysis again. In Figure 8.2 and Figure 8.3, it is noted that the theoretical performances are slightly better than the simulation results due to the asymptotic property in (8.22).

In the LMPT detector, it is assumed that the parameters σ_x^2 and σ_w^2 can be estimated in advance. In practice, however, the estimation error may exist. In Figure 8.4, the detection performance of the LMPT detector with varying normalized mean square error (NMSE) of σ_x^2 and σ_w^2 is plotted, where NMSE$(\theta) = (\theta - \hat{\theta})^2/\theta^2$, and $\hat{\theta}$ is the estimation of true value θ. It is observed from Figure 8.4 that the perturbation in the detection performance of the LMPT detector caused by small errors of the estimates of σ_x^2 and σ_w^2 is negligible.

As mentioned before, the LMPT detector does not require any reconstruction of the sparse signals. Next, the LMPT detector and the DOMP-based detector [12], which require partial reconstruction of the sparse signal, are compared in terms of the detection performance and the computational complexity. In Figure 8.5, the ROC curves of the LMPT detector and the DOMP-based detector are plotted with different number of measurements and sparsity degrees. It is clear that the LMPT detector and the DOMP-based detector have almost the same detection performance. In Table 8.1, the running time of these two detectors is compared with varying number of measurements, where $\sigma_x^2 = 5$, $\sigma_w^2 = 1$, and $p = 0.05$. All the simulations are implemented by using nonoptimized MATLAB® code on a PC (Intel Core i7 Duo CPU 5500U and 8GB RAM). It can be seen that the computational complexity of the LMPT detector is much less than that of the DOMP-based detector.

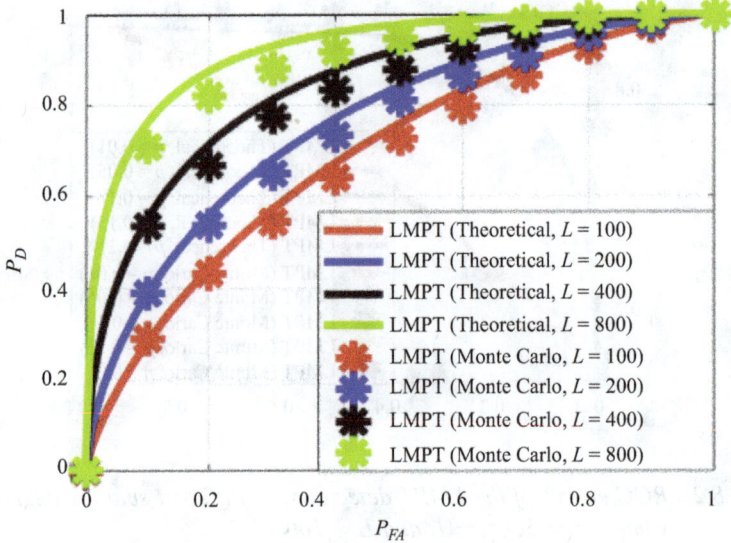

Figure 8.3   *ROC curves of the LMPT detector with different numbers of measurements, where $\sigma_x^2 = 10$, $\sigma_w^2 = 1$, and $p = 0.01$*

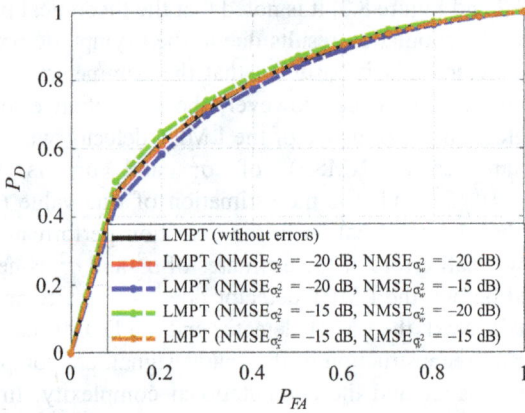

Figure 8.4   *ROC curves of the LMPT detector with different NMSEs of σ_x^2 and σ_w^2, where $L = 100$ and $p = 0.01$. The true values are $\sigma_x^2 = 10$ and $\sigma_w^2 = 1$*

8.3   The quantized LMPT detector

In this section, a communication efficient distributed detector referred to as the quantized LMPT detector is presented for the distributed detection of sparse signals. The quantized LMPT detector assumes that each radar sensor quantizes the

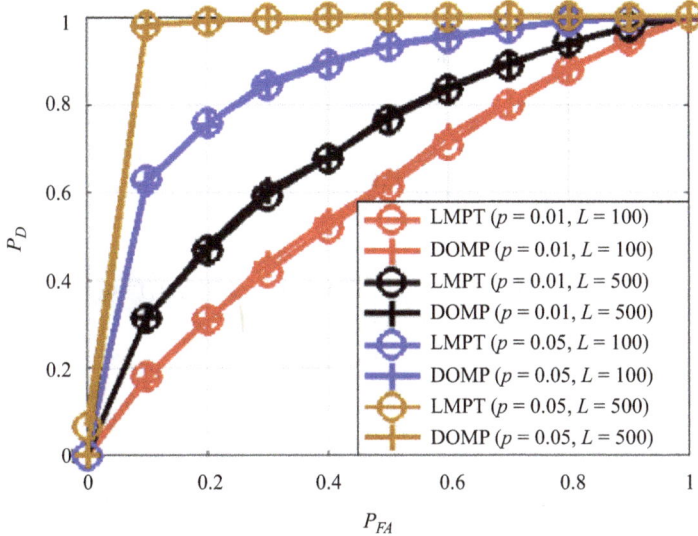

Figure 8.5 *ROC curves of the LMPT detector and the DOMP-based detector, where $\sigma_x^2 = 10$ and $\sigma_w^2 = 1$*

Table 8.1 *Running time of the LMPT detector and the DOMP-based detector for varying number of measurements*

L	Running time (s)	
	LMPT detector	**DOMP-based detector**
50	4.7×10^{-6}	3.1×10^{-3}
100	5.9×10^{-6}	6.6×10^{-3}
150	7.2×10^{-6}	1.0×10^{-2}
200	7.8×10^{-6}	1.3×10^{-2}

local data with low-bit ADC and transmits the coarsely quantized measurements instead of the high-precision measurements to the fusion center. By doing so, the quantized LMPT detector has less communication overhead in comparison with the original LMPT detector.

8.3.1 Problem formulation

Consider the problem of distributed detection of sparse stochastic signals based on quantized measurements, as illustrated in Figure 8.6. The definitions of $\mathbf{x}_l$, $\mathbf{h}_l$, and y_l are the same with that in Figure 8.1, for $l = 1, 2, \ldots, L$. The basic signal model still follows (8.1). When the communication bandwidth in the radar sensor network is limited, the local sensors have to quantize their measurements before

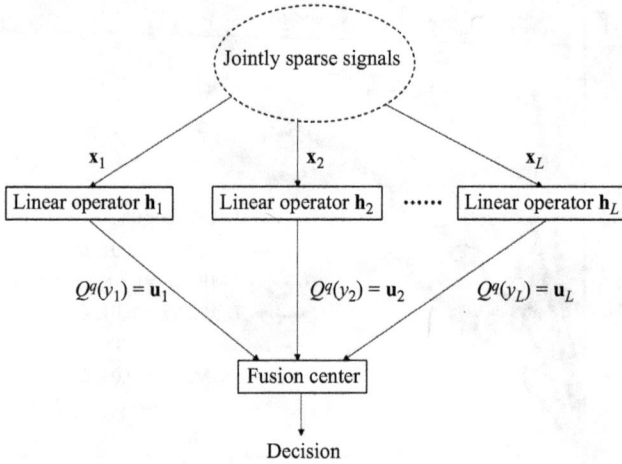

Figure 8.6   Distributed detection system with quantized data, where $Q^q(\cdot)$ denotes the q-bit quantizer

transmitting them to the fusion center. The q-bit quantizer $Q^q(\cdot)$ is used for all the measurements and defined by

$$
\mathbf{u}_l = Q^q(y_l) =
\begin{cases}
\mathbf{v}_1 & \tau_0 \leq y_l < \tau_1, \\
\mathbf{v}_2 & \tau_1 \leq y_l < \tau_2, \\
\vdots & \vdots \\
\mathbf{v}_{2^q} & \tau_{2^q-1} \leq y_l < \tau_{2^q}
\end{cases}
\tag{8.30}
$$

where $l = 1, 2, \ldots, L$, $q \in \mathbb{Z}^+$ denotes the bit depth (i.e., bits per quantized measurement), $\{\tau_m, m = 0, 1, \ldots, 2^q\}$ denote the quantization thresholds, the output $\mathbf{u}_l$ indicates the interval where y_l lies in, and $\{\mathbf{v}_k, k = 1, 2, \ldots, 2^q\}$ are the binary code words. For example, given $q = 2$, we have $\mathbf{v}_1 = $ "00", $\mathbf{v}_2 = $ "01", $\mathbf{v}_3 = $ "10" and $\mathbf{v}_4 = $ "11." Figure 8.7 illustrates the q-bit quantizer. Note that (8.30) will degenerate to the case of 1-bit quantization when $q = 1$. Assume that the same quantization thresholds $\{\tau_m, m = 0, 1, \ldots, 2^q\}$ are taken over all the local radar sensors.

The probability mass function (PMF) of $\mathbf{u}_l$ can be given by

$$
P(\mathbf{u}_l) = \prod_{i=1}^{2^q} [P(\mathbf{u}_l = \mathbf{v}_i)]^{I(\mathbf{u}_l, \mathbf{v}_i)}
\tag{8.31}
$$

for $l = 1, 2, \ldots, L$, where $I(\boldsymbol{\alpha}, \boldsymbol{\beta}) = 1$ if $\boldsymbol{\alpha} = \boldsymbol{\beta}$ and 0 otherwise. According to (8.30), $\mathbf{u}_l$ is the quantized version of the measurement y_l. Thus we have:

$$
P(\mathbf{u}_l = \mathbf{v}_i) = P(\tau_{i-1} \leq y_l \leq \tau_i)
\tag{8.32}
$$

Code words: $\mathbf{v}_1$ $\mathbf{v}_2$ $\mathbf{v}_{2^q}$

Thresholds: τ_0 τ_1 τ_2 τ_{2^q-1} τ_{2^q}

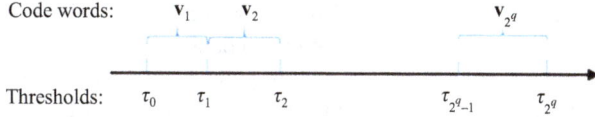

Figure 8.7   *Illustration of the q-bit quantizer, where $\{\tau_m, m = 0, 1, \ldots, 2^q\}$ are the quantization thresholds and $\{v_k, k = 1, 2, \ldots, 2^q\}$ are the binary code words*

The PDFs of $y_l|H_0$ and $y_l|H_1$ follow (8.6) and (8.7), respectively. From (8.6), (8.7), (8.31), and (8.32), we have

$$P(\mathbf{u}_l|H_0) = \prod_{i=1}^{2^q} \left[F\left(\frac{\tau_i}{\sigma_w}\right) - F\left(\frac{\tau_{i-1}}{\sigma_w}\right) \right]^{I(\mathbf{u}_l,\mathbf{v}_i)} \tag{8.33}$$

and

$$P(\mathbf{u}_l|H_1;p) = \prod_{i=1}^{2^q} \left[F\left(\frac{\tau_i}{\sqrt{p\kappa_l^2 + \sigma_w^2}}\right) - F\left(\frac{\tau_{i-1}}{\sqrt{p\kappa_l^2 + \sigma_w^2}}\right) \right]^{I(\mathbf{u}_l,\mathbf{v}_i)} \tag{8.34}$$

where $F(\cdot)$ is given in (8.27).

All the quantized measurements received at the fusion center are denoted by

$$\mathbf{U} \triangleq \{\mathbf{u}_1, \mathbf{u}_2, \ldots, \mathbf{u}_L\} \tag{8.35}$$

Under the assumption that all the L measurements are independent, we have:

$$P(\mathbf{U}) = \prod_{l=1}^{L} P(\mathbf{u}_l) \tag{8.36}$$

The fusion center makes a global decision about the absence or presence of the sparse signals based on the quantized data $\mathbf{U}$.

8.3.2   Formulation of the quantized LMPT detector

In this section, the quantized LMPT detector is derived in detail. As stated in (8.5), the problem of distributed detection of sparse stochastic signals can be viewed as the problem of one-sided and close hypothesis testing. The detection can be carried out by the logarithmic LRT:

$$\ln P(\mathbf{U}|H_1;p) - \ln P(\mathbf{U}|H_0) \underset{H_0}{\overset{H_1}{\gtrless}} \ln \eta_{q-\text{bit}} \tag{8.37}$$

where $\eta_{q-\text{bit}}$ is the decision threshold. The difference between (8.13) and (8.37) lies in that the former is based on the high-precision data $\{y_1, y_2, \ldots, y_L\}$ while the latter is based on the quantized data $\{\mathbf{u}_1, \mathbf{u}_2, \ldots, \mathbf{u}_L\}$. Since the sparsity degree p is close to zero, by taking the first-order Taylor's series expansion of $\ln P(\mathbf{U}|H_1; p)$ about zero, we have:

$$\ln P(\mathbf{U}|H_1; p) \approx \ln P(\mathbf{U}|H_1; p = 0) + p\left(\frac{\partial \ln P(\mathbf{U}|H_1; p)}{\partial p}\right)_{p=0} \tag{8.38}$$

where

$$\frac{\partial \ln P(\mathbf{U}|H_1; p)}{\partial p}$$
$$= \sum_{l=1}^{L} \frac{\partial \ln P(\mathbf{u}_l|H_1; p)}{\partial p}$$
$$= \sum_{l=1}^{L} \sum_{i=1}^{2^q} \frac{\kappa_l^2 I(\mathbf{u}_l, \mathbf{v}_i)}{2\left(p\kappa_l^2 + \sigma_w^2\right)^{3/2}} \cdot \frac{\left[\tau_{i-1}G\left(\frac{\tau_{i-1}}{\sqrt{p\kappa_l^2 + \sigma_w^2}}\right) - \tau_i G\left(\frac{\tau_i}{\sqrt{p\kappa_l^2 + \sigma_w^2}}\right)\right]}{\left[F\left(\frac{\tau_i}{\sqrt{p\kappa_l^2 + \sigma_w^2}}\right) - F\left(\frac{\tau_{i-1}}{\sqrt{p\kappa_l^2 + \sigma_w^2}}\right)\right]} \tag{8.39}$$

and

$$G(\alpha) = \frac{1}{\sqrt{2\pi}} \exp\left(-\frac{\alpha^2}{2}\right) \tag{8.40}$$

From (8.33)–(8.36), we have

$$\ln P(\mathbf{U}|H_1; p = 0) = \ln P(\mathbf{U}|H_0) \tag{8.41}$$

By substituting (8.38), (8.39), and (8.41) into (8.38), we can approximate (8.38) as

$$\left.\frac{\partial \ln P(\mathbf{U}|H_1; p)}{\partial p}\right|_{p=0} \begin{array}{c} H_1 \\ > \\ < \\ H_0 \end{array} \frac{\ln \eta_{q-\text{bit}}}{p} \tag{8.42}$$

Similar to the deduction in Section 8.2.2, the Fisher information in the case of quantized data is given by

$$\text{FI}_{q-\text{bit}}(p) = E\left[\left(\frac{\partial \ln P(\mathbf{U}|H_1; p)}{\partial p}\right)^2\right] = \sum_{l=1}^{L} \text{FI}_{q-\text{bit},l}(p) \tag{8.43}$$

where

$$\mathrm{FI}_{q\text{-bit},l}(p) = \frac{\kappa_l^4}{4\left(p\kappa_l^2 + \sigma_w^2\right)^2}$$

$$\cdot \sum_{i=1}^{2^q} \frac{\left[\dfrac{\tau_i}{\sqrt{p\kappa_l^2 + \sigma_w^2}} G\left(\dfrac{\tau_i}{\sqrt{p\kappa_l^2 + \sigma_w^2}}\right) - \dfrac{\tau_{i-1}}{\sqrt{p\kappa_l^2 + \sigma_w^2}} G\left(\dfrac{\tau_{i-1}}{\sqrt{p\kappa_l^2 + \sigma_w^2}}\right)\right]^2}{\left[F\left(\dfrac{\tau_i}{\sqrt{p\kappa_l^2 + \sigma_w^2}}\right) - F\left(\dfrac{\tau_{i-1}}{\sqrt{p\kappa_l^2 + \sigma_w^2}}\right)\right]}$$

(8.44)

for $l = 1, 2, \ldots, L$.

Multiplying both sides of (8.42) by a scale factor $1/\sqrt{\mathrm{FI}_{q\text{-bit}}(0)}$, the test statistic of the quantized LMPT detector is given by

$$T(\mathbf{U}) = \frac{\dfrac{\partial \ln P(\mathbf{U}|H_1;p)}{\partial p}\Big|_{p=0}}{\sqrt{\mathrm{FI}_{q\text{-bit}}(0)}} \underset{H_0}{\overset{H_1}{\underset{<}{>}}} \eta'_{q\text{-bit}}$$

(8.45)

where $\eta'_{q\text{-bit}} = \dfrac{\ln \eta_{q\text{-bit}}}{p\sqrt{\mathrm{FI}_{q\text{-bit}}(0)}}$.

When the value of L is large, similar to the analysis in Section 8.2.2, the test statistic $T(\mathbf{U})$ in (8.45) asymptotically follows the Gaussian distribution:

$$T(\mathbf{U}) \overset{a}{\sim} \begin{cases} \mathcal{N}(0, 1) \ H_0 \\ \mathcal{N}\left(\mu_{q\text{-bit}}, 1\right) H_1 \end{cases}$$

(8.46)

where the mean value $\mu_{q\text{-bit}}$ is given by

$$\mu_{q\text{-bit}} = p\sqrt{\mathrm{FI}_{q\text{-bit}}(0)}$$

(8.47)

One may find the similarity of (8.23) and (8.47). Note that the fisher information terms in (8.23) and (8.47) are calculated by using the high-precision measurements and quantized measurements, respectively.

From (8.46), the relationship among the probability of false alarm P_{FA}, the probability of detection P_D, and the decision threshold $\eta'_{q\text{-bit}}$ in (8.45) can be derived by

$$P_{FA} = P\left(T(\mathbf{U}) > \eta'_q | H_0\right) = 1 - F\left(\eta'_q\right)$$

(8.48)

$$P_D = P\left(T(\mathbf{U}) > \eta'_q | H_1\right) = 1 - F_{\mu_q}\left(\eta'_q\right)$$

(8.49)

and

$$\eta'_q = F^{-1}(1 - P_{FA}) \tag{8.50}$$

respectively, where $F(\eta'_q)$ and $F_{\mu_q}(\eta'_q)$ are calculated according to (8.27) and (8.28).

8.3.3   Design of the local quantizers

In Section 8.3.2, the quantized LMPT detector was presented with given quantizers. In what follows, the local quantizers are devised to achieve the near-optimal detection performance of the quantized LMPT detector. Assume that the same quantization rule, that is, the saturated quantization [32] with $\tau_0 = -\infty$ and $\tau_{2^q} = +\infty$, is carried out at all the measurements. Accordingly, these two extreme thresholds, that is, τ_0 and τ_{2^q}, will not be optimized.

It can be observed from (8.46) that increase in the mean value μ_q will improve the detection performance of the quantized LMPT detector. This inspires us to design the quantization thresholds as follows:

$$\{\hat{\tau}_m, m = 1, 2, \ldots, 2^q - 1\} = \underset{\{\tau_m, m=1,2,\ldots,2^q-1\}}{\arg\max} \quad \mu_{q-\text{bit}}, \text{ s.t. } \tau_1 < \tau_2 < \ldots < \tau_{2^q-1} \tag{8.51}$$

where $\hat{\tau}_m$ is the estimate of τ_m. The relationship of the mean value μ_q and the thresholds $\{\tau_m, m = 1, 2, \ldots, 2^q - 1\}$ can be found from (8.43), (8.44), and (8.47). Substituting (8.47) into (8.51) yields:

$$\{\hat{\tau}_m, m = 1, 2, \ldots, 2^q - 1\} = \underset{\{\tau_m, m=1,2,\ldots,2^q-1\}}{\arg\max} \quad \text{FI}(0), \text{ s.t. } \tau_1 < \tau_2 < \ldots < \tau_{2^q-1} \tag{8.52}$$

It is difficult to find the closed-form solutions of (8.52). Here solving (8.52) is accomplished by using a numerical optimization algorithm, that is, the particle swarm optimization (PSO) algorithm [33], which does not rely on the concavity property of the problem to be solved. PSO stems from the social cooperative and competitive behaviors of bird flocking and fish schooling [33], which aims at optimizing a problem by iteratively improving a set of particles. Actually, PSO is a stochastic optimization algorithm and its global convergence is guaranteed under certain assumptions [34,35]. Despite the difficulty of examining the assumptions given in [34,35], PSO is still considered as a superior optimization approach and has been successfully applied to signal processing field [27,36–38]. The selection of the parameters in PSO suggested by [27,36–38] are adopted here to solve the problem in (8.52). It is worth emphasizing that PSO is based on simple iterations with affordable computational complexity [38].

8.3.4   Comparison with the original LMPT detector

In this section, the performance loss caused by quantization is theoretically analyzed by comparing the quantized LMPT detection and the original LMPT detector.

8.3.4.1   1-bit quantization

When the communication bandwidth in the radar sensor network is extremely
limited, 1-bit data transmission may be needed. In this section, the performance
loss caused by 1-bit quantization is investigated.

In the case of 1-bit quantization, the problem of (8.52) degenerates to the
following problem:

$$\hat{\tau} = \arg\max_{\tau} \mathrm{FI}_{q-\mathrm{bit}}(0)\left(\frac{\tau}{\sigma_w}\right)$$

$$= \arg\max_{\tau} \left[\frac{\tau}{\sigma_w} \cdot G\left(\frac{\tau}{\sigma_w}\right)\right]^2 \cdot \left\{\left[F\left(\frac{\tau}{\sigma_w}\right)\right]^{-1} + \left[1 - F\left(\frac{\tau}{\sigma_w}\right)\right]^{-1}\right\}$$

$$(8.53)$$

The numerical solution of (8.53) provided by PSO converges to two points:

$$\frac{\hat{\tau}}{\sigma_w} \approx \pm 1.575 \tag{8.54}$$

The objective function in (8.53) is plotted in Figure 8.8, which is consistent
with (8.54). Since this objective function is even, the selections of $\hat{\tau} \approx 1.575\sigma_w$ and
$\hat{\tau} \approx -1.575\sigma_w$ produce the same detection performance. Without loss of generality, we select:

$$\hat{\tau} \approx 1.575\sigma_w \tag{8.55}$$

Note that the quantization threshold given in (8.55) is valid for all the local
measurements. Substituting (8.55) into (8.47), the mean value $\mu_q^{1-\mathrm{bit}}$ under 1-bit
quantization approximately achieves the maximum:

$$\mu_{1-\mathrm{bit, \, max}} = p\sqrt{\sum_{l=1}^{L} \frac{0.6084\kappa_l^4}{4\sigma_w^4}} \tag{8.56}$$

where 0.6084 is just the maximum value numerically solved by PSO shown in
Figure 8.8.

Consider a homogeneous case that $\kappa_l^2 = \kappa^2$ for $l = 1, 2, \ldots, L$, (8.23) and
(8.56) become

$$\mu_c = p\frac{\kappa^2}{\sigma_w^2}\sqrt{\frac{L}{2}} \tag{8.57}$$

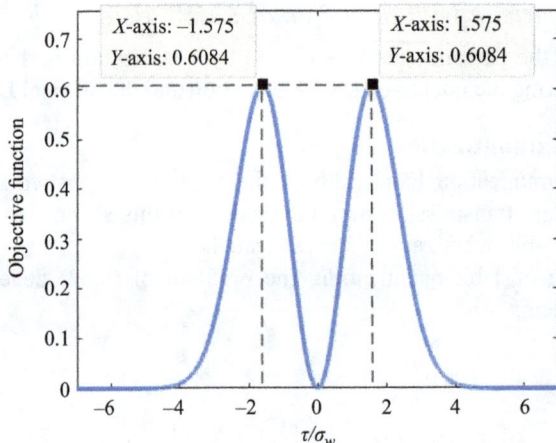

Figure 8.8   The objective function in (8.53), where the scaling factor $\kappa^4/4\sigma_w^4$ is set to 1. The maximum value is achieved at $\hat{\tau}/\sigma_w \approx \pm 1.575$

and

$$\mu_{1-\text{bit,max}} = 0.55p\frac{\kappa^2}{\sigma_w^2}\sqrt{\frac{L}{2}} \tag{8.58}$$

respectively. Therefore, to ensure that the 1-bit LMPT detector and the original LMPT detector have the same detection performance, that is, $\mu_{1-\text{bit,max}} = \mu_c$, the following condition should be satisfied:

$$\frac{L_{1-\text{bit}}}{L_c} = \left(\frac{1}{0.55}\right)^2 \approx 3.3 \tag{8.59}$$

where $L_{1-\text{bit}}$ and L_c denote the required number of measurements for the 1-bit LMPT detector and the original LMPT detector, respectively. This means that the 1-bit LMPT detector needs to increase the number of measurements to $3.3L$ to achieve the same performance as the original LMPT detector with L measurements. Such a 1-bit LMPT detector is still more communication efficient than the original LMPT detector since the former only needs to transmit a total number of $3.3L$ bits, while the latter transmits L high-precision measurements (e.g., $64L$ bits if the 64-bit quantizer is taken for the high-precision data).

8.3.4.2   Multibit quantization

Compared to 1-bit quantization, multilevel quantization incurs less loss of detection performance of the quantized LMPT detector. A question naturally arises: can multibit LMPT detector with the optimal local quantizers achieve similar detection performance as the original LMPT detector?

Table 8.2   Designed quantization thresholds for 2-bit LMPT detector

2-bit LMPT		
Quantization thresholds obtained by the PSO algorithm	$\hat{\tau}_1$	$-2.263\sigma_w$
	$\hat{\tau}_2$	$-1.325\sigma_w$
	$\hat{\tau}_3$	$1.623\sigma_w$

Table 8.3   Designed quantization thresholds for 3-bit LMPT detector

3-bit LMPT		
Quantization thresholds obtained by the PSO algorithm	$\hat{\tau}_1$	$-2.694\sigma_w$
	$\hat{\tau}_2$	$-1.787\sigma_w$
	$\hat{\tau}_3$	$-1.057\sigma_w$
	$\hat{\tau}_4$	$-0.980\sigma_w$
	$\hat{\tau}_5$	$1.565\sigma_w$
	$\hat{\tau}_6$	$2.154\sigma_w$
	$\hat{\tau}_7$	$2.915\sigma_w$

It is natural that the detection performance of the multibit LMPT detector converges to that of the original LMPT detector when the bit depth q tends to infinity, that is, $\lim_{q \to +\infty} \mu_{q-\text{bit}} = \mu_c$. In Table 8.2 and Table 8.3, the quantization thresholds obtained by the PSO algorithm for 2-bit and 3-bit quantization are listed, respectively. Substituting these designed quantization thresholds into (8.47), we have:

$$\mu_{2-\text{bit,max}} = 0.85\mu_c \tag{8.60}$$

and

$$\mu_{3-\text{bit,max}} = 0.95\mu_c \tag{8.61}$$

Thus, it can be concluded that the detection performances of the 3-bit LMPT detector and the original LMPT detector are very close when the designed quantization thresholds are taken at all the local sensors. In other words, for the problem of the detection of sparse stochastic signals, it is unnecessary to use finely quantized measurements to compensate for the performance loss caused by quantization.

8.3.5   Simulation results

In this section, simulation results are provided to corroborate the theoretical analysis and demonstrate the performance of the quantized LMPT detector. The length of sparse signals is $N = 1,000$. All the linear operators $\{\mathbf{h}_l, l = 1, 2, \ldots, L\}$ are sampled from standard normal distribution and normalized so that $\|\mathbf{h}_l\|_2 = 1$, for $l = 1, 2, \ldots, L$. The definition of SNR is the same as that in Section 8.2.3.

8.3.5.1 Effect of the designed quantizers

We first evaluate the effect of the threshold of 1-bit quantization given in (8.54) on the performance of distributed detection. In Figure 8.9, four scenarios with different parameters are taken into account and 10,000 Monte Carlo trials are performed to examine the design of the 1-bit quantization threshold. The sparsity degree is set to $p = 0.05$, and the probability of false alarm is $P_{FA} = 0.05$. It is observed that the threshold provided in (8.54) leads to the optimal probability of detection.

For multibit quantization, the problem in (8.52) is a multivariable optimization problem. We take the 2-bit quantizer solved from (8.52) as an example to show its superiority over the uniform quantizers, which have been widely investigated in existing literature [39,40]. In Figure 8.10, the ROC curves of the 2-bit LMPT detector generated by the uniform quantizers and the quantizers designed according to (8.52) are plotted. The range of quantization thresholds of the uniform quantizer is set to $[-5\sigma_w, 5\sigma_w]$. In Figure 8.10, the solid lines represent the theoretical performance, while the star and diamond marks denote the average performance of the Monte Carlo simulations. As shown in Figure 8.10, the 2-bit LMPT detector with

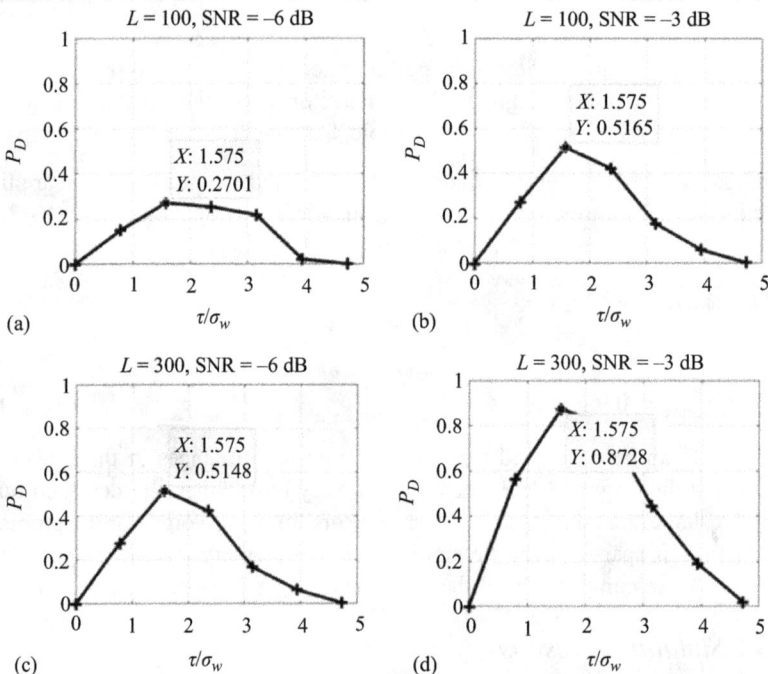

Figure 8.9 *Probability of detection of the 1-bit LMPT detector versus varying threshold. The sparsity degree is $p = 0.05$ and the probability of false alarm is $P_{FA} = 0.05$: (a) $L = 100$, $SNR = -6$ dB, (b) $L = 100$, $SNR = -3$ dB, (c) $L = 300$, $SNR = -6$ dB, and (d) $L = 300$, $SNR = -3$ dB*

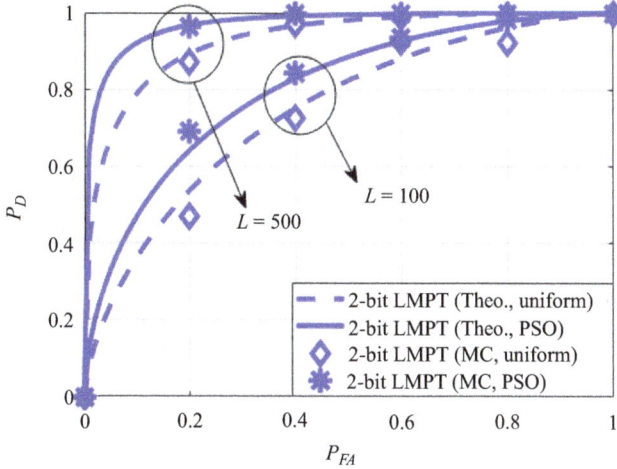

Figure 8.10   ROC curves of the 2-bit LMPT detector with the uniform quantization and the designed quantization thresholds, where p = 0.05 and SNR = −6 dB. Two scenarios are considered, that is, L = 100 and L = 500

the designed quantization thresholds provides better performance than that based on uniform quantizers. The reason is that the uniform quantizers do not consider the relationship between the quantization thresholds and the detection performance.

8.3.5.2   Performance of the quantized LMPT detector

In Figure 8.11, the ROC curves of the 1-bit LMPT detector and the original LMPT detector are plotted with different number of measurements, where the number of the measurements of the 1-bit LMPT detector is set to be 3.3 times larger than that of the original LMPT detector in all the Monte Carlo trials. As illustrated in Figure 8.11, the 1-bit LMPT detector with $3.3L$ measurements and the original LMPT detector with L measurements approximately provide the same detection performance. This is consistent with our theoretical analysis given in Section 8.3.4.

In Figure 8.12, the ROC curves of the LMPT detector based on high-precision data and quantized data with different parameters are plotted. It is clear that the theoretical and simulated ROC curves are consistent with each other. One can see that there is an obvious performance gap between the 1-bit LMPT detector and the original LMPT detector when they have the same number of measurements. The increase of the bit depth leads to a gain in the detection performance. The performance loss caused by quantization can be negligible when the bit depth increases to 3, which is consistent with our theoretical analysis given in Section 8.3.4.

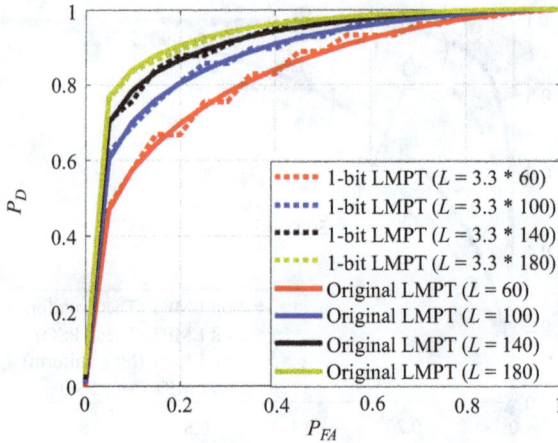

Figure 8.11   ROC curves of the 1-bit LMPT detector and the original LMPT detector with different number of measurements, where p = 0.01 and SNR = −5 dB

8.4   Conclusion

In this chapter, the problem of the detection of sparse stochastic signals with radar sensor networks was studied. The BG distribution was imposed on the sparse signals and accordingly the problem of distributed detection of sparse signals was converted into the problem of close and one-sided hypothesis testing. The original LMPT detector was presented to detect sparse signals from high-precision measurements without any requirement of signal reconstruction. Simulation results demonstrated that to achieve the same detection performance, the original LMPT detector has a much lower computational burden than the DOMP-based detector. We further presented the quantized LMPT detector to solve the problem of the distributed detection of sparse signals with quantized measurements. To ensure the optimal detection performance, a method for the design of the quantizers at the local sensors was presented. Theoretical analysis of the performance of both original and quantized LMPT detectors was consistent with the simulation results. Simulation results also demonstrated that (1) the 1-bit LMPT detector with $3.3L$ measurements approximately achieves the same detection performance as the original LMPT detector with L high-precision measurements; and (2) the detection performance of the 3-bit LMPT detector is very close to that of the original LMPT detector.

Note that the independence between the measurements is assumed in this chapter for mathematical simplicity and the coherent measurements are not considered. If a part of measurements is coherent, the detection performance is expected to be higher than that obtained from independent measurements since the coherence processing produces larger SNR gain.

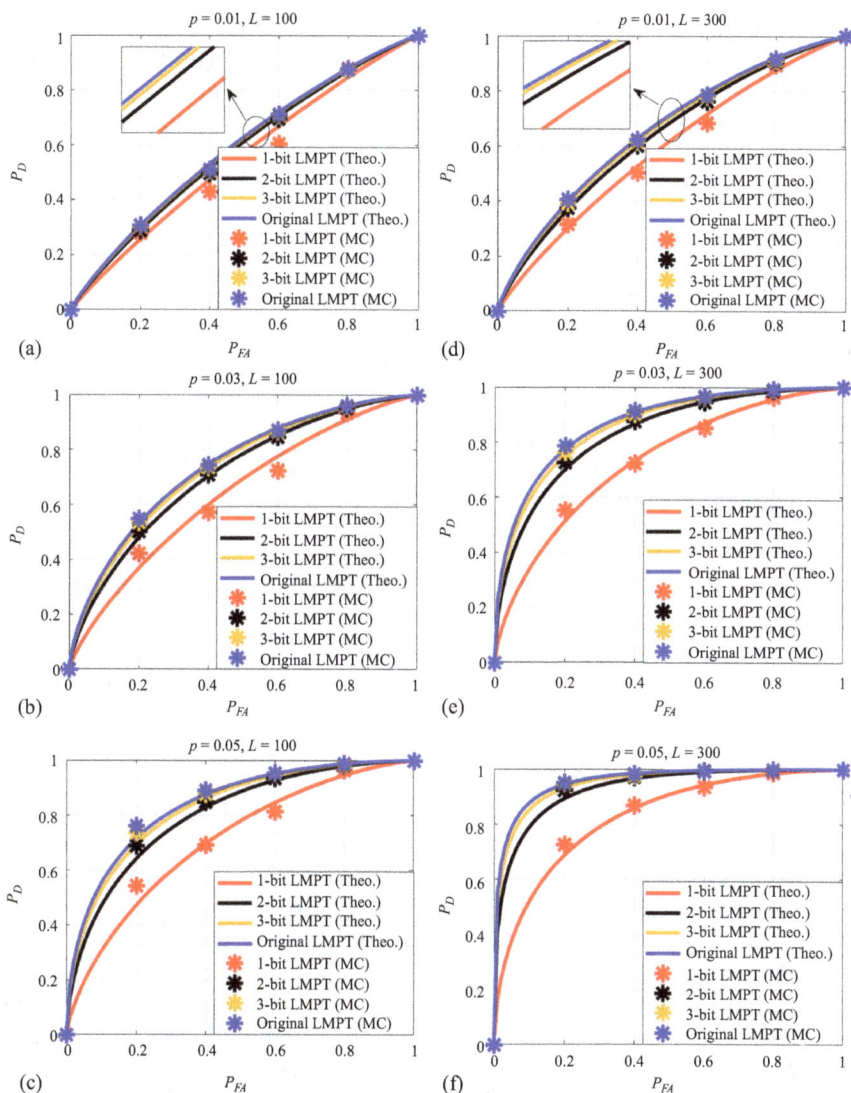

Figure 8.12 ROC curves of the original LMPT detector and the quantized LMPT detector, where $\sigma_x^2 = 5$, $\sigma_w^2 = 1$: (a) $p = 0.01$, $L = 100$, (b) $p = 0.03$, $L = 100$, (c) $p = 0.05$, $L = 100$, (d) $p = 0.01$, $L = 300$, (e) $p = 0.03$, $L = 300$, and (f) $p = 0.05$, $L = 300$

Here the LMPT detectors were derived under the assumption that the fusion center collects all the measurements from the local sensors and makes global decision. These LMPT detectors can also be extended to the case that no fusion center is available in the radar sensor network. In this case, decentralized detection of sparse signals can be achieved by employing the LMPT strategy at each local radar sensor by sharing the measurements with its neighboring sensors. Note that the performance loss may be incurred by decentralized detection since each sensor only collaborates with its neighbors and the access to the data collected by all the sensors in the network is time-consuming.

References

[1]   Baker C. J. and Hume A. L. "Netted radar sensing." *IEEE Aerospace and Electronic Systems Magazine*. 2003; 18(2): 3–6.

[2]   Liu S., Hagiwara H., Shoji R., Tamara H., and Okano T. "Radar network system to observe and analyze Tokyo Bay vessel traffic." *IEEE Aerospace and Electronic Systems Magazine*. 2004; 19(11): 3–11.

[3]   Thomä R. S., Andrich C., Del Galdo G., *et al.* "Cooperative passive coherent location: A promising 5G service to support road Safety." *IEEE Communications Magazine*. 2019; 57(9): 86–92.

[4]   Schmitt A. and Collins P. "Demonstration of a network of simultaneously operating digital noise radars." *IEEE Antennas and Propagation Magazine*. 2009; 51(2): 125–130.

[5]   Matthias W. "Passive WLAN radar network using compressed sensing." *Proceedings of IET International Conference on Radar Systems (Radar 2012)*. Glasgow, UK: IET; 2012, pp. 1–6.

[6]   Colone F., Martelli T., Bongioanni C., Pastina D., and Lombardo P. "WiFi-based PCL for monitoring private airfields." *IEEE Aerospace and Electronic Systems Magazine*. 2017; 32(2): 22–29.

[7]   Griffiths H. D. and Baker C. J. *An introduction to passive radar*. Norwood, MA: Artech House, 2017.

[8]   Hariri A. and Babaie-Zadeh M. "Compressive detection of sparse signals in additive white gaussian noise without signal reconstruction." *Signal Processing*. 2017; 131: 376–385.

[9]   Davenport M. A., Boufounos P. T., Wakin M. B., and Baraniuk R. G. "Signal processing with compressive measurements." *IEEE Journal of Selected Topics in Signal Processing*. 2010; 4(2): 445–460.

[10] Zayyani H., Haddadi F., and Korki M. "Double detector for sparse signal detection from one-bit compressed sensing measurements." *IEEE Signal Processing Letters*. 2016; 23(11): 1637–1641.

[11] Duarte M., Davenport M., Wakin M., and Baraniuk R. G. "Sparse signal detection from incoherent projections." *Proceedings of the IEEE International Conference on Acoustics, Speech and Signal Processing (ICASSP)*. Toulouse, France: IEEE; 2006, pp. 305–308.

[12] Wimalajeewa T. and Varshney P. K. "Sparse signal detection with compressive measurements via partial support set estimation." *IEEE Transactions on Signal and Information Processing over Networks*. 2017; 3(1): 46–60.

[13] Zhao W. and Li G. "On the detection probability of sparse signals with sensor networks based on distributed subspace pursuit." *Proceedings of IEEE China Summit and International Conference on Signal and Information Processing (ChinaSIP)*. Chengdu, China: IEEE; 2015, pp. 324–328.

[14] Li G., Zhang H., Wimalajeewa T., and Varshney P. K. "On the detection of sparse signals with sensor networks based on Subspace Pursuit." *Proceedings of IEEE Global Conference of Signal Information Processing (GlobalSIP)*. Atlanta, GA, USA: IEEE; 2014, pp. 438–442.

[15] Soussen C., Idier J., Brie D., and Duan J. "From Bernoulli-Gaussian deconvolution to sparse signal restoration." *IEEE Transactions on Signal Processing*. 2011; 59(10): 4572–4584.

[16] Korki M., Zhang J., Zhang C., and Zayyani H. "Iterative Bayesian reconstruction of non-IID block-sparse signals." *IEEE Transactions on Signal Processing*. 2016; 64(13): 3297–3307.

[17] Ziniel J. and Schniter P. "Dynamic compressive sensing of time-varying signals via approximate message passing." *IEEE Transactions on Signal Processing*. 2013; 61(21): 5270–5284.

[18] Wu Q., Zhang Y. D., Amin M. G., and Himed B. "Multi-task Bayesian compressive sensing exploiting intra-task dependency." *IEEE Signal Processing Letters*. 2015, 22(4): 430–434.

[19] Ramirez D., Via J., Santamaria I., and Scharf L. L. "Locally most powerful invariant tests for correlation and sphericity of Gaussian vectors." *IEEE Transactions on Information Theory*. 2013; 59(4): 2128–2141.

[20] Xiao Y. H., Huang L., Xie J., and So H. C. "Approximate asymptotic distribution of locally most powerful invariant test for independence: Complex case." *IEEE Transactions on Information Theory*. 2018; 64(3): 1784–1799.

[21] Wang X., Li G., and Varshney P. K. "Detection of sparse signals in sensor networks via locally most powerful tests." *IEEE Signal Processing Letters*. 2018; 25(9): 1418–1422.

[22] Kay S. M. *Fundamentals of statistical signal processing: Volume II: Detection theory*. Upper Saddle River, NJ: Prentice-Hall, 1998.

[23] Le B., Rondeau T. W., Reed J. H., and Bostian C. W. "Analog-to-digital converters." *IEEE Signal Processing Magazine*. 2005; 22(6): 69–77.

[24] Wang T., Zhang W., Maunder R. G., and Hanzo L. "Near-capacity joint source and channel coding of symbol values from an infinite source set using Elias gamma error correction codes." *IEEE Transactions on Communications*. 2014; 62(1): 280–292.

[25] Nadendla V. S. S. and Varshney P. K. "Design of binary quantizers for distributed detection under secrecy constraints." *IEEE Transactions on Signal Processing*. 2016; 64(10): 2636–2648.

[26] Fang J., Liu Y., Li H., and Li S. "One-bit quantizer design for multisensor GLRT fusion." *IEEE Signal Processing Letters*. 2013; 20(3): 257–260.

[27]   Gao F., Guo L., Li H., Liu J., and Fang J. "Quantizer design for distributed GLRT detection of weak signal in wireless sensor networks." *IEEE Transactions on Wireless Communications*. 2015; 14(4): 2032–2042.

[28]   Wang X., Li G., and Varshney P. K. "Detection of sparse stochastic signals with quantized measurements in sensor networks." *IEEE Transactions on Signal Processing*. 2019; 67(8): 2210–2220.

[29]   Fulvio G., Antonio De M., and Lee P. *Waveform design and diversity for advanced radar systems*. London, UK: IET Publisher, 2012. Chapter 7.

[30]   Papoulis A. and Pillai S. U. *Probability, random variables, and stochastic processes*. New York, NY, USA: Tata McGraw-Hill Education; 2002.

[31]   Kay S. M. *Fundamentals of statistical signal processing: Volume I: Estimation theory*. Upper Saddle River, NJ: Prentice-Hall, 1993.

[32]   Yang Z., Xie L., and Zhang C. "Variational Bayesian algorithm for quantized compressed sensing." *IEEE Transactions on Signal Processing*. 2013; 61(11): 2815–2824.

[33]   Kennedy J. and Eberhart R. C. "Particle swarm optimization." *Proceeding of IEEE International Conference of Neural Networks*. Perth, WA, Australia: IEEE; 1995. pp. IV:G1942–IV:G1948.

[34]   Jiang M., Luo Y. P., and Yang S. Y. "Stochastic convergence analysis and parameter selection of the standard particle swarm optimization algorithm." *Information Processing Letter*. 2007; 102(1): 8–16.

[35]   Hui Q. and Zhang H. "Global convergence analysis of swarm optimization using paracontraction and semistability theory." *Proceeding of American Control Conference (ACC)*. Boston, MA, USA: IEEE; 2016, pp. 2900–2905.

[36]   Cheung R. C. Y., Aue A., and Lee T. C. M. "Consistent estimation for partition-wise regression and classification models." *IEEE Transactions on Signal Processing*. 2017; 65(14): 3662–3674.

[37]   Duarte C., Barner K. E., and Goossen K. "Design of IIR multi-notch filters based on polynomially-represented squared frequency response." *IEEE Transactions on Signal Processing*. 2016; 64(10): 2613–2623.

[38]   Bayram S., Gezici S., and Poor H. V. "Noise enhanced hypothesis-testing in the restricted Bayesian framework." *IEEE Transactions on Signal Processing*. 2010; 58(8): 3972–3989.

[39]   Ohno S., Shiraki T., Tariq M. R., and Nagahara M. "Mean squared error analysis of quantizers with error feedback." *IEEE Transactions on Signal Processing*. 2017; 65(22): 5970–5981.

[40]   Zhang Z., Zhang L., Hao F., and Wang L. "Periodic event-triggered consensus with quantization." *IEEE Transactions on Circuits and Systems II: Express Briefs*. 2016; 63(4): 406–410.

Chapter 9
Summary and perspectives

9.1 Summary

The motivation for writing this book came from an experiment, in which direct applications of some compressed sensing (CS) algorithms to the data collected by some real radar systems failed to achieve satisfactory performance. This experiment made us realize that something beyond the simple sparsity is necessary for signal processing in various radar tasks. Moreover, it is found that a number of books on the fundamentals of CS but few books are devoted to explaining how to apply the CS-based methods to solve the practical problems in the radar area.

This book aims to introduce the advanced sparsity-driven models and methods that were designed for radar tasks such as detection, imaging, and classification, mainly based on the author's publications in the last decade. Besides the theoretical analysis, a number of simulations and experiments on real radar data were provided through this book to intuitively illustrate the effect of the advanced sparsity-driven models and methods.

9.1.1 Sparsity-driven radar detection

The sparsity-driven radar detection was discussed in Chapters 3 and 8.

- In Chapter 3, the two-level block sparsity model was imposed on the signal received by an airborne antenna array radar and applied to space-time adaptive processing (STAP). By enhancing both the cluttered sparsity of the angle-Doppler spectrum and the joint sparsity of multiple training snapshots collected from multiple range resolution cells, the covariance matrix of the ground clutter can be accurately estimated from the small number of training snapshots and accordingly the performance of moving target detection can be improved. The superiority of this model was demonstrated by the experiments on the radar data collected by the DARPA Mountain-top program.
- In Chapter 8, the distributed detector based on the locally most powerful test (LMPT) was presented for sparse signal detection with radar networks. The LMPT detector is computationally efficient since it does not require the reconstruction of sparse signals. In order to decrease the communication overhead within the network, the quantized LMPT detector was further developed under the assumption that each local sensor first quantizes the

measurements with a low-bit analog-to-digital converter (ADC) and then transmits the coarsely quantized data to the fusion center. It was demonstrated that (1) the detection performance loss induced by 1-bit quantization can be compensated by increasing the number of measurements by 3.3 times, and (2) the detection performance loss caused by 3-bit quantization can be neglected if the designed quantizes are taken at all the local sensors.

9.1.2   Sparsity-driven radar imaging

The sparsity-driven radar imaging was discussed in Chapters 2, 3, 4, 5, and 6.

- In Chapter 2, the hybrid matching pursuit (HMP) algorithm was presented for enhancing the quality of radar imaging, by combining the strength of the orthogonal matching pursuit (OMP) in the basis-signal selection and the strength of subspace pursuit (SP) in basis-signal re-evaluation. Another algorithm referred to as look-ahead hybrid matching pursuit (LAHMP) was further presented for refining the basis-signal selection per iteration by evaluating the effect of basis-signal selection on the recovery error in next iteration. The experiments on through-wall radar data showed that both of these two algorithms can improve the radar image quality at the cost of an increase of the computational complexity.

- In Chapter 3, the two-level block sparsity model was presented to enhance the quality of multichannel radar imaging. This model aims to promote both of the clustered sparsity of the single-channel radar image and the joint sparsity of the radar images generated at multiple receiving channels. Compared to only enforcing the clustered sparsity or the joint sparsity, the two-level block sparsity model not only ensures the high-resolution radar image of extended targets but also significantly suppresses the artifacts outside the target area. The effectiveness of the two-level block sparsity model was demonstrated by experiments on through-wall radar data.

- In Chapter 4, the parametric sparse representation (PSR) method was presented to eliminate the influence of the model uncertainty on radar imaging. The model uncertainty may come from the unknown motion of the targets, the undesired trajectory error of the radar platform, and the unknown characteristics of the complicated propagation. The key idea of PSR is to formulate the basis-signal dictionary as a function of the parameters describing the model uncertainty so that the dictionary is adjustable during the process of radar image formation. Two approaches for solving both of the sparse radar image and the parametric dictionary, that is, alternating iterations and parameter searching, were introduced. Examples of applying PSR to synthetic aperture radar (SAR) refocusing of moving targets, SAR motion compensation, and inverse synthetic aperture radar (ISAR) imaging of maneuvering aircrafts were demonstrated.

- In Chapter 5, the Poisson disk sampling and the iterative shrinkage thresholding like (IST-like) algorithm were combined to simultaneously achieve high-resolution and wide-swath in the scenario of single-antenna SAR imaging. The Poisson disk sampling strategy was taken to enlarge the pulse interval and accordingly the range swath without Doppler aliasing, and the IST-like algorithm was developed for suppressing the nonstructured noise induced by the stochastic pulse sampling. The experimental results on space-borne SAR data demonstrated that the range swath width can be increased by at least 1.5 times without obvious sacrifice of SAR imaging quality.

• In Chapter 6, two algorithms of sparsity-driven radar imaging with coarsely quantized data, that is, the parametric quantized iterative hard thresholding (PQIHT) and the enhanced-binary iterative hard thresholding (E-BIHT), were presented. Coarse quantization may be demanded for lowering hardware cost and saving communication overhead. In order to mitigate the influence of the quantization error, the original PSR framework and the two-level block sparsity model were extended to the case of coarsely quantized data. The experimental results on SAR data showed that these two algorithms can reconstruct satisfactory images of moving targets and stationary targets, respectively, from 1-bit quantized data.

9.1.3   Sparsity-driven radar classification

The sparsity-driven radar classification was discussed in Chapter 7.

• In Chapter 7, two sparsity-aware algorithms of micro-Doppler analysis were presented for radar classification of rigid-body and nonrigid-body targets, respectively. For rigid-body targets, the problem of micro-Doppler-based classification can be attributed to the problem of estimation of the micro-Doppler parameters such as the Doppler repetition period, the Doppler amplitude, and the initial phase. This goal was achieved by using the PSR method, and the pruned orthogonal matching pursuit (POMP) algorithm was further designed to reduce the computational complexity. For nonrigid-body targets such as dynamic hand gestures, their radar signals are generally diffi-cult to be formulated as an analytical form. In this case, the time-frequency locations and the corresponding reflectivity coefficients of the dominant components were extracted via the Gabor decomposition and fed into the modified-Hausdorff-distance-based nearest neighbor (NN) classifier for recognition. Experimental results on simulated data and real radar data demonstrated the effectiveness of these algorithms in terms of the time-frequency resolution and the recognition accuracy.

9.2   Perspectives

9.2.1   Structured models

Beyond the simple sparsity, structured models are more conducive to describing the intrinsic characteristics of signals. Some structured models such as the block sparsity, the joint sparsity, the tree-structured sparsity, and the low-rank property have been exploited in a wide range of applications [1]. In various radar tasks, it is worth investigating more sophisticated models to represent the intrinsic structures of radar signals case by case. The structure of the radar signal may be dependent on waveform. For example, for analyzing chirp signals, the linear-shaped sparsity in the time-frequency domain is more accurate than the general block sparsity. The structured model may also be related to the environment. For instance, when multipath radar signals are exploited, besides the joint sparsity of the direct-path and indirect-path signals, the strong correlation between their waveforms also holds. In summary, the design of the structured models for various radar tasks should be application-dependent, and the specifically designed models can be expected to work better than the general sparse models if there is a good match between the assumed and the actual scenarios.

9.2.2   Practical databases

Although a number of sparsity-driven algorithms of radar signal processing have been developed, the statistical analysis on the sparse solutions is still inadequate. Statistical analysis of the results of sparse signal processing is necessary for a number of radar tasks, for example, constant false alarm rate (CFAR) detection in a radar image reconstructed by sparsity-driven algorithms. However, it is not easy to theoretically deduce the probability density function (PDF) and the moment statistics of the sparse solutions based on the statistical knowledge of measured data and a priori probability distribution of clutter and noise, because reconstruction of sparse signals is basically an inverse problem. A feasible way is to establish practical radar databases and perform a large number of experiments in various clutter environments. As the MSTAR program [2] has greatly promoted the development of radar target recognition methods, huge databases are expected to be capable of effectively evaluating the sparsity-driven algorithms for radar detection, imaging and classification, and reliably providing statistical knowledge of their sparse solutions.

9.2.3   Combination of sparsity and data-driven algorithms

The signal processing algorithms developed in recent decades can be generally divided into two kinds, that is, model-based and data-driven. The classical CS algorithms are model-based, and the state-of-the-art deep learning algorithms are data-driven. It is promising to combine the strengths of these two kinds of algorithms. Learning from the measurement data can adjust the predefined sparse signal model. A good example is dictionary learning [3], in which the dictionary and the sparse solution are iteratively updated so that the learned dictionary better represents the measurement data. Introducing model and structure into deep learning algorithms can make the learning process interpretable and well-trained with a relatively small amount of training data [4]. As for the combination of the model-based and data-driven algorithms, an enduring challenge lies in the theoretical guarantee and the generalization ability of the algorithms.

9.2.4 Generalized sparsity of heterogeneous data

When radar and other kinds of sensors collaborate together to observe the same region of interest or the same group of targets, the fusion of heterogeneous data is regarded as an effective way to improve the performance and the robustness. There is no doubt that separate sparse representations of heterogeneous data have different forms and meanings. A question naturally arises: is it possible to bridge the gap between the sparse representations of heterogeneous data? This task is to some extent equivalent to defining a generalized sparsity of heterogeneous data. Before answering this question, we first look at the recently developed image style transfer (IST) [5,6], in which a natural image can be rendered into specific artistic styles. The key idea of IST is to first extract the content features and the style features via the deep convolutional neural networks (DCNN) and then generate the synthetic image by combining the content features of the natural image and the style features of the artistic painting. Inspired by this interesting work, heterogeneous remote sensing images are fused based on IST for earthquake damage assessment in [7], where the optical and SAR remote sensing images are deemed to have the same semantic content but different styles. The results in [7] can give inspiration for defining generalized sparsity of heterogeneous data, that is, finding the relationship between the semantic content features of heterogeneous data and separately representing their individual style features.

References

[1] Duarte M. F. and Eldar Y. C. "Structured compressed sensing: From theory to applications." *IEEE Transactions on Signal Processing*. 2011; 59(9): 4053–4085.

[2] Hummel R. "Model-based ATR using synthetic aperture radar." *Proceedings of IEEE International Radar Conference*. Washington, DC, USA, 2000. pp. 856–886.

[3] Ivana T. and Pascal F. "Dictionary learning." *IEEE Signal Processing Magazine*. 2011; 28(2): 27–38.

[4] Eldar Y. C. "From compressed sensing to deep learning: Tasks, structures, and models." *Plenary Speech at the 45th International Conference on Acoustics, Speech, and Signal Processing (ICASSP)*. Barcelona, Spain: IEEE; May 5, 2020.

[5] Gatys L. A., Ecker A. S., and Bethge M. "Image style transfer using convolutional neural networks." *Proceedings of IEEE Conference on Computer Vision and Pattern Recognition*. Honolulu, HI, USA, 2016. pp. 2414–2423.

[6] Justin J., Alexandre A., and Li F.-F. "Perceptual losses for real-time style transfer and super-resolution." *Proceedings of European Conference on Computer Vision*. Amsterdam, The Netherlands: Springer; 2016. pp. 694–711.

[7] Jiang X., Li G., Liu Y., Zhang X.-P., and He Y. "Change detection in heterogeneous optical and SAR remote sensing images via deep homogeneous feature fusion." *IEEE Journal of Selected Topics in Applied Earth Observations and Remote Sensing*. 2020; 13: 1551–1566.

Index